Proceedings of the First International Workshop on the

Analysis of Multi-Temporal Remote Sensing Images

SERIES IN REMOTE SENSING

Series Editor: Arthur P. Cracknell *(Univ. Dundee)*

Published

Vol. 1: Soft Computing in Remote Sensing Data Analysis
Eds: E. Binaghi, P. A. Brivio & A. Rampini

Series in Remote Sensing Vol. 2

Proceedings of the First International Workshop on the

Analysis of Multi-Temporal Remote Sensing Images

Multitemp 2001

University of Trento, Italy

13-14 September 2001

Editors

Lorenzo Bruzzone

University of Trento, Italy

Paul Smits

Joint Research Centre, European Commission, Italy

World Scientific

New Jersey • London • Singapore • Hong Kong

Published by

World Scientific Publishing Co. Pte. Ltd.

P O Box 128, Farrer Road, Singapore 912805

USA office: Suite 1B, 1060 Main Street, River Edge, NJ 07661

UK office: 57 Shelton Street, Covent Garden, London WC2H 9HE

British Library Cataloguing-in-Publication Data
A catalogue record for this book is available from the British Library.

European Commission – Joint Research Centre
EUR 20240 EN

ANALYSIS OF MULTI-TEMPORAL REMOTE SENSING IMAGES
Proceedings of Multitemp 2001

ISBN 981-02-4955-1

Printed in Singapore by Uto-Print

Foreword

The development of effective methodologies for the analysis of multi-temporal data is one of the most important and challenging issues that the remote-sensing community should face in the next years. The importance and timeliness of this issue are directly related to the ever-increasing quantity of multi-temporal data provided by the numerous remote-sensing satellites that orbit around our planet. The synergistic use of multi-temporal remote-sensing data and advanced analysis methodologies results in the possibility to solve complex problems related to the monitoring of the Earth surface and atmosphere at different scales. In this context, important examples of applications are change detection, assessment of land-cover dynamics, monitoring of natural resources, etc. However, the advances in the methodologies for the analysis of multi-temporal data have been significantly under-illuminated with respect to other topics (e.g., classification). At the same time, the link between the end-users' needs and the scientific community seems to be weak. For these reasons, the first international workshop on the *Analysis of Multi-temporal Remote Sensing Images* (*MultiTemp-2001*) has been organized. The workshop was held at the University of Trento, Trento, Italy (September 13–14, 2001).

This volume contains the proceedings of *MultiTemp-2001*. The workshop addressed the scientific developments in the field of the analysis of multi-temporal remote-sensing images in connection with real applications and end-users' requirements. Its main aim was to assess the state of the art of both methodologies and applications related to the analysis of multi-temporal remote-sensing images and geospatial data, as well as to provide a forum for researchers in this subject area to exchange views and report their latest results. This aim was successfully accomplished, thanks to the contribution of researchers from different communities (including image analysis, pattern recognition, computer science, forestry, management of natural resources, and ecology) representing academia, industry and governmental organizations. The proceedings contain 49 papers organized in three thematic sections dealing with *Image Analysis and Algorithms*, *Monitoring and Management of Natural Resources*, and *Land-Cover Dynamics*.

The workshop was organized and sponsored by the University of Trento (School of Telecommunication Engineering), Italy, and the Joint Research Center (JRC) of the European Commission, in cooperation with the IEEE Geoscience and Remote Sensing Society (IEEE GRS-S). It was supported by the Italian Association for Remote Sensing and the ITC-IRST, Trento, Italy.

We are grateful to all people who helped us in the organization of MultiTemp-2001. First of all, our thanks are due to the Program Chairs (Gianni Vernazza and Alessandro Annoni) and to the members of the Scientific Committee, who greatly contributed to select the papers presented at the workshop. Thanks are also due to the invited speakers: David Goodenough (Natural Resources Canada), Roger King (Mississippi State University, USA), and Ross Lunetta (US Environmental Protection Agency, USA). Last but not least, we are grateful to Paola Bodio and Elisabetta Nones for the logistic organization of the workshop, and to Roberto Cossu for supporting the local organization and for maintaining the MultiTemp 2001 website.

Lorenzo Bruzzone and Paul Smits
General Workshop Chairs

Trento, September 2001

General Workshop Chairs

Lorenzo Bruzzone (University of Trento, Italy)

Paul Smits (Joint Research Centre, Italy)

Program Chairs

Gianni Vernazza (University of Genoa, Italy)

Alessandro Annoni (Joint Research Centre, Italy)

Scientific Committee

Jon Atli Benediktsson (University of Iceland, Iceland)

Alessandro Brivio (National Research Council, Italy)

Chi Hau Chen (University of Massachusetts, USA)

Pol Coppin (University of Leuven, Belgium)

Mario Gomarasca (National Research Council, Italy)

Roger King (Mississippi State University, USA)

David Landgrebe (Purdue University, USA)

Jacqueline Le Moigne (NASA, USA)

Ross Lunetta (US Environmental Protection Agency, USA)

Maria Petrou (University of Surrey, UK)

Paul Rosin (Cardiff University, UK)

Robert Schowengerdt (University of Arizona, USA)

Sebastiano Serpico (University of Genoa, Italy)

Ashbindu Singh (United Nations Environment, USA)

Anne Schistad-Solberg (University of Oslo, Norway)

James Tilton (NASA, USA)

John Townshend (University of Maryland, USA)

Urs Wegmüller (Gamma Remote Sensing, Switzerland)

Contents

Monitoring and Management of Natural Resources

Land-cover Dynamics

Image Analysis and Algorithms

DIGITAL CHANGE DETECTION METHODS IN NATURAL ECOSYSTEM MONITORING: A REVIEW

POL COPPIN*, ERIC LAMBIN**, INGE JONCKHEERE* AND BART MUYS*

**Geomatics and Forest Engineering Group, Department of Land Management, Katholieke Universiteit Leuven, Vital Decosterstraat 102, B-3000 Leuven, Belgium, tel 32 16 329740, fax 32 16 329760*

***Department of Geography, Université Catholique de Louvain, 3 Place Pasteur, B-1348 Louvain-la-Neuve, Belgium, tel 32 10 474477, fax 32 10 472877*

Email corresponding author: pol.coppin@agr.kuleuven.ac.be

Natural ecosystems are in a state of permanent flux at a variety of spatial and temporal scales all around the world. Causes of these fluxes can be natural as well as anthropogenic, or may combine the two. Moreover, all scientific evidence clearly points to the fact that impacts of, for example, global change, on land surface attributes are not uniformly distributed on the face of the earth either. Techniques based on multitemporal multispectral satellite-acquired data have demonstrated potential as a means to detect, identify, map, and monitor natural ecosystem changes, irrespective of their causal agents. This review paper, which summarizes the methods and the results of digital change detection in the optical/infrared domain, has as its primary objective a synthesis of the state-of-the-art today. We approach digital change detection from three angles. First, the different perspectives from which the variability in ecosystems and the change events have been approached are summarized. Change detection between pairs of images (bi-temporal) as well as between time profiles of imagery-derived indicators (trends), and, where relevant, the appropriate choices for digital imagery acquisition timing and change interval length definition, are discussed. Second, pre-processing routines to either establish a more direct linkage between remote sensing data and biophysical phenomena, or to temporally mosaic imagery and extract time profiles, are reviewed. Third, the actual change detection methods themselves are categorized in an analytical framework and critically evaluated. Ultimately, we highlight how some of these methodological aspects are being improved upon as this review is being written, and we summarize the new developments that can be expected in the near future.

1 Introduction

The fact that sustainability has become a primary objective in present-day natural ecosystem management, has as one of its consequences the continuous need for accurate and up-to-date resource data. What is more, where it concerns large-area processes, any in-depth understanding of the changes has to be based on an accurate monitoring of land surface attributes over at least a few decades. At such regional scales, monitoring poses a number of methodological challenges. The lack of quantitative, spatially explicit and statistically representative data on natural ecosystem change has left the door open to simplistic representations, e.g., "marching deserts". While it is possible to find local examples of such extreme changes, empirical studies in grassland, savannas and open forest ecosystems generally revealed the predominance over secular land cover conversions of inter-annual climatic variability, ecosystem resilience and complex land-cover change trajectories.

Until vertical aerial photos became available in the 1930's, natural resource managers were totally dependent upon information obtained on the ground. Aerial photography immediately proved to be a valuable asset, greatly facilitating the acquisition of much of the required knowledge. The launch of Landsat-1 in 1972 marked the beginning of the satellite remote sensing for renewable resources applications era. Because of their synoptic and repetitive data acquisition capabilities, satellite-based sensors hold the potential to detect, identify, and map changes that are important to ecosystem managers. While limitations in data storage, data processing, and quantitative analysis capabilities initially hampered the successful use of satellite data, advanced processing algorithms and computer hardware now make it feasible to monitor on a continuous or periodic basis most areas of our planet. Going into the twenty-first century, recent developments in sensor technology and initiatives such as the Global Monitoring of Environment and Security (GMES), have paved the way for fully operational monitoring systems.

Digital change detection essentially comprises the quantification of temporal phenomena from multi-date imagery that is most commonly acquired by satellite-based multispectral sensors. The scientific literature reveals, however, that digital change detection is a difficult task to perform. An interpreter analysing large-scale aerial photography will almost always produce more accurate results with a higher degree of precision [1]. Nevertheless, visual change detection is difficult to replicate because different interpreters produce different results. Furthermore, visual detection incurs substantial data acquisition costs. Apart from offering consistent and repeatable procedures, digital methods can also more efficiently incorporate features from the non-optical parts of the electromagnetic spectrum.

This paper reviews the state-of-the-art of digital change detection in the optical/infrared domain with emphasis on natural ecosystems. It is a thorough actualisation of a five-year-old previous effort [2] that was largely confined to bi-temporal change detection methodologies (comparing the same area at two points in time). The present review has been expanded to now also encompass the emerging field of multi-temporal trend analysis (comparing the same area over longer time intervals with multiple imagery, e.g., over the length of a growing season), also often referred to as change monitoring based on time profiles.

2 Ecosystem change and multi-temporal imagery

2.1 Change

Natural ecosystems are continuously changing, where change is defined as "an alteration in the surface components of the vegetation cover" [3] or as "a spectral/spatial movement of a vegetation entity over time" [4]. The rate of change can either be dramatic and/or abrupt, as exemplified by fire; or subtle and/or gradual, such as biomass accumulation. Change can therefore be seen as a categorical variable (class) or in a continuum. Authors generally distinguish between land-cover *conversion*, i.e. the complete replacement of one cover type by another, and land-cover *modification*, i.e. more subtle changes that affect the character of the land cover without changing its overall classification. Land-cover modifications are generally more prevalent than land-cover conversions. Some natural ecosystem modifications are human-induced, including deforestation for land-use conversion. Others have natural origins resulting from, for example, flooding and disease epidemics.

Various classifications of change in natural ecosystems have been proposed. A problem with many of these categorical schemes is that the change classes are often not mutually exclusive. Colwell *et al.* [5] therefore suggested a more hierarchical framework, whereby the classification is mutually exclusive and totally exhaustive. It, however, still falls short by not providing a link between change event and causal agent. Hobbs [6] focused more on ecological aspects and differentiated between seasonal vegetation responses, inter-annual variability, and directional change. The latter may be caused by intrinsic vegetation processes (e.g. succession), land-use conversion, other human-induced changes (e.g., pollution stress) and alterations in global climate patterns (e.g., global warming). Khorram *et al.* [7] concentrated on the spatial environment in which the change occurs: "Some changes may affect entire areas uniformly and instantaneously, while others may take the form of slow advances or retreats of boundaries between classes, and still other changes may have very complex spatial textures." In the spatial context, they proposed four types of change whereby polygons either (1) become a different category, (2) expand, shrink, or alter shape, (3) shift position, or (4) fragment or coalesce.

The ability of any system to detect and monitor change in natural ecosystems depends not only on its capability to adequately deal with the initial static situation, but also on its capacity to account for variability at one scale, e.g., seasonal, while interpreting changes at another, e.g., directional [6]. Moreover, the ability to "detect" is a function of the "from" and "to" classes, the spatial extent, and the context of the change [7].

Not all detectable changes, however, are equally important. On the other hand, it is also probable that some changes of interest will not be captured very well, or at all, by any given system. Of particular interest to the natural ecosystem scientist and/or manager are first and foremost vegetation disturbances caused by short-term natural phenomena such as insect infestation, fire and flooding, and changes resulting from human activities, e.g., resource exploitation and land use conversion. While the former are likely to be temporary and may in some cases even be self-correcting, evidence of the latter generally remains much longer. Secondary, but therefore not less important, are the ecosystem changes that ensue from alterations in larger-scale processes, such as global warming etc. In this case, changes in trends are analysed, rather than actual change events. The proper understanding of the nature of the change and the principles that enable its detection and categorization usually encompass more sophistication than the simple detection of the change event itself.

In summary, the main challenges facing natural ecosystem change monitoring from space come from the requirements to: (i) detect modifications in addition to conversions (e.g., quantify forest cover degradation due to selective logging or fires); (ii) monitor rapid and abrupt changes in addition to the progressive and incremental changes (e.g., assess the impact of a flood, drought or fire, versus a progressive expansion of agriculture); (iii) separate interannual variability from secular trends, given the shortness of the available time series - 20 to 30 years - (e.g., assess dry land degradation); (iv) understand and correct for the scale dependence of statistical estimates of change derived from remote sensing data at different spatial resolutions; and (v) match the temporal sampling rates of observations of processes to the intrinsic scales of these processes (e.g., monitor rapidly evolving processes such as floods or biomass burning).

2.2 *Imagery Acquisition*

Change detection for natural ecosystem monitoring generally requires phenological conditions to be comparable, be it on a two-point time scale (bi-temporal change detection), or on a continuous time scale (trend analysis). However, in the latter case, it is also possible that these conditions themselves are the objects that are being monitored, e.g., in natural variability monitoring.

2.2.1 Bi-temporal change detection

The appropriate selection of imagery acquisition dates is as crucial to a bi-temporal change detection method as is the choice of the sensor(s), change categories, and change detection algorithms. The problem has two dimensions: the calendar acquisition dates and the change interval length (temporal resolution).

Anniversary dates or anniversary windows (annual cycles or multiples thereof) are often used because they minimize discrepancies in reflectance caused by seasonal vegetation fluxes and sun angle differences. However, even at anniversary dates, or within anniversary windows, phenological disparities due to local precipitation and temperature variations may present real problems. Forest ecosystem dynamics in the temperate regions illustrate this clearly. All else being equal, the reflectance of tree leaves in the visible part of the EMR spectrum is higher in spring and autumn than in the middle of the growing season. Changes in the near infrared part of the spectrum are less distinct. The stage of change at a particular time in spring and autumn depends on the site, tree species and varying genotypes of the same species. Local seasonal effects are especially confusing during leaf-out and autumn coloration. Hame [8] therefore concluded that for bi-temporal change detection, summer and winter are the best seasons because of their phenological stability. What is more, selecting the summer, or the driest period of the year for the locale, will enhance spectral separability, yet minimize spectral similarity, because of excessive surface wetness prevailing during other periods of the year [9]. However, the optimal selection of the season for bi-temporal forest cover change detection data acquisition remains a topic of contention in the pertinent literature.

For most documented studies, the periodicity of the data acquisition seems to have been determined according to the availability of satellite data of acceptable quality. While visually analysing Landsat-1 multispectral scanner (MSS) data for forest cover regeneration assessment, Aldrich [10] concluded that a minimum time interval of three years was required to detect non-forest to forest changes. Colwell *et al.* [5] judged a two-year separation between MSS scenes insufficient to reliably map re-vegetating areas in South Carolina. They advised a periodicity of at least five years. Gregory *et al.* [11] reported that in southern Oklahoma three separate classes of clear cuts have been identified from processed Landsat MSS data: those created less than six years before image acquisition, those that dated from six to 15 years before data capture, and those older than 15 years. Park *et al.* [12], again using Landsat MSS data, suggested a one-year interval to detect forest to non-forest (successional herbs, urban or agricultural development) changes, a three to five year interval to monitor non-forest to successional shrubs stage, and another five to ten years to detect the consecutive establishment of a forest cover. Coppin & Bauer [13] tested two-, four-, and six-year intervals for canopy change detection and found that a two-year cycle was optimal to study aspen establishment and storm damage in the Great Lakes region with Landsat Thematic Mapper (TM) imagery. Their four- and six-year cycles performed best for human-induced and natural canopy disturbances such as thinning, cutting, and dieback.

2.2.2 Trend analysis

To circumvent the problem of the selection of optimal imagery acquisition dates, some investigators have approached natural ecosystem monitoring by comparing seasonal development curves or profiles. These require time series of remotely sensed indicators of relevant land surface attributes which, in turn, are constructed from daily imagery acquisitions provided by such sensors as the NOAA AVHRR, SPOT VEGETATION, SeaWIFS, MODIS, ATSR, etc. Change detection based on such profiles has been proven appropriate for regional studies of largely climate-driven land surface attribute changes [14, 15], of phenology modifications [16], and of inter-annual net primary production variability [17].

One of the advantages of profile-based techniques lays in the fact that the issue of influence of phenology on change detection performance is resolved, because data are collected throughout the growing season. As such, changes inherently linked to seasonality can be separated from other changes. A serious disadvantage, however, is that, until today, only coarse- spatial- resolution sensors provide the high temporal frequency of observations that is necessary to establish time profiles, This drastically limits the change categories that can be detected and monitored. Lawrence and Ripple [18] have nevertheless been able to monitor the natural ecosystem disturbances caused by a volcanic eruption by calculating change curves from a time series of Landsat TM data acquired at high temporal frequency.

3 Data pre-processing for change detection

The primary challenge in deriving accurate natural ecosystem change information is representative of the standard remote sensing problem: maximization of the signal-to-noise ratio. Inherent noise will affect the change detection capabilities of a system or even create unreal change phenomena. Causes of such unreal changes can be, among others, differences in atmospheric absorption and scattering due to variations in water vapour and aerosol concentrations of the atmosphere at disparate moments in time, temporal variations in the solar zenith and/or azimuth angles, and sensor calibration inconsistencies for separate images. Pre-processing of satellite images prior to actual change detection is essential and has as its unique goals the establishment of a more direct linkage between the data and biophysical phenomena, the removal of data acquisition errors and image noise, and the masking of contaminated and/or irrelevant scene fragments.

The pre-processing of multi-date imagery, when absolute comparisons between different dates or periods are to be carried out, is much more demanding than in the single-date case. It commonly comprises a series of sequential operations, including (but not necessarily in this order) calibration to radiance or at-satellite reflectance, atmospheric correction or normalization, image registration, geometric correction, mosaicing, sub-setting and masking (e.g., for clouds, water, irrelevant features). Often these procedures are accompanied by a data transformation to vegetation indices that are known to exhibit a strong positive relationship between up-welling radiance and the vegetative cover of natural ecosystems. The principal advantages of vegetation indices over single-band radiometric responses are their ability to considerably reduce the data volume for processing and analysis, and their inherent capability to provide information not available in any single band. However, no single vegetation index can be expected to totally summarize the information in multidimensional spectral data space.

Wallace & Campbell [19] aptly stated that adequate indices can be found for different purposes and that indices derived for one analysis may be inappropriate in another context.

3.1 Bi-temporal change detection

Of the various aspects of pre-processing for bi-temporal change detection, there are two outstanding requirements: multi-date image registration, and radiometric rectification. It should be evident that accurate spatial registration of the multi-date imagery is absolutely essential to digital change detection. In a study on the impact of mis-registration on change detection simulations of MODIS data run with spatially degraded Landsat MSS images, Townshend *et al.* [20] clearly demonstrated that to achieve errors of only 10% in vegetation index values, registration accuracies of 0.2 pixels or less were required. However, change detection capabilities are intrinsically limited by the spatial resolution of the digital imagery. Further, residual mis-registration at the below-pixel level commonly degrades the areal assessment of the change events somewhat, specifically at the change/no-change boundaries. This within-pixel shift is inherent to any digital change detection technique [21].

The second critical requirement for successful change detection is that a common radiometric response is required for quantitative analysis of one or more image pairs acquired on different dates. For many change detection applications, absolute radiometric correction is unnecessary, but variations in solar illumination conditions, in atmospheric scattering and absorption, and in detector performance need to be normalized or, in other words, radiometric properties of a subject image need to be adjusted to those of a reference image. Duggin & Robinove [22] strongly insisted, we believe correctly, on the calibration of raw sensor data to meaningful physical units prior to any multi-temporal analysis. Only through reliable radiometric calibration can a researcher be confident that observed spatial or temporal changes are real differences and not artefacts introduced by differences in the sensor calibration, atmosphere, and/or sun angle [23].

Collins and Woodcock [24] examined three levels of radiometric pre-processing: no pre-processing, matched digital counts, and matched reflectances (full radiometric correction). They considered some form of radiometric correction essential, but the intermediate level of radiometric pre-processing (matched digital counts) in which several invariant features were used to calibrate a regression equation to predict time-2 digital numbers from those at time-1 (normalization) was found to be as accurate and easier to apply than the full radiometric correction. Songh et al. [25] demonstrated that one relative and seven absolute atmospheric correction algorithms all improved final data analysis when tested for change detection purposes. Contrary to expectations, the more complicated algorithms did not necessarily lead to improved performance. The authors recommended either simple dark object subtraction with or without a Rayleigh atmospheric correction (absolute correction), or atmospheric normalization (relative correction).

Atmospheric normalization can be full-scene-based and is then called simple regression normalization or SR [26], or pseudo-invariant-feature-based which is labeled as PIF. Hall *et al.* [27] have developed such a PIF radiometric normalization technique that corrects or rectifies images from common scenes through use of sets of landscape elements whose reflectance is nearly constant over time. The technique provides a relative calibration and does not require sensor calibration or atmospheric turbidity data, although correction to absolute surface reflectance can be accomplished if sensor calibration coefficients, an atmospheric correction algorithm (model), and atmospheric turbidity data are available. The technique resulted in data accurate to within 1% absolute reflectance in the visible and near infrared bands. Similar procedures have been used by Caselles & Garcia [28], Conel [29] and Coppin et al. [30]. Elvidge et al. [31] developed an automatic scatter gram-controlled regression (ASCR) normalization technique. It uses pixels from no-change regions identified from scatter grams, instead of from whole images as in the SR method or from spectral/spatial features as in the PIF procedure. More sophisticated approaches, which use atmospheric models for correction, but derive the atmospheric parameters from the satellite imagery, include, Ahern *et al.* [32], Chavez [33], and Hill & Sturm [34]. All of these studies make it clear that only when all sources of variation but the surface cover can be adjusted for (absolute calibration) or normalized to a common standard (relative calibration), will it be possible to detect and identify changes in natural ecosystems from multi-date imagery.

The debate around the use of filtering as a pre-processing technique for change detection has not been settled. Riordan [35] applied a modification of Nagao's and Matsuyama's edge-preserving image-smoothing algorithm to reduce minor variations in radiance values and to allow the comparison of relatively homogeneous groups of pixels. He judged the result ineffective. For tropical deforestation assessment, Singh [36] tried both image smoothing and edge enhancements, but did not see any improvement in the final change detection accuracy with either method. Baraldi & Parmiggiani [37], however, suggested the application of edge-preserving image-smoothing filters previous to image analysis in order to enhance the homogeneity of the spectral response of a thematic class and at the same time to eliminate noise effects. Bruzzone and Pietro [38] reported positive results when exploiting the spatial-contextual information contained in the neighborhood of each pixel to reduce the effects of noise and hence to increase change detection accuracy. In addition, the definition of an adaptive pixel neighborhood or parcel allowed for a more precise location of the borders of changed areas. To overcome low frequency / high contrast problems in Landsat TM imagery covering sand sheet desert environments, Kwarteng and Chavez [39] advocated the use of high pass spatial filters with relatively large kernel sizes, with a 50 percent add-back option and followed by edge enhancement, before submitting the data to actual change detection.

Some researchers advocate the use of texture features for change detection. Most consulted literature sources explicitly state, however, that texture information must only be used in conjunction with spectral data, and that both sources are complimentary to each other (e.g., [40]). Reed [41] cautioned the user community about the integration of texture measures in data analysis procedures. Classification accuracy is improved only under certain conditions. In his research, the addition of textural features, derived from a transformed vegetation index map of TM data, degraded the classification accuracy in distinguishing spectrally similar vegetation covers in all but one case. He conceded, however, that the choice of texture index (angular second moment and contrast) and window size (7x7 pixels) might have influenced the outcome significantly. Smits and Annoni [42] computed texture distance measures based on the texture indices contrast and homogeneity. They report mis-detection and false alarm rates of only 15% in urban expansion change detection and proved their method robust for mis-registration.

3.2 Trend analysis

As time-profile-based change detection methods work with data acquired on a large number of observation dates, adequate pre-processing is of utmost importance. Just as with bi-temporal approaches, pre-processing includes the geometric registration of successive images, but this time at the sub-pixel level, and radiometric corrections. However, as one is dealing with a series of images, data artefacts that are inherent to the type of imagery used and that render the comparison of data observations or measurements at different times difficult, must be removed, and a process of temporal compositing (mosaicing) must be executed. With respect to the former, many present-day high-temporal-frequency sensors have wide filed-of-views, which necessitates a correction for directionality (BRDF) effects. For example, changes in illumination conditions due to a drift in equator crossing time of the AVHRR sensors result in a viewing angle difference for off-nadir observations [43]. And alterations in AVHRR sensor responsitivity over time constitute another major source of image noise [43, 44].

The fact that controversy still rages over wether previous trend analyses detected real trends in the Earth's climate system (e.g., [16]), or simply resulted from a combination of calibration residuals and satellite-orbit drift [43], illustrates how critical the artefact issue still is. In addition, nature itself can render trend analysis much more difficult: the eruptions of Mount Pinatubo in mid 1991 caused significant noise in the global AVHRR time series due to the varying presence of aerosols in the atmosphere. With respect to temporal compositing, a range of procedures has been proposed for wide-angle sensors. As they are not directly related to actual change detection, the reader is referred to the relevant specialist literature.

4 Change detection methods

4.1 *Bi-temporal algorithms*

4.1.1 Background

All digital change detection is affected by spatial, spectral, temporal, and thematic constraints. The type of method implemented can profoundly affect the qualitative and quantitative estimates of the disturbance [45]. Even in the same environment, different approaches may yield different change maps. The selection of the appropriate method therefore takes on considerable significance. Most digital change detection methods are based on per-pixel classifiers and change information contained in the spectral-radiometric domain of the images. They combine both the procedures for change extraction (change detection algorithm) and those for change separation/labelling (change classification routine). Change extraction and change separation / labelling are preliminaries to any change modelling. In other words, predicting with a statistical or ecological model using independent variables where, when, and/or why change occurs, requires prior detection, measurement, and categorization of natural ecosystem change patterns.

Statistical and/or spatial decision rules that are derived from a heuristic understanding of the change event often constitute the backbone of the separation/labelling exercise. These rules, however, are not change-detection-algorithm-specific. In other words, they can be applied irrespective of the algorithm that generated the change data. They comprise the complete range of pattern recognition procedures available to the image analyst, from edge enhancement and simple thresholding (visual or statistically-based), to supervised and unsupervised classification, to segmentation and spatial analysis rules.

A wide variety of digital change detection algorithms have been developed over the last two decades. They basically can be summarized in two broad categories to which different reviewers have attached definitions that vary in complexity and, to a certain extent, in coverage. Malila [46] recognized the categories as change measurement (stratification) methods versus classification approaches. Pilon *et al.* [47] amplified the description of the first category to "enhancement approaches involving mathematical combinations of multi-date imagery which, when displayed as a composite image, show changes in unique colors". Singh [36] changed the focus slightly by centering the definitions more on a temporal scale: simultaneous analysis of multitemporal data versus comparative analysis of independently produced classifications for different dates. Other scientists [3, 48] have employed multi-class schemes.

Whatever combination of change detection algorithm and classification routine is applied, it is obvious that a wide assortment of alternatives exist and that all have varying degrees of flexibility and availability. As already stated, change classification routines are not specific to change detection. The overview that follows is therefore restricted to change detection algorithms only. All change detection algorithms that have been found documented in the literature by mid 2001 can be grouped into nine distinctly different categories plus one heterogeneous group encompassing hybrid methods and less frequently implemented or more esoteric algorithms. The first six are those most frequently used for monitoring vegetative canopies, while the other algorithms are less common and/or remain in an experimental stage. However, the order in which the algorithms are presented does not imply any ranking nor qualitative judgement. Moreover, because the algorithms are not necessarily independent of the data sources for which their implementation has been documented, examples typical for natural ecosystem monitoring are given.

4.1.2 Post-classification comparison

Post-classification comparison is also commonly referred to as "delta classification". It involves independently produced spectral classification results from each end of the time interval of interest, followed by a pixel-by-pixel or segment-by-segment comparison to detect changes in cover type. By properly coding the classification results, a complete matrix of change is obtained, and change classes can be defined by the analyst.

The principal advantage of delta classification lies in the fact that the two dates of imagery are separately classified, thereby minimizing the problem of radiometric calibration between dates. By choosing the appropriate classification scheme, the method can also be made insensitive to a variety of types of transient changes in selected terrain features that are of no interest [5]. However, the accuracy of the post-classification comparison is totally dependent on the accuracy of the initial classifications. The final accuracy very closely resembles that resulting from the multiplication of the accuracies of each individual classification and may be considered intrinsically low. The difficulty thus lies in securing completely consistent, analogous and highly accurate target identifications for each iteration. Mis-classification and mis-registration errors that may be present in the original images are compounded and results obtained using post-classification comparison are therefore frequently judged unsatisfactory [49].

A significant example of the use of the post-classification comparison approach is the work of Hall *et al.* [50] in which Landsat images acquired in 1973 and 1983 were classified into five forest successional classes (clearings, regeneration, broadleaf, conifer and mixed). Application of a PIF normalization (see 3.1) enabled classification of the 1973 image using 1983 ground data for classifier training. Following the two classifications a matrix of class changes over the 10-year interval was constructed and the transition rates between classes were calculated.

Xu & Young [51] preceded their post-classification comparison by a manual segmentation of the images according to ground features and characteristics of the scene. They then classified all segments separately for each date via a supervised maximum likelihood pattern recognition routine. They concluded that this approach, sometimes referred to as "pre-stratified delta classification", enabled them to avoid some obvious errors in classification (e.g., pixels classified as built-up areas on areas known to be moor lands in southeast Scotland).

4.1.3 Composite analysis

By using combined registered data sets, or corresponding subsets of bands, collected under similar conditions with respect to natural ecosystem phenology but from different years, classes where vegetative canopy change is occurring would be expected to have statistics significantly different from those where no change has occurred, and could be identified as such. The method can incorporate multistage decision logic and is sometimes referred to as "spectral / temporal change classification", "multi-date clustering", or "spectral change pattern analysis". While this technique necessitates only a single classification, it is a very complex one, in part because of the added dimensionality of two dates of data. In numerous cases it requires many classes and many, often redundant features when no discriminant analysis has preceded the process. It furthermore demands prior knowledge of the logical inter-relationships of the classes and should only be used when the researcher is intimately familiar with the study area [26]. Burns & Joyce [9] found the method to produce only change in forest cover *per se* without providing accurate information on the character of the change. Schowengerdt [52] remarked that, since spectral and temporal features have equal status in the combined data set, they can not be easily separated in the pattern recognition process. As a consequence, class labelling may be difficult. Researchers who have used this technique for natural ecosystem change detection include Colwell *et al.* [5], Hall *et al.* [53] and Sader [54].

4.1.4 Univariate image differencing

Univariate image differencing is probably the most widely applied change detection algorithm. It involves subtracting one date of original or transformed (e.g., vegetation indices, albedo, etc.) imagery from a second date that has been precisely registered to the first. With "perfect" data, this would result in a data set in which positive and negative values represent areas of change and zero values represent no change.

Banner & Lynham [55] used a multitemporal vegetation index difference based on calculated NDVI's for MSS data sets. They then density-sliced the difference vegetation index image. They found the method impractical for forest cutover delineation due to the sensitivity of the NDVI to grass growth and the development of other vegetation in the clear cuts, but useful for monitoring vegetative competition within the cutovers. Analogously, Lyon et al. [56] implemented NDVI differencing and found it the superior vegetation change detection technique for monitoring deforestation and loss of vegetation. Nelson [48] delineated forest canopy changes due to Gypsy Moth defoliation in Pennsylvania more accurately with vegetation index differencing than with any other single band difference or band ratioing (see 4.1.5). Mukai *et al.* [57] computed normalized difference channels for MSS5 and MSS6 to detect areas infested by Pine Bark Beetle in Japan. They were able to distinguish three classes of infestation with increasing band-5 and decreasing band-7 pixel values; light, moderate, and heavy.

Hame [58] suggested histogram matching and Yasuoka et al. [59] band-to-band normalization before differencing TM data so as to yield bands with comparable means and standard deviations and to reduce scene-dependent effects. In comparative analyses of the six reflective TM difference channels, Hame [58] as well as Fung [60] found that the TM3 difference channel contained the highest information content for vegetative cover monitoring.

Coppin & Bauer [21] suggested a standardization of the differencing algorithm (difference divided by the sum) to minimize the occurrence of identical change values depicting different change events.

Serneels et al. [61] applied double univariate image differencing in a spatial-contextual approach to separate anthropogenic changes from climate variability in a savanna environment. The technique computed and combined changes at pixel and landscape scales. First, univariate image differencing was applied to pairs of smoothed (101 X 101 pixels low-pass filter) vegetation index images to delineate landscape-scale changes. Second, local-scale patterns were detected via pixel-level image differencing between the two original (unsmoothed) full-resolution images. Finally, the latter change image was subtracted from the former, resulting in a change image wherein all pixels that behaved differently over time at both scales had the highest change values.

Cohen et al. [62] expanded bi-temporal image differencing in a multi-temporal context. They comparatively assessed two approaches: merged versus simultaneous image differencing. The first was based on an unsupervised classification, repeated five times, using five sequential date-pairs of difference Landsat images between 1972 and 1993, and requiring the merging of the results from five separate time intervals into a single change map. The second involved a single unsupervised classification of the full sequential difference image set. Both methods gave consistent and comparable results, with simultaneous image differencing being considerably more cost-effective to implement.

4.1.5 Image ratioing

Image ratioing is one of the simplest and quickest change detection methods. Data are ratioed on a pixel-by-pixel basis. A pixel that has not changed will yield a ratio value of one. Areas of change will have values either higher or lower than one.

To represent the range of the ratio function in a linear fashion and to encode the ratio values in a standard eight or sixteen bit code common to many PC-based image analysis software packages, a two-level (for values smaller than or greater than zero) normalizing function can be applied to all ratio values not equal to zero [26]. Because of the non-Gaussian bimodal distribution of ratioed multiple-date images, Riordan [38] criticized the ratio change detection algorithm in combination with an empirical threshold definition as being statistically invalid. The areas delineated on either side of the distribution mode are not equal. Consequently the standard deviation cannot be used for threshold definition.

Howarth & Wickware [49] combined MSS5 and MSS7 ratios in a single color composite. They found that, while the band-5 ratio emphasized changes in water level due to flooding, the band- 7 ratio assigned the brightest pixel values to areas where changes in vegetation cover were dominant. They were not able, however, to make a quantitative assessment of the changes.

4.1.6 Bi-temporal linear data transformation

Various linear data transformation techniques can be applied to two-date imagery that has been stacked in 2n-dimensional space (where n is the number of input bands per image). They concentrate information pertaining to statistically minor modifications in the state of the natural ecosystem (minor as contrasted to the entirety of the image scene) in orthogonal components, producing uncorrelated differences. The most important linear transformation is the one based on principal component analysis or PCA, with tasselled cap [63] and multi-variate alteration detection or MAD [64] being less frequently implemented.

The exact nature of the principal components derived from multitemporal data sets is difficult to ascertain without a thorough examination of the eigen-structure of the data and a visual inspection of the combined images (eventually via multidimensional temporal feature space analysis, see 4.1.10). To avoid drawing faulty conclusions, the analysis should not be applied as a change detection method without a thorough understanding of the study area [65]. The link between vegetative canopy change and tasselled cap transforms, on the other hand, appears to be more solid, an observation supported by Collins and Woodcock [66].

Richards [67] applied PCA to two-date MSS imagery to monitor brushfire damage and vegetation regrowth over extensive areas in Australia. Provided the major portion of the variance in the multitemporal sequence was associated with correlated (constant, unchanged) land cover, areas of localized change were enhanced in some of the lower components, particularly principal components three and four. Ingebritsen & Lyon [68] did exactly the same thing to detect and monitor vegetation changes around a large open-pit uranium mine in Washington and a wetland area in Nevada. Under the assumptions that the two original images both had an intrinsic dimensionality of two (first two principal components, rest primarily noise), that these dimensions were related to soil brightness and vegetation greenness, and that the change in land cover and/or vegetation condition exceeded some threshold value, four meaningful principal components resulted. They were stable brightness, stable greenness, change in brightness (somewhat analogous to an albedo difference signature), and change in greenness (somewhat similar to a near infrared / red ratio difference image). The latter component proved to be well-related to change in vegetative cover and insensitive to variations in slope and aspect.

According to Fung & LeDrew [69], standardized PCA gave more accurate results because its PC's were found to be better aligned along the object of interest: change. Singh & Harrison [70] concurred, citing a substantive improvement in signal-to-noise ratio and image enhancement by using standardized variables in the PCA. On the subject of subsetting the image to derive multitemporal PC's, the authors stated that, although even lower principal components can detect some land cover changes, the statistics extracted from data subsets are not recommended for change detection due to the great variability and uncertainty of the unextracted part of the data. They furthermore found the multitemporal PCA too scene-dependent and suffering from the serious drawback that no prior information was available regarding the nature of the components before actual processing.

Fung [65] reported on a comparative analysis of a multidate standardized PCA and a multidate tasselled cap transformation of a multitemporal Landsat TM data set. The PCA rotation was based on the merged 12-band TM image. Three components were found associated with change: PC3 with changes in soil brightness, PC5 with changes in NIR reflectance and thus vegetative vigor, and PC6 with changes in the contrast between the middle infrared bands and the PAR (photo synthetically active radiation) bands, thus changes in wetness. The usefulness of these lower-order multidate PC's to highlight localized change was confirmed by Lee *et al.* [71]. They ran a principal factor analysis on the transformed image and found the lower-order PC's to contain significantly more unique variance. The tasselled cap transformation was carried out as follows: The spectral bands of the two dates were assigned the tasselled cap coefficients as derived by Crist & Cicone [63] with positive coefficients for the first date and negative coefficients for the second. The derived vectors were not orthogonal and consequently were subjected to a Gram-Schmidt transformation. The process generated output vectors that were effectively orthogonal to each other. Three change tasselled cap images were produced detailing differences in greenness, brightness, and wetness. The greenness change image gave the highest classification accuracy among all resulting PCA and tasselled cap change-related images. Fung [65] clearly advocated the use of the tasselled cap transformation. Under this algorithm, the inherent data structure could be clearly depicted and the derived variables were physically based and independent of scene content. Twelve TM input bands could moreover be reduced to two, maximum three, significant change bands. While Hame [8] concluded that the original TM bands accomplished more than the transformed features (bi-temporal PC's or differences between paired PC's based on either the covariance matrix or the correlation matrix) to separate change classes in Finland, Coppin & Bauer [21] found the second principal component of vegetation index band pairs to be an excellent indicator of change in the temperate forest cover in the north-central U.S.A. Kwarteng and Chavez [39] used a similar selective PCA approach to successfully detect and map surface changes dealing with urban development, vegetation growth, and coastal wetland and sand sheet surface differences.

Collins and Woodcock [66] have developed a multitemporal generalization of the tasselled cap transformation. Application of the technique produced multitemporal analogs of the brightness, greenness, and wetness---the three primary dimensions of the tasselled cap transformation---and a component measuring change. Nielsen et al. [64] proposed the MAD, which is an extension of the traditional canonical correlations analysis and is invariant to linear scaling of the input data. As such MAD is insensitive to, for example, differences in sensor gain settings, or to linear radiometric and atmospheric correction schemes. MAD, in combination with a posterior maximum auto-correlation factor (MAF) transformation, gave significantly better results than PCA of simple differenced data in the detection of coherent patterns of spatial change in urbanization zones in Australia. Moreover, the MAD transformation provides a way of combining different data types that may be useful in historical change detection studies.

4.1.7 Change vector analysis

Change vector analysis or CVA is a multi-variate change detection technique that processes the full dimensionality (spectral + temporal) of the image data and produces two outputs: change magnitude and change direction. A major advantage is its capability to concurrently analyse change in all data layers as opposed to selected bands.

The first automated CVA change detection algorithm that took furthermore account of spatial scene characteristics was developed at the Environmental Research Institute of Michigan in the late seventies. It consolidated a tasselled cap transformation to greenness-brightness, an image-segmentation to spatially contiguous pixel groups or "blobs", and a characterization of the movement of the individual segments in spectral space in terms of magnitude and direction [46].

Colwell *et al.* [5] applied the algorithm in Kershaw County, South Carolina and found that, to be effective, it required absolutely precise image registration and normalization (changes in brightness needed to be scaled to be approximately equal to changes in greenness to avoid elliptical change thresholds) and a considerable amount of operator interaction. A forest mask had to be created, and parameters had to be set to control the formation of blobs and to threshold the change vectors for change, with respect to their magnitude (defining change from no change) as well as their direction (labelling the type of change). While the relative utility of the technique to assess the type of change was not clear, CVA performed well with respect to automated change detection.

Prior to computing change as a vector or distance in multispectral space, Yokota & Matsumoto [72] applied a preliminary transformation to all raw digital numbers (new value = old value minus the average brightness over the six TM bands) to "accentuate" the disturbances. They then calculated a multi-band Euclidean distance measure as the spectral differentiation norm. This measure was computed as the square root of the sum of the squares of the pixel value differences between the two dates over the six bands.

Lambin & Strahler [73] applied similar principles of segmentation and vector movement characterization to multitemporal vegetation indices, surface temperatures, and spatial texture data. Combined with principal component analysis of the change vectors, the technique proved to be effective in detecting and categorizing different land cover changes operating at different time scales in West Africa. Johnson and Kasischke [74] amply illustrated the capability of CVA in general to be an effective technique to capture all changes and not just *a priori* defined change events.

4.1.8 Image regression

A mathematical model that best describes the fit between two multi-date images of the same area can be developed through stepwise regression. The algorithm assumes that a pixel at time-2 is linearly related to the same pixel at time-1 in all bands of the EMR spectrum acquired by the sensor. This implies that the spectral properties of a large majority of the pixels have not changed significantly during the time interval [75]. The dimension of the residuals is an indicator of where change occurred. The regression technique accounts for differences in mean and variance between pixel values for different dates. Simultaneously, the adverse effects from divergences in atmospheric conditions and/or sun angles are reduced.

The critical part of the method is the definition of threshold values or limiting dimensions for the no-change pixel residuals. When Burns & Joyce [9] applied the technique to each pair of spectral MSS bands for land cover change detection via a third-degree polynomial linear equation, they found the green band (MSS4) to perform better than the other band pairs, however, still with relatively low accuracy. Singh [71], on the other hand, reported the highest change detection accuracy for tropical forest change detection with the regression method and the MSS5 band. A couple of years later [36], he reviewed that statement and concluded that the regression method performed only marginally better than uni-variate image differencing techniques in detecting tropical forest cover changes. This conclusion was confirmed in a later study in urban environment [76].

4.1.9 Multi-temporal spectral mixture analysis

The increased data dimensionality associated with high-spectral resolution (HSR) data gave rise to spectral mixture analysis or SMA, based on the premise that

HSR image elements are composed of multiple "pure" spectral signatures or end-members. If a linear mixing model is assumed, the overall reflectance of the image element may be computed from the reflectances of the composing end-members, weighted by their respective surface proportions. As a consequence, image element changes over time are directly mirrored in modifications of the end-member proportions, especially where more subtle natural ecosystem changes are concerned. Because end-members represent the spectra of known natural ecosystem components (e.g., canopy, soil, shadow, etc.), they have the advantage of providing physically-based standardized measures of fractional abundance.

Multi-temporal SMA or MSMA algorithms were implemented with good results by Adams et al. [77] and Roberts et al. [78] to monitor the physical nature of land cover changes in the Brazilian Amazon with Landsat TM imagery. The first group used reference end-members (derived from field or laboratory spectra of known components), while the second implemented an iterative process whereby reference library spectra were manipulated using image end-members in order to derive candidate reference end-members for green vegetation, soil, shade and non-photosynthetic vegetation. The major advantage of MSMA when compared to other bi-temporal change detection algorithms lays in its capability to recover natural ecosystem change signals at much finer event scales, e.g., thinning in forest ecosystems.

4.1.10 Multidimensional temporal feature space analysis

This change detection algorithm uses image overlay as a digital enhancement technique for on-screen change delineation. It comprises the one-step combination of a maximum of three individual bands (more steps are possible) which, when displayed as a composite image via the blue, green and red color guns of a cathode ray tube (CRT), portray changes in unique colors. The multidimensional temporal feature space analysis method seems to be most appropriate in natural environments where changes are relatively subtle. However, it provides the analyst with little information regarding the nature of the change [79]. Its use is mostly restricted to the creation of binary change masks to eliminate no-change areas before further analysis with any of the other algorithms, or to the visual definition of training areas related to particular change phenomena.

Banner & Lynham [55] obtained a better forest clear-cut identification and boundary delineation displaying MSS5 of time-1 via the blue CRT color gun and MSS5 of time-2 via the red color gun, than with multitemporal vegetation index differencing (see 4.5). Areas of change were highlighted in one of the two gun colors, while areas of no change appeared gray after adjustment of the color balance for the composite image. Hall *et al.* [53] applied the same technique with MSS7 data, however, assigning the red color gun to the MSS7 of time-1 and the green one to the MSS7 of time-2. They were able to connect the variations in red hues to qualitative variations in aspen tree defoliation in Alberta, Canada. Areas without defoliation appeared yellow on the composite image. Rencz [80], however, remarked that the fast vegetative regeneration in cutover areas precluded the use of any of the MSS reflective infrared bands for clear cut monitoring with this approach. Because of its relative insensitivity to this phenomenon, the use of red-band overlays was advised.

Werle *et al.* [81] created composite multitemporal images to visually monitor clear-cut and regeneration areas on Vancouver Island, Canada, assigning blue to TM3 for time-1, green to TM3 for time-2, and red to TM4 for time-2 on the CRT. In a multiseasonal assessment of regeneration cover in burned forest in Alberta, Canada, Kneppeck & Ahern [82] applied another color combination: blue to TM4 for summer, green to TM4 for fall, and red to TM3 for summer. Alwashe & Bokhari [83] merged TM bands 2, 4, and 5 of two different acquisition dates via an intensity-hue-saturation (IHS) transform to represent directly the bitemporal variations within one single image product. Vegetation differences showed up in distinctly different colors.

4.1.11 Hybrid and less frequently implemented algorithms

Only a few authors have documented the combined use of different change detection algorithms (hybrid schemes) in an attempt to minimize commission errors. Pilon et al. [79] applied such a hybrid scheme to change detection in semi-arid north-western Nigeria, performing post-classification comparison in areas where change was detected by other algorithms.

At about the same time CVA was developed, change detection procedures called inner product analysis and correlation analysis [59] were implemented in single instances. For both, the difference between the multispectral vectors of a pixel at two points in time was expressed as the cosine of the angle between them, but the correlation analysis also took into account the means of the multispectral vectors. While the inner product method was found to be sensitive to mis-registration, mixed pixels, sensor gain and offset fluctuations, and changes in absolute radiance, the correlation method substantially reduced these effects.

In 1985, Rencz proposed a multitemporal biomass index for forest clear cut monitoring, using the ratio of an MSS7 and MSS5 difference at time-2 over an MSS7 and MSS5 sum at time-1. The results had a very limited usefulness. The error was concentrated in the omission of forest cutovers in which there remained a relatively large number of unfelled residual hardwoods.

In theory, no-change areas can be treated as having slowly varying background gray levels. These variations can be approximated by a background image, for example, a low-pass filtered variant of the original. A subtraction of such an approximation from the image potentially could be used to create a new data set that accentuates change phenomena. Singh [36, 71] used the technique in tropical deforestation monitoring with only modest results.

Ridd and Liu [76] explored a chi square transformation encompassing the six reflective Landsat TM bands to create a single change image. The method is applicable only if the scenes are relatively unchanged. A disadvantage is that change related to specific spectral directions may not be readily identified.

4.1.12 Comparison and evaluation of methods

The literature indicates that natural ecosystem changes can be monitored via a variety of detection methods, with most providing positive results. With the exception of Singh's 1989 paper, it has taken until the mid nineties, however, to see comparisons and evaluations of alternative approaches documented.

Sunar [84] studied the equivalence of multidimensional temporal feature space analysis, univariate image differencing, PCA, and composite analysis, in order to detect and delineate land cover changes in ecosystems under intense development pressure. He could define no optimal method, as each algorithm had its own merits with respect to production ease, information content and interpretability. Similarly, Coppin and Bauer [21], Ridd and Liu [76] and Cohen and Fiorella [85] could make no conclusive statements regarding the superiority or inferiority of univariate image differencing versus in the first case selective PCA, in the second case image regression, and in the third case composite analysis.

When comparative conclusions are made, often univariate image differencing is cited as the, or one of the, preferred change detection algorithms.

Singh [71] nevertheless found univariate image differencing more effective than bi-temporal PCA for tropical deforestation monitoring. Muchoney and Haack [86] tested four methods on multi-temporal SPOT data: merged PCA, image differencing, spectral-temporal change classification (composite analysis), and post-classification comparison, in order to identify changes in hardwood defoliation caused by gypsy moth advances. Defoliation was most accurately detected by the image differencing and PCA approaches. Michener and Houhoulis [87] cross-referenced composite analysis, PCA, and univariate image differencing for monitoring vegetation responses to extensive flooding in southwest Georgia. Univariate NDVI differencing most accurately identified vegetation changes in their multi-temporal SPOT data set. Comparable conclusions were reached by Macleod and Congalton [88] when comparing post-classification comparison to univariate image differencing and PCA algorithms: image differencing performed significantly better in monitoring alterations in aquatic vegetation ecosystems encompassing submerged eelgrass. In a pilot study on North American Landscape Characterization or NALC land cover change detection, Yuan and Elvidge [89] systematically tested and evaluated 75 change detection methods using both visual and statistical procedures. Initial results again suggested image differencing outperforming the other algorithms. While Sohl [90] found specific quantitative values of change most accurately and easily provided by an enhanced image differencing algorithm, he cited the CVA as excelling at providing rich qualitative detail about the nature of the change. Mas [91], on the other hand, obtained the highest accuracy in land cover change detection in a coastal zone of Mexico using post-classification comparison. In single band analysis, selective PCA even outperformed image differencing due to its capacity to more efficiently remove inter-image variability left-over after radiometric normalization. Li and Yeh [92] indicated a clear superiority of PCA over post-classification comparison in urban expansion change detection.

Collins and Woodcock [24] have compared three multitemporal linear data transformation algorithms: PCA, tasselled cap, and Gramm-Schmidt orthogonalization. Tasselled cap and PCA both gave better results than the Gramm-Schmidt technique. However, the authors recommended the Kauth-Thomas approach because it identified change in a more consistent and interpretable manner.

4.2 Trend analysis

Trend analysis requires the comparison of temporal development curves, also called time trajectories or time profiles, of different relevant indicators, and this for successive growing seasons or years. The inherent high temporal frequency in data acquisition not only expedites the detection of ecosystem modifications, but also greatly facilitates the characterization of phenological variations in ecosystem status. When the time trajectory of one or several remotely-sensed indicators for a particular pixel departs from the "normal" (or average, or optimal, depending on the objectives of the study), a seasonal or inter-annual change event or process is detected [73]. In order to bring this about, several investigators computed simple anomalies in time profiles of vegetation indices. For example, Myneni et al. [16] and Plisnier et al. [93] calculated a change parameter as the vegetation index value for a given month minus the average value for the time series, divided by the standard deviation. The use of signal processing techniques such as Fourier analysis to examine frequency distributions of high-temporal-resolution AVHRR data has also been documented. The algorithm showed definite capabilities in determining seasonal and sub-seasonal variability in Brazilian Amazon forest cover [94]. Standardized PCA of regional to continental scale time series from wide-angle sensors also proved a powerful technique to separate changes that were taking place at different time frequencies, e.g., decadal time-scale changes in productivity, seasonal changes, sensor-related value drifts, and vegetation-index-value variations related to the El Nino Southern Oscillation or ENSO phenomenon [95, 96].

Trend analysis methods have been implemented with different wide-field-of-view high-temporal-resolution sensors (AVHRR, SeaWIFS, VEGETATION), and different indicators (vegetation indices, surface temperature, spatial structure data). With the availability of still relative short time series (ten to a couple of years), mostly climate-driven fluctuations in surface conditions of natural ecosystems and natural disaster (drought, flood, fire) impacts have been detected [17, 97]. In addition to allowing an unambiguous detection of many abrupt changes in different ecosystems, trend analysis has proven sensitive to more subtle alterations in primary productivity, vegetation phenology, ecosystem dynamics and seasonality, often more so than classical bi-temporal approaches. Even in a multi-stage context, the latter techniques may suffer from an obvious under-sampling at the time scale. This phenomenon has proven especially problematic when dealing with abrupt and often brief ecological events such as fire, flooding, dry spells and vegetation stress. However, given the typical coarse spatial resolution and the large area coverage of trend analysis imagery, validation with independent data sets is a major challenge for natural ecosystem monitoring.

When cross-referencing trend analysis and bi-temporal change detection methods, Borak et al. [98] found statistically significant relationships between AVHRR time-profile-based change metrics and classical change indicators computed from Landsat and SPOT imagery for several African study sites.

5 New developments

The focus of the above review has been on operational data enhancement and data analysis procedures for natural ecosystem change detection with digital imagery, especially satellite-acquired. There are, however, a number of recent and expected advancements that will increase the accuracy and effectiveness of change detection with digital satellite data. Further improvements in hard- (sensor and computer) and software systems, in ecosystem management models, and in change detection algorithms and models, can all be expected to improve the capability of satellite remote sensing for natural ecosystem monitoring.

Since the mid 1990's we have witnessed the development and launch of an unprecedented number of satellite sensing systems. It now appears that by 2005 there will be 15-20 commercial and government satellites acquiring data at spatial resolutions of one to 30 meters. In addition to the technical improvements, the 1992 U.S. Land Remote Sensing Act now also provides for the distribution of the data at the cost of reproduction. Access to space-based digital information has thus become much better and continues to do so.

Although geographic information systems (GIS) are not a new development, it is only in the last decennium that they have gained widespread acceptance as a practical tool for natural ecosystem management. The incorporation of GIS technology in digital change detection methods enables the delivery of disturbance maps, derived from any descriptive change model, in a timely fashion at scales that are consistent with ecosystem management objectives. Today, most image processing systems are integrated with, or at least compatible with, GIS systems, and classifications of remotely sensed data are commonly viewed as inputs to geographic information systems. At the same time increasing attention is being given to using GIS data layers as ancillary inputs to classification of remote sensing data. Further improvements in image analysis and display systems, and in the integration of graphical user interfaces, data base management systems, statistical analysis (including spatial statistics) and process modelling subroutines, along with advanced GIS and image analysis functions, are to be expected.

Artificial intelligence or knowledge-based expert systems offer further opportunities. They provide a way to integrate other features of vegetative cover categories besides spectral change information, thereby overcoming some of the limitations of the traditional statistical classifiers. Such change category recognition methods make use of existing or prior knowledge of the scene content (e.g. original ecosystem status, location, size, relationship with other cover types, shape, etc.) to guide and assist the classification, which follows spatial reasoning lines. As such, they parallel the interpretative procedures employed by a photo-interpreter much more closely. The methods assume that there are similar and identifiable characteristics that each cover type possesses. Although artificial intelligence approaches to natural ecosystem change detection have largely remained in a conceptual design stage [99], researchers developing ecological models have started incorporating inputs from remote sensing and GIS techniques to analyse spatial patterns and processes [100, 101]. Land cover and change determined from remotely sensed imagery are integral components of such models.

In recent years, the use of machine learning algorithms, among which artificial neural networks and decision tree classifiers, has gained considerable attention as an alternative to conventional approaches such as the maximum likelihood classification [102, 103]. Increased classification accuracy is often cited as the primary reason for developing and applying these techniques. However, machine-learning algorithms can also be computationally very complex and require a considerable number of training samples. Documented implementation of machine learning algorithms in the change detection environment is rather scarce and of recent dates, including the use of multi-layer perceptrons [104, 105], of learning vector quantization [106], and of decision tree classifiers such as, for example, Quinlan's C5.0 [106]. The first study also provided evidence that the neural network approach made use of the same spectral signals and scene characteristics as bi-temporal linear data transformation algorithms, e.g., the Gramm-Schmidt orthogonalization. This suggests that the non-linearities in the relationship between the spectral inputs and, in this case forest mortality patterns as the change event, accounted for the improved results using the neural network (with back-propagation training). While perceptron procedures were reported as the most difficult to replicate, tree classifiers were the easiest to use, and learning vector quantization the best performer where it concerned change detection accuracy [106]. Although the implicit use of the time dependency of the data sets represents a major advantage of machine learning algorithms in bi-temporal change analysis, it is not clear yet if the benefits of the higher accuracy outweigh the cost of the additional training data.

Lastly, we can look forward to the development of improved and new change detection algorithms and models. Bruzzone and Serpico conceptualised the unsupervised SMI or "selective use of multispectral information" algorithm [107] to reduce the effect of bi-temporal registration noise, and a supervised non-parametric iterative technique based on the "compound classification rule" for minimum error [108] to improve change detection accuracy. To provide a richer information base on class membership and its dynamics, Foody and Boyd [109] suggested running the post-classification comparison algorithm with fuzzy classifications instead of with conventional "hard" ones. The effect was especially positive when the ecosystem changes were operating at a scale finer than the spatial resolution of the sensor. Morisette et al. [110] explored the use of generalized linear models or GLM's for enhancing standard methods of satellite-based land-cover change detection. GLM's were shown to be helpful in examining different change metrics and useful by applying the resulting model throughout the image to get a probability of change estimate as well as pixel-specific estimates of the variability of change estimate. Smits and Myers [111] investigated a method that can be used to characterize and understand the spatial behavior of change by decomposing the change intensity image into a tree of entities called echelons. They indicated that such a tree could be extremely helpful in discovering connections between changes. Just recently, Hazel [112] advocated the use of object-level change detection (OLCD) instead of the traditional pixel-based algorithms, and Yamamoto and Hanaizumi [113] proposed three-dimensional segmentation for temporal change detection.

6 Conclusions

The data-gathering capabilities of space-borne remote sensors have generated great enthusiasm over the prospect of establishing remote sensing-based systems for the continuous monitoring of natural ecosystems. Although Aldrich's prediction in 1975 of the accuracy of satellite remote sensing for monitoring forest change was not quickly or easily achieved, today it is well established that remote sensing imagery, particularly digital data, can be used to monitor and map changes in natural ecosystems. Recent results (e.g., [13, 66]) have demonstrated that it is feasible to develop automated forest cover monitoring methodologies. When the inherent limitations of digital approaches are appropriately dealt with, pre-processing is adequately incorporated, and optimal change detection algorithms are selected, then Aldrich's prediction can be met.

Although all of the possible change detection methods have not been applied to the same data for evaluation, consideration of over a 100 change detection studies leads to the following conclusions:

1. Vegetation indices are more strongly related to changes in the scene than the responses of single bands.
2. Precise registration of multi-date imagery is a critical prerequisite of accurate change detection. However, residual mis-registration at the

below-pixel level somewhat degrades areal assessment of change events at the change/no-change boundaries.

3. Some form of image matching or radiometric calibration is recommended to eliminate exogenous differences, for example due to differing atmospheric conditions, between image acquisitions. The goal should be that following image rectification, all images should appear as if they were acquired with the same sensor, while observing through the atmospheric and illumination conditions of the reference image.
4. Image differencing and linear transformations appear to perform generally better than other bi-temporal change detection methods. Differences among the different change maps and their accuracies are undoubtedly related to the complexity and variability in the spatial patterns and spectral-radiometric responses of natural ecosystems, as well as to the specific attributes of the methods used.
5. Patterns of seasonal and interannual variations in land surface attributes, which can be driven by climatic variability (e.g., ENSO), natural disasters (e.g., fires, floods), land use changes (e.g., deforestation) or global climate change (e.g., climate warming), can be detected using high temporal frequency data from wide-field-of-view sensors, provided that great care has been taken to remove sensor-related artifacts in time series and that appropriate profile-based change detection method is applied. This research area is still in its infancy compared to the more classic bi-temporal change detection techniques used with medium to fine spatial resolution remote sensing data.
6. The capability of using remote sensing imagery for change detection will be enhanced by improvements in satellite data that will become available over the next several years, and by the integration of remote sensing and GIS techniques, along with the use of supporting methods such as expert systems and ecosystem simulation models.

Analysis of the literature provides ample evidence to support the conclusion that multi-date satellite imagery can be effectively used to detect and monitor changes in natural ecosystems. At the same time we agree with the observation of Collins & Woodcock [24] that one of the challenges confronting the remote sensing research community is to develop an improved understanding of the change detection process on which to build an understanding of how to match applications and change detection methods.

References

1. Edwards, G. Image segmentation, cartographic information and knowledge-based reasoning: getting the mixture right. *Proc. IGARSS'90 Sym.*, University of Maryland, College Park, Maryland (1990) pp. 1641-1644.
2. Coppin, P.R. and Bauer, M.E., Digital change detection in forest ecosystems with remote sensing imagery. *Remote Sensing Reviews* **13** (1996) pp. 207-234.
3. Milne, A.K. Change direction analysis using Landsat imagery: a review of methodology. *Proc. IGARSS'88 Symp. (ESA SP-284)*, Edinburgh, Scotland, (1988) pp. 541-544.
4. Lund, H.G. Change: now you see it - now you don't! *Proc. Int. Conf. on Renewable Resource Inventories for Monitoring Changes and Trends*, Oregon State University, Corvallis (1983) pp. 211-213.
5. Colwell, J.E., Davis, G. and Thomson, F. *Detection and measurement of changes in the production and quality of renewable resources*. USDA Forest Service Final Report (1980) No. 145300-4-F, ERIM, Ann Arbor, Michigan.
6. Hobbs, R.J. Remote sensing of spatial and temporal dynamics of vegetation. In: *Remote Sensing of Biosphere Functioning* (R.J. Hobbs and H.A. Mooney, eds). New York: Springer Verlag, (1990) pp. 203-219.
7. Khorram S., Biging, G.S., Chrisman, N.R., Colby, D.R., Congalton, R.G., Dobson, J.E., Ferguson, R.L., Jensen, J.R. and Mace, T.H. *Accuracy Assessment of Land Cover Change Detection.* Report 101, Computer Graphics Center, Raleigh, North Carolina (1994).
8. Hame, T.H. Interpretation of forest changes from satellite scanner imagery. In: *Satellite Imageries for Forest Inventory and Monitoring; Experiences, Methods, Perspectives, Research Notes No. 21*, Department of Forest Mensuration and Management, University of Helsinki, Helsinki, Finland (1988) pp. 31-42.
9. Burns, G.S. and Joyce, A.T. Evaluation of land cover change detection techniques using Landsat MSS data. *Proc. 7th Pecora Symp.*, Sioux Falls, South Dakota (1981) pp. 252-260.
10. Aldrich, R.C. Detecting disturbances in a forest environment. *Photogramm. Eng. Remote Sens.* **41** (1975) pp. 39-48.
11. Gregory, M.S., Walsh, S.J. and Vitek, J.D. Mechanics of monitoring forest clear cuts and their regeneration. *Proc. 7th Int. Symp. on Machine Processing of Remotely Sensed Data with Special Emphasis on Range, Forest, and Wetlands Assessment*, Purdue University, West Lafayette, Indiana (1981) pp. 520-527.
12. Park, A.B., Houghton, R.A., Hicks, G.M. and Peterson, C.J. Multitemporal change detection techniques for the identification and monitoring of forest

disturbances. *Proc. 17th Int. Symp. on Remote Sensing of Environment*, Ann Arbor, Michigan, (1983) pp. 77-97.

13. Coppin, P.R. and Bauer, M.E., The potential contribution of pixel-based canopy change information to stand-based forest management in the northern U.S. *J. Environ. Management* **44** (1995) pp. 69-82.
14. Lambin, E.F. and Ehrlich, D., Land-cover changes in sub-Saharan Africa (1982-1991): Application of a change index based on remotely sensed surface temperature and vegetation indices at a continental scale. *Remote Sens. Environ.* **61** (1997) pp. 181-200.
15. Kawabata, A., Ichii, K. and Yamaguchi, Y. Global monitoring of interannual changes in vegetation activities using NDVI and its relationships to temperature and precipitation. *Int. J. Remote Sensing* **22** (2001) pp. 1377-1382.
16. Myneni, R.B., Keeling, C.D., Tucker, C.J., Asrar, G. and R.R. Nemani. Increased plant growth in the northern high latitudes from 1981 to 1991. *Nature* **386** (1997) pp. 698-702.
17. Behrenfeld, M.J., Randerson, J.T., McClain, C.R., Feldman, G.C., Los, S.O., Tucker, C.J., Falowski, P.G., Field, C.B., Frouin, R., Esaiais, W.E., Kolber, D.D. and Pollack, H., Biospheric Primary Production during an ENSO transition. *Science* **291** (2001) pp. 2594-2597.
18. Lawrence, R.L. and Ripple W.J. Calculating change curves for multitemporal satellite imagery: Mount St. Helens 1980-1995. *Remote Sens. Environ.* **67** (1999) pp. 309-319.
19. Wallace, J.F. and Campbell, H. Analysis of remotely sensed data. In: *Remote Sensing of Biosphere Functioning* (R.J. Hobbs and H.A. Mooney, eds). New York: Springer Verlag (1989) pp. 297-304.
20. Townshend, J.R.G., Justice, C.O., Gurney, C. and McManus, J. The impact of misregistration on change detection. *IEEE Trans. Geosci. Remote Sens.,* **30** (1992) pp. 1054-1060.
21. Coppin, P.R. and Bauer, M.E., Processing of multitemporal Landsat TM imagery to optimize extraction of forest cover change features. *IEEE Trans. Geosci. Remote Sens.* **32** (1994) pp. 918-927.
22. Duggin, M.J. and Robinove, C.J. Assumptions implicit in remote sensing data acquisition and analysis. *Int. J. Remote Sens* **11** (1990) pp. 1669-1694.
23. Robinove, C.J. Computation with physical values from Landsat digital data. *Photogramm. Eng. Remote Sens.* **48** (1982) pp. 781-784.
24. Collins, J.B. and Woodcock, C.E. An assessment of several linear change detection techniques for mapping forest mortality using multitemporal Landsat TM data. *Remote Sens. Environ.* **56** (1996) pp. 66-77.
25. Songh, C., Woodcock, C.E., Seto, K.C., Lenney M.P. and Macomber, S.A., Classification and change detection using Landsat TM data: when and how to correct atmospheric effects? *Remote Sens. Environ.* **75** (2001) pp. 230-244.
26. Jensen, J.R. Urban change detection mapping using Landsat digital data. *The American Cartographer,* **81** (1983) pp. 127-147.

27. Hall, F.G., Strebel D.E., Nickeson, J.E, and Goetz, S.J. Radiometric rectification: Toward a common radiometric response among multi-date, multi-sensor images. *Remote Sens. Environ.* **35** (1991a) pp. 11-27.
28. Caselles, V. and M.J. Lopez Garcia. An alternative simple approach to estimate atmospheric correction in multitemporal studies. *Int. J. Remote Sens.* **10** (1989) pp. 1127-1134.
29. Conel, J.E. Determination of surface reflectance and estimates of atmospheric optical depth and single scattering albedo from Landsat Thematic Mapper data. *Int. J. Remote Sens.* **11** (1990) pp. 783-828.
30. Coppin, P., Nackaerts, K, Queen, L and Brewer K., Operational monitoring of green biomass change for forest management. *Photogramm. Eng. Remote Sens.* **67** (2001) pp. 603-611
31. Elvidge, C.D., Yuan, D., Rifgeway, D.W. and Lunnetta, R.S., Relative radiometric normalization of Landsat multispectral scanner (MSS) data using automatic scattergram-controlled regression. *Photogramm. Eng. Remote Sens.* **61** (1995) pp. 1255-1260
32. Ahern, F.J., R.J. Brown, J. Cihlar, R. Gauthier, J. Murphy, R.A. Neville, and P.M. Teillet. Radiometric correction of visible and infrared remote sensing data at the Canada Center for Remote Sensing. In: *Remote Sensing Yearbook 1988/89* (A. Cracknell and L. Hayes, eds.). London: Taylor and Francis, (1988) pp. 101-127.
33. Chavez, P.S. Radiometric calibration of Landsat Thematic Mapper multispectral images. *Photogramm. Eng. Remote Sens.,* **55** (1989) pp.1285-1294.
34. Hill, J. and B. Sturm. Radiometric correction of multitemporal Thematic Mapper data for use in agricultural land cover classification and vegetation monitoring. *Int. J. Remote Sens.* **12** (1991) pp. 1471-1491.
35. Riordan, C.J. Change detection for resource inventories using digital remote sensing data. *Proc. Workshop on In-Place Resource Inventories: Principles and Practices*, University of Maine, Orono, Maine (1981) pp. 278-283.
36. Singh, A. Digital change detection techniques using remotely-sensed data. *Int. J. Remote Sens.*, **10** (1989) pp. 989-1003.
37. Baraldi, A. and Parmiggiani, F. Classification methods on multispectral SPOT images. *Proc. IGARSS'89 Symp*, Vancouver, Canada (1989) pp. 1203-12078.
38. Bruzzone L. and Prieto D.F., An adaptive parcel-based technique for unsupervised change detection. *Int. J. Remote Sensing* **21** (2000) pp. 817-822.
39. Kwarteng, A.Y. and Chavez, P.S., Change detection study of Kuweit City and environs using multi-temporal Landsat Thematic Mapper data. *Int. J. Remote Sensing* **19** (1998) pp. 1651-1662.
40. He, D.C. and Wang, L. Texture feature extraction from texture spectrum. *Proc. IGARSS'90 Symp.*, Washington, D.C. (1990) pp. 1987-1990.

41. Reed, B.C. Using textural measures to distinguish spectrally similar vegetation. *Techn. Pap. ACSM-ASPRS'88 Conv. Vol.4*, St. Louis, Missouri, (1988) pp. 40-49.
42. Smits, P.C. and Annoni, A. Towards specification-driven change detection. *IEEE Trans. Geosci. Remote Sens.* **38** (2000) pp. 1484-1488.
43. Gutman, G.G. On the use of long-term global data of land reflectances and vegetation indices derived from the advanced very high resolution radiometer. *J. Geophys. Res.* **104** (1999) pp.6241-6255.
44. Cihlar, J., Chen, J.M., Li, Z., Huang, F., Latifovic, R. and Dixon, R. Can interannual land surface signal be discerned in composite AVHRR data? *J. Geophys. Res.* **103** (1998) pp.23,163-23,172.
45. Colwell, J.E. and Weber, F.P. Forest change detection. *Proc. 15th Int. Symp. on Remote Sensing of Environment*, Ann Arbor, Michigan (1981) pp. 839-852.
46. Malila, W.A. Change vector analysis: an approach detecting forest changes with Landsat. *Proc. 6th Int. Symp. on Machine Processing of Remotely Sensed Data*, Purdue University, West Lafayette, Indiana (1980) pp. 326-335.
47. Pilon, P.G., Howarth, P.J. and Adeniyi, P.O. Improving the detection of human- induced change in West Africa's semi-arid zone using multitemporal Landsat MSS imagery. *Proc. 21st Int. Symp. on Remote Sensing of Environment*, Ann Arbor, Michigan, (1987) pp. 797-804.
48. Nelson, R.F. Detecting forest canopy change due to insect activity using Landsat MSS. *Photogramm. Eng. Remote Sens.* **49** (1983) pp. 1303-1314.
49. Howarth, P.J. and Wickware, G.M. Procedures for change detection using Landsat. *Int. J. Remote Sens.* **2** (1981) pp. 277-291.
50. Hall, F.G., Botkin, D.B., Strebel, D.E., Woods, K.D., and Goetz, S.J. Large-scale patterns of forest succession as determined by remote sensing. *Ecology* **72** (1991b) pp. 628-640.
51. Xu, H. and Young, J.A. Monitoring changes in land use through integration of remote sensing and GIS. *Proc.IGARSS'90 Symp.*, University of Maryland, College Park, Maryland (1990) pp. 957-960.
52. Schowengerdt, R.A. *Techniques for Image Processing and Classification in Remote Sensing.* New York: Academic Press (1983) p. 199.
53. Hall, R.J., Crown, P.H. and Titus, S.J. Change detection methodology for aspen defoliation with Landsat MSS digital data. *Can. J. Remote Sens.* **10** (1984) pp. 135-142.
54. Sader, S.A. Remote sensing investigations of forest biomass and change detection in tropical regions. In: *Satellite Imageries for Forest Inventory and Monitoring; Experiences, Methods, Perspectives,* Research Notes No. 21, Department of Forest Mensuration and Management, University of Helsinki, Helsinki, Finland (1988) pp. 31-42.
55. Banner, A. and Lynham, T. Multi temporal analysis of Landsat data for forest cutover mapping - a trial of two procedures. *Proc. 7th Can. Symp. on Remote Sensing*, Winnipeg, Canada (1981) pp. 233-240.

56. Lyon, J., Yuan, D., Lunetta, R. and Elvidge C. A change detection experiment using vegetation indices. *Photogramm. Eng. Remote Sens.* **64** (1998) pp. 143-150.
57. Mukai, Y., Sugimura, T., Watanabe, H. and Wakamori, K. Extraction of areas infested by pine bark beetle using Landsat MSS data. *Photogramm. Eng. Remote Sens.* **53** (1987) pp. 77-81.
58. Hame, T.H. Satellite image aided change detection. In: *Remote Sensing-Aided Forest Inventory, Research Notes No. 19*, Department of Forest Mensuration and Management, University of Helsinki, Helsinki, Finland (1986) pp. 47-60.
59. Yasuoka, Y., Detection of land-cover changes from remotely sensed images using spectral signature similarity. *Proceedings of the 9th Asian Conference on Remote sensing,* November 1988, pp. G-3-1-G-3-6.
60. Fung, T. An assessment of TM imagery for land cover change detection. *IEEE Trans. Geosci. Remote Sens.* **28** (1990) pp. 681-684.
61. Serneels, S., Said, M. and Lambin, E.F. Land-cover changes around a major East African wildlife reserve: the Mara ecosystem, *Int. J. Remote Sens.* (2001) *in press.*
62. Cohen, W.N., Fiorella, M., Gray, J., Helmer, E. and Anderson K., An efficient and accurate method for mapping forest clear cuts in the Pacific Northwest using Landsat imagery. *Photogramm. Eng. Remote Sens.* **64** (1998) pp. 293-300.
63. Crist, E.P. and Cicone, R.C. Application of the tasseled cap concept to simulated Thematic Mapper data. *Photogramm. Eng. Remote Sens.* **50** (1984) pp. 343-352.
64. Nielsen, A., Conradsen K and Simpson J., Multivariate alteration detection (MAD) and MAF post processing in multispectral bitemporal image data: new approaches to change detection studies. *Remote sens. Environ.* **64** (1998) pp. 1-19.
65. Fung, T. and LeDrew, E. Application of principal components analysis to change detection. *Photogramm. Eng. Remote Sens.* **53 (**1987) pp. 1649-1658.
66. Collins, J.B. and Woodcock, C.E. Change detection using the Gramm-Schmidt transformation applied to mapping forest mortality. *Remote Sens. Environ.* **50** (1994) pp. 267-279.
67. Richards, J.A. Thematic mapping from multitemporal image data using the principal components transformation. *Remote Sens. Environ.* **16 (**1984) pp. 35-46.
68. Ingebritsen, S.E. and Lyon, R.J. Principal components analysis of multitemporal image pairs. *Int. J. Remote Sens*. **6** (1985) pp. 687-696.
69. Singh, A. and Harrison, A.. Standardized principal components. *Int. J. Remote Sens.* **6** (1985) pp. 883-896.
70. Lee, J.K., Lulla, K.P. and Mausel, P.W. Data structure characterization of multispectral data using principal component and principal factor analysis. *Geocarto* **2** (1989) pp. 43-47.

71. Singh, A. Change detection in the tropical forest environment of North-eastern India using Landsat. *Remote Sensing and Tropical Land Management*. New York: John Wiley and Sons Ltd. (1986) pp. 237-254.
72. Yokota, T. and Matsumoto, Y. Detection of seasonal and long-term changes in land cover from multitemporal Landsat MSS data. *Proc. IGARSS'88 Symp. ESA SP-284*, Edinburgh, Scotland (1988) pp. 215-218.
73. Lambin, E.F. and Strahler, A.H. Change-vector analysis in multitemporal space: a tool to detect and categorize land-cover change processes using high temporal-resolution satellite data. *Remote Sens. Environ.* **48** (1994a) pp. 231-244.
74. Johnson, R.D. and Kasischke, E.S., Change vector analysis: a technique for the multispectral monitoring of land cover and condition. *Int. J. Remote Sensing* **19** (1998) pp. 411-426.
75. Vogelmann, J.E. Detection of forest change in the Green Mountains of Vermont using multispectral scanner data. *Int. J. Remote Sens.*, **9** (1988) pp. 1187-1200.
76. Ridd, M.K. and Liu, J., A comparison of four algorithms for change detection in an urban environment. *Remote Sens. Environ.* **63** (1998) pp. 95-100.
77. Adams, J.B., Sabol, D.E., Kapos, V., Almeida Filho, R., Robers, D.A., Smith, M.O. and Gillespie, A.R., Classification of multispectral images based on fractions of end members: applications to land-cover change in the Brazilian Amazon. *Remote Sens. Environ* **12** (1995) pp. 137-154.
78. Roberts, D.A., Batista, G.T., Pereira, J., Waller, E.K. and Nelson, B.W., Change identification using multitemporal spectral mixture analysis: applications in eastern Amazonia. In: *Remote Sensing Change detection: Environmental monitoring Applications and Methods* (C. Elvidge and R. Lunetta, eds.), Ann Arbor Press, Ann Arbor MI, pp. 137-161.
79. Pilon, P.G., Howarth, P.J., Bullock, R.A. and Adeniyi, P.O., An enhanced classification approach to change detection in semi-arid environments. *Photogramm. Eng. Remote Sens.* **54** (1988) pp.1709-1716.
80. Rencz, A.N. Multitemporal analysis of Landsat imagery for monitoring forest cutovers in Nova Scotia. *Can. J. Remote Sens.* **11** (1985) pp. 188-194.
81. Werle, D., Lee, Y.J. and Brown, R.J. The use of multi-spectral and radar remote sensing data for monitoring forest clear cut and regeneration sites on Vancouver Island. *Proc. 10th Canadian. Symp. on Remote Sensing*, Edmonton, Canada (1986) pp. 319-329.
82. Kneppeck, I.D. and Ahern, F.J. Stratification of a regenerating burned forest in Alberta using Thematic Mapper and C-SAR images. *Proc. IGARSS'89 Symp.*, Vancouver, Canada (1989) pp. 1391-1396.
83. Alwashe, M.A. and Bokhari, A.Y.. Monitoring vegetation changes in Al Madinah, Saudi Arabia, using Thematic Mapper data. *Int. J. Remote Sens.* **14** (1993) pp.191 197.

84. Sunar, F., An analysis of changes in a multi-date data set: a case study in the Ikitelli area, Istanbul, Turkey. *Int. J. Remote Sens.* **19** (1998) pp. 225-235.
85. Cohen, W.B., and Fiorella, M., Comparison of methods for detecting conifer forest change with Thematic Mapper imagery. In: *Remote Sensing Change detection: Environmental monitoring Applications and Methods* (C. Elvidge and R. Lunetta, eds.), Ann Arbor Press, Ann Arbor MI, pp. 89-102.
86. Muchoney, D.M. and Haack, B.N. Change detection for monitoring forest defoliation. *Photogramm. Eng. Remote Sens.* **60** (1994) pp. 1243-1251.
87. Michener, W.K. and Houhoulis, P.F., Detection of vegetation changes associated with extensive flooding in a forested ecosystem. *Photogramm. Eng. Remote Sens.* **63** (1997) pp. 1363-1374.
88. Macleod, R.B. and Congalton, R.G. A quantitative comparison of change-detection algorithms for monitoring eelgrass from remotely sensed data. *Photogramm. Eng. Remote Sens.* **64** (1998) pp. 207-216.
89. Yuan, D. and Elvidge, C., NALC land cover change detection pilot study: Washington D.C. area experiments. *Remote Sens. Environ.* **66** (1998) pp. 166-178.
90. Sohl, T.L., Change analysis in the United Arab Emirates: An investigation of techniques. *Photogramm. Eng. Remote Sens.* **65** (1999) pp. 475-484.
91. Mas, J.F., Monitoring land-cover change: A comparison of change detection techniques. *Int. J. Remote Sensing* **20** (1999) pp. 139-152.
92. Li, X. and Yeh A.. Principal component analysis of stacked multi-temporal images for the monitoring of rapid urban expansion in the Pearl River Delta. *Int. J. Remote Sensing* **19** (1998) pp. 1501-1518.
93. Plisnier, P-D., Serneels, S. and Lambin, E.F.. Impact of ENSO on East African ecosystems: Multivariate analysis based on climatological and remote sensing data, *Global Ecol. & Biogeog. Letters* **9** (2000) pp. 481-497.
94. Andres, L., Salas, W.A. and Skole, D. Fourier analysis of multi-temporal AVHRR data applied to a land cover classification. *Int. J. Remote Sens.* **15** (1994) pp. 1115-1121.
95. Eastman, J.R. and Fulk, M. Long sequence time series evaluation using standardized principal components. *Photogramm. Eng. Remote Sens.* **59** (1993) pp. 991-996.
96. Young, S.S. and Wang, C.Y. Land-cover change analysis of China using global-scale Pathfinder AVHRR Landover (PAL) data, 1982-92 (2001).
97. Lupo, F., Reginster, I. and Lambin, E.F. Monitoring land-cover changes in West Africa with Spot VEGETATION: Impact of natural disasters in 1998-1999. *Int. J. Remote Sens., in press.*
98. Borak, J.S., Lambin, E.F. and Strahler, A.H. The use of temporal metrics for land-cover change detection at coarse spatial scales, *Int. J. Remote Sens*ing **21** (2000) pp.1415-1432.

99. McRoberts, R.E., Schmoldt, D.L. and Rauscher, H.M. Enhancing the scientific process with artificial intelligence: forest science applications. *Art. Intelligence Appl.* **5** (1991) pp. 5-26.
100. Ustin, S.L., Smith, M.O., and Adams, J.B. Remote sensing of ecological processes: A strategy for developing and testing ecological models using spectral mixture analysis. In: *Scaling of Physiological Processes: Leaf to Globe* (J.R. Ehleringer and C.B. Field, eds.). San Diego: Academic Press (1993) pp. 339-357.
101. Mladenoff, D.J. and Host, G.E. Ecological Perspective: Current and potential applications of remote sensing and GIS to ecosystem analysis. *Remote Sensing and GIS in Ecosystem Management*, (V.A. Sample, ed.), Washington, D.C.: Island Press (1994) pp. 218-242.
102. Benediktsson, J.A., Swain, P.H., and Ersoy, O.K. Neural network approaches versus statistical methods in classification of multisource remote sensing data. *IEEE Trans. Geosci. Remote Sens.* **28** (1990) pp. 540-552.
103. Bischof, H., Schneider, W., and Pinz, A.J. Multispectral classification of Landsat images using neural networks. *IEEE Trans. Geosci. Remote Sens.* **30** (1992) pp.482-490.
104. Gopal, S. and Woodcock, C.E., Remote sensing of forest change using artificial neural networks. *IEEE Trans. Geosci. Remote Sens.* **34** (1996) pp. 398-404.
105. Dai Long X. and Khorram S., Remotely sensed change detection based on artificial neural networks, *Photogramm. Eng. Remote Sens.* **65** (1999) pp. 1187-1194.
106. Chang Cheung-Wai, J., Chan Kwok-Ping and Yeh Gar-On, A., Detecting the nature of change in an urban environment: a comparison of machine learning algorithms. *Photogramm. Eng. Remote Sens.* **67** (2001) pp. 213-225.
107. Bruzzone, L. and Serpico, S.B., Detection of changes in remotely-sensed images by the selective use of multi-spectral information. *Int. J. Remote Sensing* **18** (1997) pp. 3883-3888.
108. Bruzzone, L and Serpico, S.B., An iterative technique for the detection of land-cover transitions in multitemporal remote-sensing images. *IEEE Trans. Geosci. Remote Sens.* **35** (1997) pp. 858-867.
109. Foody, G.M. and Boyd, D.S., Detection of partial land cover change associated with the migration of inter-class transitional zones. *Int. J. Remote Sens.*, **14** (1999) pp. 2723-2740.
110. Morisette, J.T., Khorram, S. and Mace, T., Land-cover change detection enhanced with generalized linear models. *Int. J. Remote Sensing* **20** (1999) pp. 2701-2721.
111. Smits, P.C. and Myers, W.L., Echelon approach to characterize and understand spatial structures of change in multitemporal remote sensing imagery. *IEEE Trans. Geosci. Remote Sens.* **38** (2000) pp. 2299-2309.

112. Hazel, G.G., Object-level change detection in spectral imagery. *IEEE Trans. Geosci. Remote Sens.* **39** (2001) pp. 553-561.
113. Yamamoto, T. and Hanaizumi, H., A change detection method for remotely sensed multispectral and multitemporal images using 3-D segmentation. *IEEE Trans. Geosci. Remote Sens.*, **39** (2001) pp. 976-985.

AUTOMATED AND PRECISE IMAGE REGISTRATION PROCEDURES

U. WEGMÜLLER, C. WERNER, T. STROZZI, AND A. WIESMANN
Gamma Remote Sensing, Thunstrasse 130, CH-3074 Muri BE, Switzerland
Tel: +41 31 9517005, Fax: +41 31 9517008, e -mail: wegmuller@gamma-rs.ch

In this contribution image registration methods are presented and their performance is discussed based on examples. Automated and precise image registration is feasible in many cases. Key elements of the presented registration techniques are the determination of a large number of registration offsets throughout the images, the calculation of a quality measure for each of these estimates, the derivation of a registration function and its quality, and the resampling of the images to the common reference geometry. The registration accuracies achieved clearly meets the "registration error < 0.2 pixel" requirement of SAR interferometry. Application examples include multi-temporal SAR imagery, multi-sensor SAR imagery, optical imagery, registration between SAR and optical data, and between satellite imagery and other spatial map type data, and between SAR imagery and a simulated data set calculated based on a DEM.

1 Introduction

Multi-temporal analysis is one of the most important techniques of remote sensing. Change detection and monitoring applications are directly depending on multi-temporal observations - for other applications as thematic mapping or the retrieval of bio- and geophysical parameters, multi-temporal data allows to enhance the performance by reducing signal noise through multiple independent observations. The use of multi-sensor data, for example SAR and optical data, is another way to increase the information content. Whatever the motivation for the multi-temporal or multi-sensor analysis is, a very important processing step is the registration of the data to a common geometry.

In this contribution two image registration methods are presented. Both methods were originally developed in the context of SAR interferometry. One of the two methods is much more generally applicable. Interferometry requires SAR image registration at sub-pixel accuracy (< 0.2 pixel) in order not to reduce scene coherence. As control point selection is particularly tedious in SAR images because of the image speckle, there was a clear demand for automated and precise registration algorithms. Key elements of the presented registration techniques are the automated determination of a large number of registration offsets throughout the image, the calculation of a quality measure for each of these estimates, the derivation of a registration function and its quality, and the resampling of the images to the common reference geometry. The two methods differ in the registration offset estimation method used. The "coherence optimization method" calculates small interferograms and optimizes its coherence, the "intensity cross-correlation method" estimates the registration offsets based on the cross-correlation function of corresponding sections of real valued intensity images. The first method requires coherence between at least small sections of the images and is therefore limited to multi-temporal SAR images with a minimum level of coherence. The second method depends on scene contrast and features and is therefore much more widely applicable. The intensity cross-correlation method was found to be very robust for the registration of multi-temporal SAR data of a single sensor. It was also successfully applied to register SAR data from different sensors, optical data, and SAR images with optical images and with landuse inventory data. SAR images were also successfully registered with simulated images calculated based on a digital elevation model, as used to automate terrain corrected SAR geocoding.

2 Concept

The basic concept of the presented image registration procedures consists of 3 main steps. In a first step the geometry used for the registration is selected and all images are brought into this geometry in an initial registration step. In this step it is important to consider the imaging geometry as good as possible; appropriate geometric models for the imaging process are preferred to parametric approaches based on tie points. Repeat-pass SAR images, for example, can be registered in slant range, ground range or map geometry, but a SAR image in slant range geometry should never be registered to an image in ground-range or map geometry. To register images of different sensors a map geometry is usually selected. This means that each image will be geocoded prior to the fine-registration step.

The second step is the derivation of the fine registration function using an automated technique. The steps used in the derivation of this fine registration function are the selection of image chips, the determination of registration offsets for these chips, and a regression analysis on the registration offsets to derive the fine registration function. This step will be discussed in more detail in Section 3.

In a third step the fine registration function is used to resample the data to the common reference geometry. In order to avoid multiple resampling of the data the fine registration function can be used to "refine" the initial registration function, so that the two transformations are conducted in a single resampling step, which allows to minimize problems introduced by the resampling.

3 Registration offset estimation

The key step in the derivation of the fine registration function is the automated estimation of reliable registration offsets. To meet the reliability requirement, quality control is conducted for each single offset estimate. Lower quality estimates have to be rejected and, in addition, a relatively large number of reliable estimates is required to minimize statistical errors. Fine registration functions with few free parameters to be determined in the regression analysis are preferred.

To meet these requirements a large number of image locations are automatically selected for the registration offset estimation. In our implementation the selected m x n locations are on a regular grid. For each selected location a registration offset is estimated and for each estimate the quality is determined. The two main methods used for the registration offset determination are the intensity cross-correlation method and the coherence optimization method which are both described in more detail below. The estimated offsets and the corresponding quality measures are the input to the regression analysis. Estimates with a quality measure above a specified threshold, are used to determine the parametric fine registration function using a least squares technique. In the case of images which are already in the same image geometry bi-linear functions are usually sufficient, but in principle more complicated functions could also be used. In addition to the registration function the standard deviation of the estimates from this function is calculated and used for quality control and to iteratively reject estimates which deviate more than a specified multiple (for example 3 times) of the standard deviation from the registration function.

Very often more than one hundred estimates with a standard deviation below half a pixel are used in the final determination of the registration function, resulting in a very small statistical error of the registration function of less than a tenth of a pixel.

3.1 Intensity cross-correlation method

For the determination of an individual registration offset a small section of one real valued image is selected and cross-correlated with the other real valued image. The maximum correlation found, or more precisely the maximum of the two dimensional correlation function determined from the correlation values in the proximity of the correlation maximum, determines the offset estimate. The width of the correlation function determines the quality, with narrow correlation functions corresponding to high quality and wide correlation functions to low quality. The intensity cross-correlation method depends on sufficient image contrast. The size of the image sections used for the intensity cross-correlation method should be larger than the typical dimension of features in the image. In practice we typically use image chips of 32 x 32 pixels to 256 x 256 pixels.

3.2 Coherence optimization method

The coherence optimization method is limited to "interferometric SAR data". With interferometric SAR data we mean SAR acquisitions with a significant level of coherence. As in the intensity cross-correlation methods small section of one image is selected. But instead of optimizing the intensity cross-correlation with the other image the degree of coherence of small interferograms is optimized. The maximum coherence found, or more precisely the maximum of the two dimensional coherence function determined from the coherence values in the proximity of the coherence maximum, determines the offset estimate. The width of this function determines the quality of the estimate, with narrow functions corresponding to high quality and wide functions to low quality. The coherence optimization method depends on sufficient coherence at least for parts of the images. The size of the image sections used for the coherence optimization method can be small. In practice, we use image chips of 4 x 4 pixels to 16 x 16 pixels. In order to be able to calculate the small interferograms and determine its average coherence level single look complex (SLC) data are required.

3.3 Optimization

The presented registration methods offer some flexibility for optimization for a specific example. Several parameters, as for example the size of the image chips used to determine the individual offset estimates, quality thresholds to reject estimates as unreliable, the number of locations included in the analysis, may be adapted to the specific case. To achieve a high quality is only one criteria considered in such an optimization. The robustness of the methodology and its efficiency are also very important. Typically, the objective is not to achieve the highest possible quality but to reliably achieve a high enough quality in a short time.

The size of the image chips used has a strong effect on the quality of the individual estimate as well as on the computational efficiency. The selected chip size needs to be a minimal size to achieve the required quality. This minimal size depends on the data type and the specific example. A good measure of the accuracy of the individual estimates is found in the standard deviation of these estimates from the parametric registration function. Using larger chips also increases the chance for reliable estimates and therefore the robustness of the method. The number of chips selected needs to be high enough to achieve a sufficiently high number of accepted estimates. The spatial distribution of these estimates is another important factor for the accuracy of the registration.

The computational effort depends not only on the size and number of the image chips used but also on the size of the search region in the second image. This search region needs to be large enough that it contains the section corresponding to the search chip, but it should be as small as possible to reduce the computational effort. A good solution is a guided search. An initial registration function is used to pre-determine an expected location. This initial registration functions is typically determined in a first run with fewer, but larger image chips and search regions. The main run is then conducted for a larger number of locations but with smaller chips and search regions.

With our implementation (Werner et al., [1]), a registration is conducted for both methods without operator interaction in a reasonably short time (~ few minutes) on normal PCs or workstations.

Oversampling of the image chips is a possibility to further increase the accuracy of the individual offset estimates. This is important mainly if the use of the individual estimates goes beyond the determination of a parametric function. An example for such use is the determination of surface deformations based on the deviation of the individual offset from an average registration function (Werner et al., [2], Strozzi et al. [3]).

4 Examples

4.1 SAR interferometry

The registration requirements in SAR interferometry are high. Registration errors larger than about 0.2 pixel cause significant additional phase noise and reduce the coherence level. For this case both presented methods are applicable, allowing to validate their accuracy by cross-comparison of the results. Both methods were successfully applied to thousands of image pairs, very often as part of fully automated processing sequences. Typically, more than 100 individual estimates are determined with a standard deviation from the parametric registration function between 0.05 and 0.3 pixels.

The coherence optimization method is applied with small estimator window sizes. Therefore, the spatial resolution of the estimates is higher. This is only a relevant advantage, though, if there are inhomogeneities with higher spatial frequencies, as caused for example in the case of uncompensated terrain effects or surface displacement as in the case of moving glaciers or earthquakes. The coherence optimization method does not depend on image contrast, so it also works in the case of very homogeneous surfaces. On the other hand it depends on a sufficiently high level of coherence, which strongly reduces the number of accepted estimates for longer acquisition intervals and data with larger spatial baselines. In general we prefer the intensity cross-correlation method for its higher robustness.

4.2 Multi-temporal SAR

The main difference between the multi-temporal SAR case and SAR interferometry is that coherence cannot be expected, in general. Therefore, mainly the intensity cross-correlation method is used, which works again very well for repeat-pass ERS, JERS, and Radarsat data. The accuracy achieved is only slightly below the one in interferometric pairs. Reasons for some reduction of the accuracy are longer acquisition intervals, which lead to more temporal change in the scene) and larger differences in the initial geometries (as the data selection has not to be restricted to small baselines as in the interferometry case), which results in a higher spatial dependence of the registration offsets. Again, many individual estimates with a small standard deviation from the parametric registration function (0.05 and 0.5 pixels) are typically found.

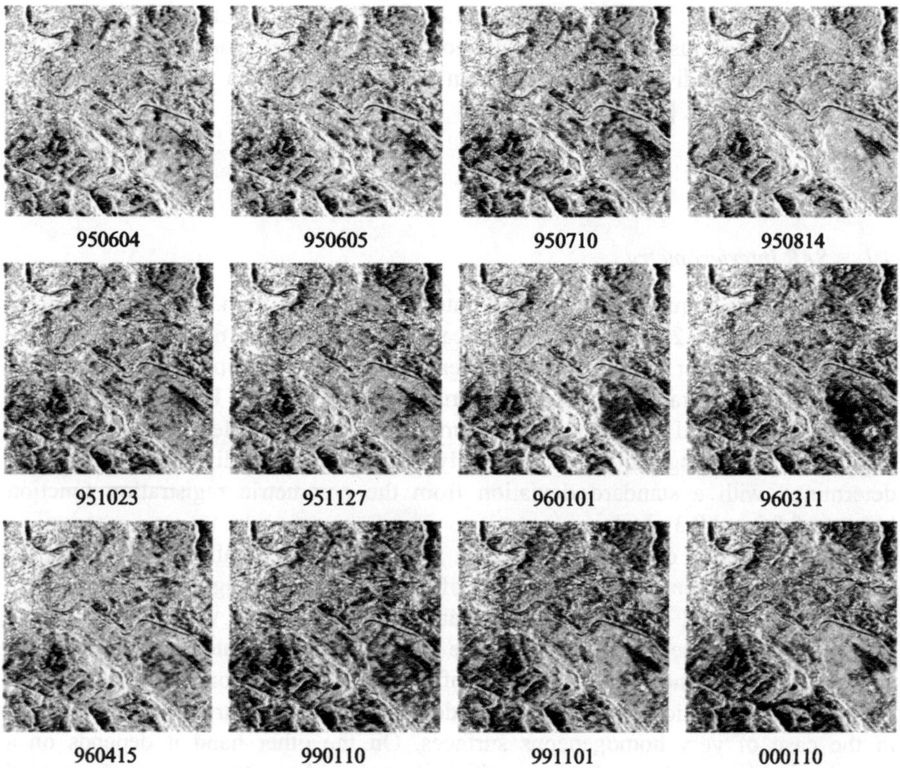

Figure 1 Section (320x320 pixels; approx. 6.4 km x 6.4 km) of the twelve repeat-orbit multi-temporal ERS SAR images over Berne (upper left corner with its "old town" within Aare river loop), which were used for the registration experiment.

To investigate the robustness of the registration of multi-temporal SAR images a small registration experiment was conducted on twelve ERS SAR backscatter images data (5-look intensity images) acquired in repeat-orbits over Berne, Switzerland (see Figure 1). This time series includes pairs with very short (Tandem pair with 1 day difference) to very long (4.5 years) acquisition intervals. The selected images cover different seasons and include one image (960101) with wet snow coverage which significantly changes the backscatter levels and image contrast. The intensity cross-correlation method with an oversampling factor of 2 was used for the image registration.

With a correlator size of 64x64 pixels a consistent high registration accuracy was achieved. From the 1024 cross-correlation analyses resulted 741 to 960 accepted estimates with standard deviations of the range and azimuth offset estimates from the regression functions between 0.041 and 0.123 pixel. This means that each image pair was successfully registered at high accuracy. As expected the standard deviations are the smallest for the Tandem pair, and larger for intervals of 35 days to several years. There is not a uniform time dependence, scene parameters as for example the wet snow cover on 960101 are very important factors. The detailed results are shown in Table 1.

Table 1: Registration of multi-temporal ERS. For 1024 evenly distributed locations the intensity cross-correlation approach with a 64 x 64 pixel window and an oversampling factor of 2 was used to estimate individual registration offsets. The number of valid estimates (first row), the standard deviation of the range offset in pixels (second row) and the standard deviation of the azimuth offset in pixels (third row), both relative to the determined regression function, are indicated. The image pairs are defined by its acquisition dates (indicated in the format yymmdd).

	950605	950710	950814	951023	951127	960101	960311	960415	990510	991101	000110
950604	960 0.041 0.043	950 0.075 0.069	874 0.082 0.074	917 0.085 0.076	923 0.089 0.075	832 0.114 0.099	844 0.084 0.078	914 0.098 0.086	854 0.092 0.084	824 0.097 0.088	784 0.117 0.088
950605		955 0.076 0.074	863 0.085 0.073	911 0.084 0.079	924 0.087 0.078	850 0.121 0.108	830 0.089 0.083	903 0.091 0.087	864 0.092 0.087	808 0.098 0.092	790 0.123 0.086
950710			902 0.085 0.078	897 0.070 0.067	852 0.107 0.091	813 0.091 0.081	792 0.082 0.077	887 0.094 0.080	897 0.104 0.088	773 0.121 0.087	741 0.082 0.074
950814				905 0.078 0.079	891 0.081 0.077	831 0.103 0.091	860 0.080 0.087	913 0.088 0.076	872 0.088 0.082	864 0.086 0.090	822 0.106 0.088
951023					922 0.083 0.076	877 0.100 0.088	895 0.077 0.073	907 0.079 0.072	918 0.088 0.082	859 0.097 0.085	801 0.111 0.090
951127						861 0.096 0.098	897 0.073 0.078	918 0.075 0.073	841 0.097 0.085	864 0.091 0.082	773 0.085 0.082
960101							810 0.091 0.102	869 0.092 0.092	845 0.121 0.098	792 0.121 0.105	782 0.086 0.097
960311								898 0.070 0.075	825 0.092 0.084	831 0.090 0.094	752 0.092 0.098
960415									875 0.094 0.083	844 0.098 0.083	789 0.099 0.085
990510										917 0.079 0.072	880 0.107 0.084
991101											884 0.087 0.078

To evaluate the influence of the correlator window size the experiment was repeated with window sizes of 16x16, 32x32, 64x64, 128x128, and 256x256 pixels. During the registration we kept track of the number of valid offset estimates found and of the standard deviation of the estimated range and azimuth offsets from the regression functions. It was found that the registration always succeeded for this data set, even with the smallest window size. For the small window sizes the acceptance rates were significantly lower. In spite of this or more precisely because of this the standard deviations remained quite low, below 0.2 pixel. For the small window sizes significantly more estimates were accepted for the Tandem data. The reasons are that there is very little temporal change for the Tandem pair and that the speckle are very similar for the Tandem data due to the high degree of coherence. Figure 2 shows the values for the Tandem pair, the average of two 35 day interval pairs, and the average of the 3 pairs with acquisition intervals of several years. The corresponding azimuth offset estimation standard deviations are very similar to the values in range direction. For the largest windows the processing efficiency gets very low. As a good compromise for efficient and robust registration a window size of 64x64 or 128x128 can be recommended.

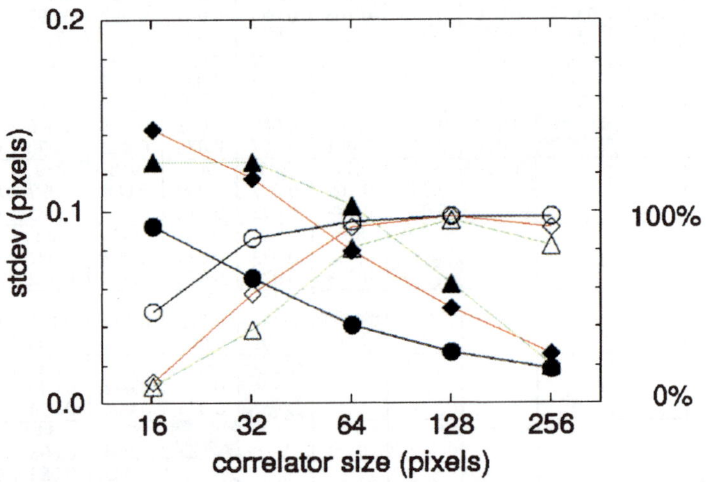

Figure 2 Offset estimate acceptance rate (opaque symbols) and range offset estimation standard deviation (filled symbols) as a function of the estimator window size, for an ERS Tandem pair (circle), 35 day pairs (diamond), and multi-year pairs (triangle) for intensity cross-correlation method applied to 5-look ERS backscatter images.

4.3 Multiple SAR sensor

The registration of SAR images from different sensors differs from the registration of multi-temporal images acquired with just one sensor in the sense that imaging parameters as the frequency, incidence angle, track orientation, and spatial resolution, differ. As mentioned in the concept section the images need to be brought in the same geometry. Typically, a map geometry is selected in this case. Each of the images is first geocoded to the map geometry. Terrain corrected geocoding is, of course, more accurate than just ellipsoid corrected geocoding and is therefore preferred if a digital elevation model (DEM) is available. Once in the same geometry the automated fine registration can be applied.

In such a way a large number of JERS and ERS scenes were successfully registered (Wiesmann et al., [4]). The accuracy achieved depends on the accuracy of the geocoding step as well as on the correspondence of specific features within the images of the two sensors. For the example JERS / ERS everything worked fine, in spite of the very difficult appearance of the images. The standard deviations of the fine registration offsets were usually between 0.5 and 1.0 pixel.

The same technique is also recommended for the registration of images acquired by the same SAR sensor but at different incidence angles, as necessary when working with Radarsat or ENVISAT ASAR images of different modes or in ascending and descending orbits.

4.4 SAR geocoding

The intensity cross-correlation method is also used in the frame of the automated SAR geocoding methodology described by Wegmüller et al., [5]. In this case a simulated SAR backscattering image, calculated based on a DEM and an assumed incidence angle dependence of the backscattering coefficient, is used as the image in the reference geometry. This simulated image does not contain any dependence on the surface type. In spite of this limitation automated fine registration with a real SAR image is feasible as soon as there is some topography in the scene. In this case somewhat larger chips (128 x 128 to 1024 x 1024) are used for the estimation of the individual offsets (Wegmüller et al., [5]).

4.5 Optical and SAR data

The number of optical data registration examples investigated so far with our implementation is much smaller than for the SAR data, simply because of the focus of our work on SAR data. Examples with SPOT and Landsat TM data confirmed nevertheless, that the intensity cross-correlation method is also applicable to the registration of optical data.

To assess the potential more quantitatively we used a data set over Switzerland consisting of a Landsat TM image, channels 2, 3, and 4, an ERS SAR backscatter image, an ERS Tandem coherence image, and a landuse inventory. All the data sets were geocoded to the same map projection with a pixel spacing of 25m and a total size of 2000x2000 pixels. The landuse inventory data is based on the Swiss "Areal Statistik" which has an original pixel spacing of 100m. The data were oversampled to 25m. In addition, the inventory was reduced to the three classes forest and water, urban, and others. In the optical and in the SAR data forests and water appear in dark gray levels. Therefore, a dark gray level was attributed to the forest and water class, an intermediate level to the others class and a bright level to the urban class. To the SAR backscattering logarithmic scaling was applied. To the optical channels and the coherence image linear scaling was applied. The resulting images which were used for the SAR – optical registration analysis are shown in Figure 3.

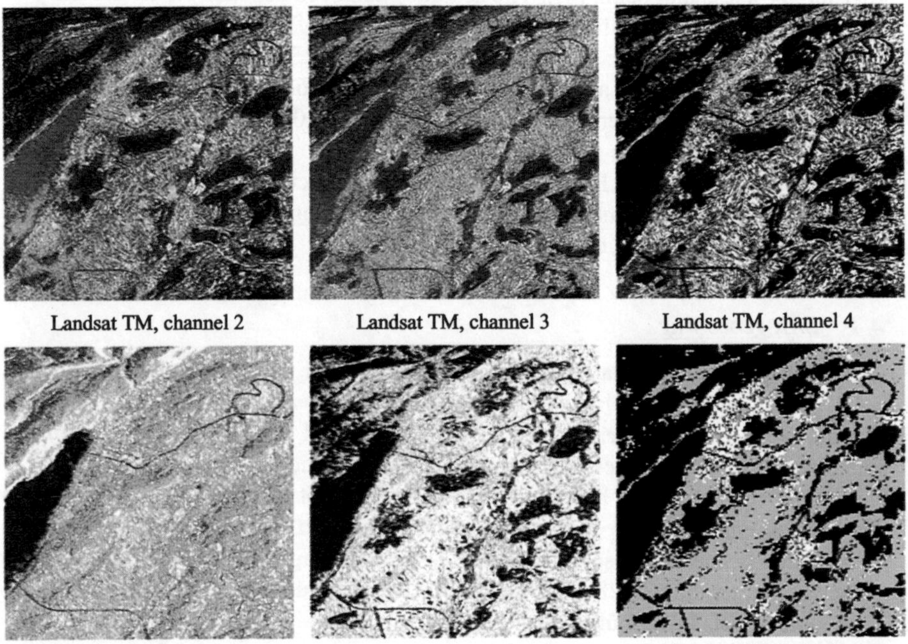

Landsat TM, channel 2 — Landsat TM, channel 3 — Landsat TM, channel 4

ERS SAR backscattering coefficient. — ERS Tandem coherence — Landuse inventory data (dark: forest and water, bright: urban, intermediate: other classes)

Figure 3 Section (640 x 640pixels,16*km* x 16*km*) of geocoded images used for the SAR – optical registration analysis.

Table 2: SAR – optical data registration experiment. Easting (diamonds) and Northing (triangles) offset estimation standard deviations versus estimator window size, for Landsat TM image, channels 2, 3, and 4, an ERS SAR backscatter image, an ERS Tandem coherence image, and a landuse inventory data set.

	ERS σ^0	ERS coherence	Inventory
TM.2	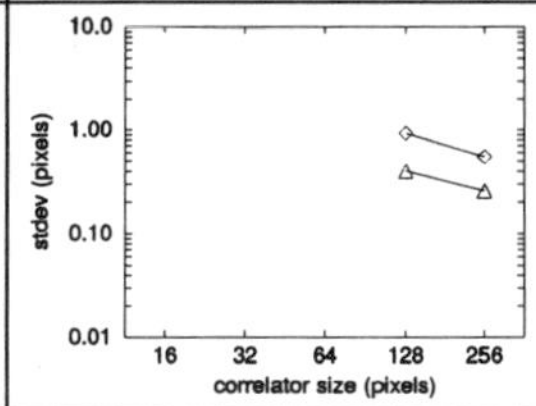	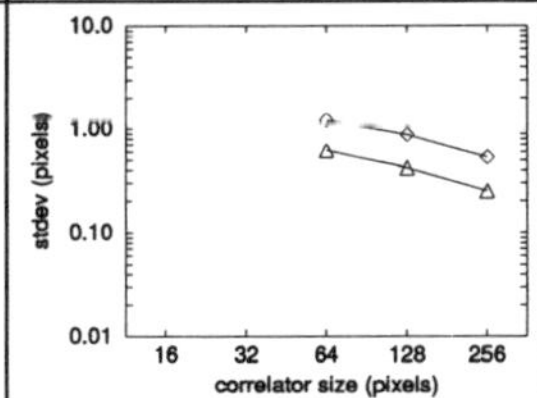	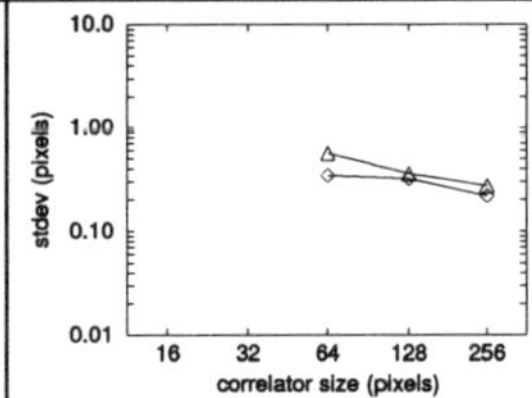
TM.3	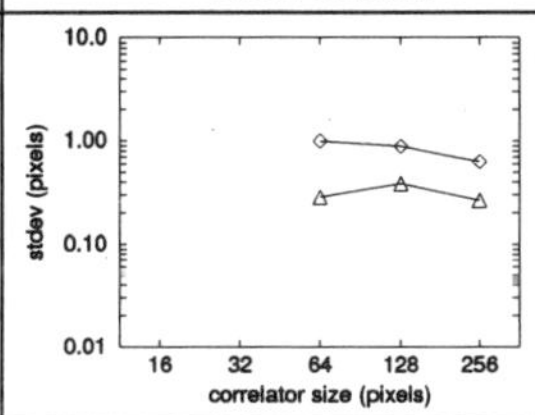	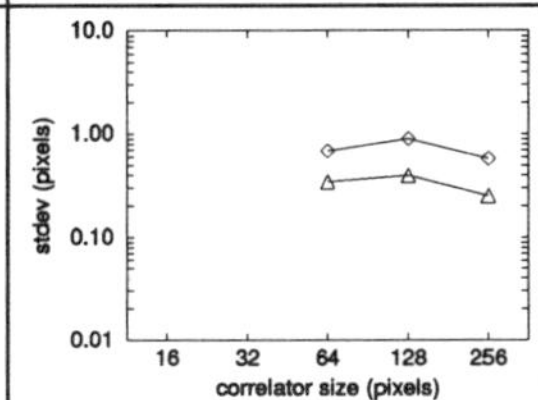	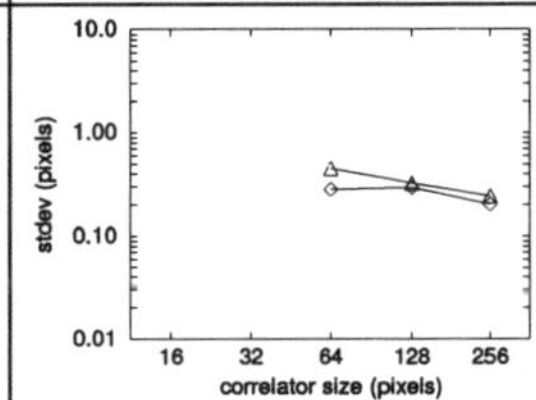
TM.4	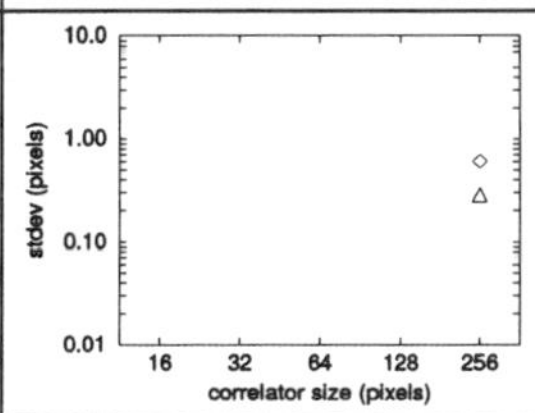	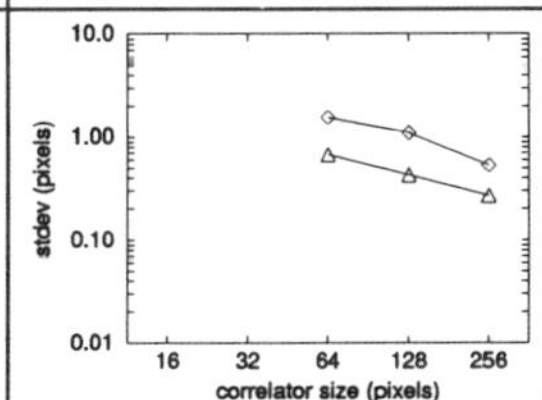	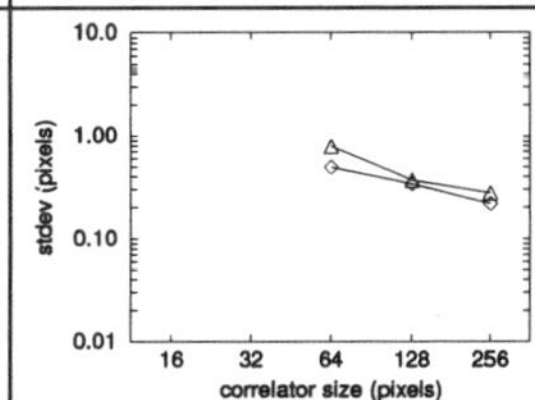
ERS σ^0		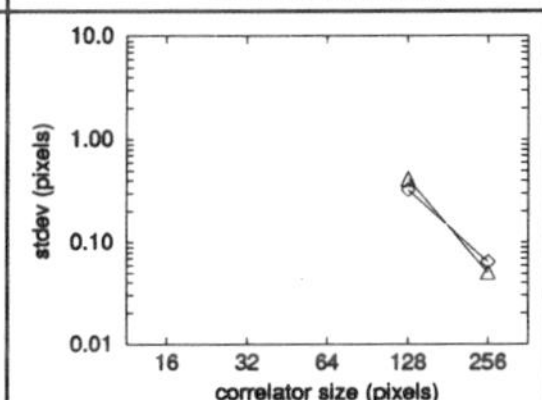	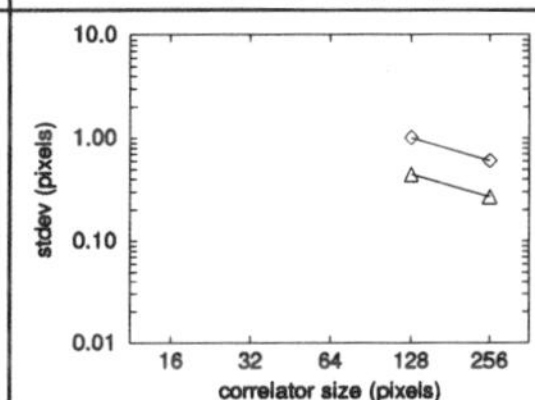
ERS Coher-encc			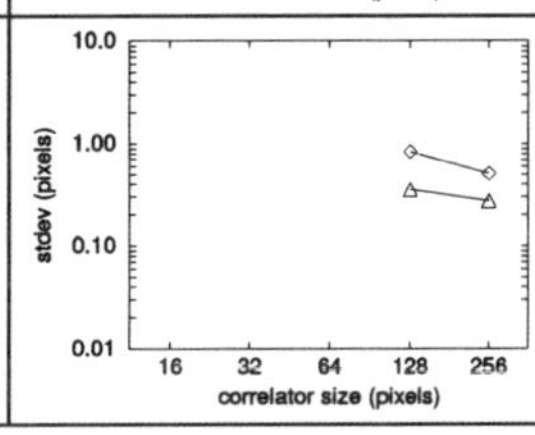

Table 3: SAR – optical data registration experiment, as Table 2, but for Landsat TM image, channels 2, 3, and 4.

TM.2 with TM.3	TM.2 with TM.4	TM.3 with TM.4
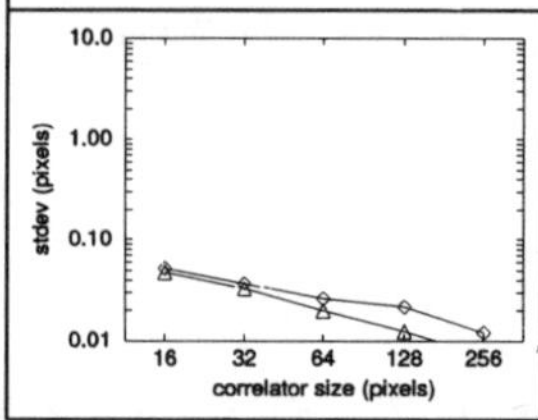	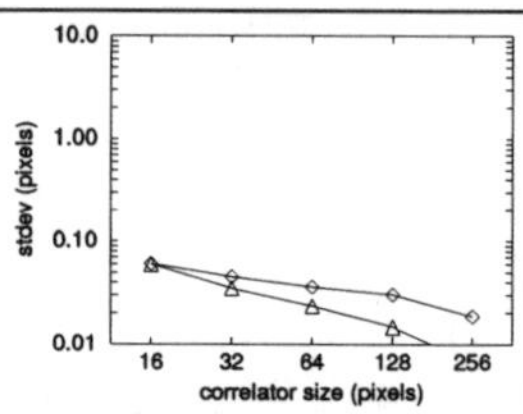	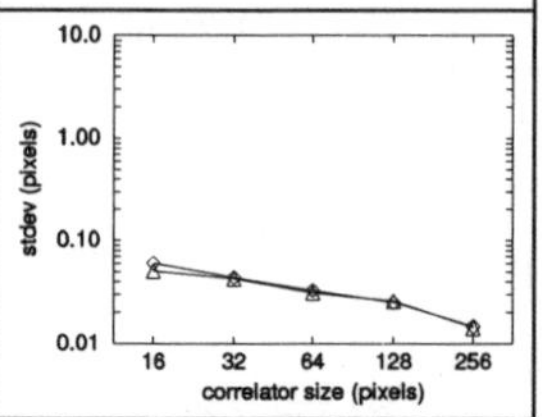

The intensity cross-correlation method with an oversampling factor of 2 was used for the image registration. To evaluate the influence of the correlator window size the registration was repeated with window sizes of 16x16, 32x32, 64x64, 128x128, and 256x256 pixels. The standard deviation of the estimated registration offsets in Easting and Northing from the regression functions are shown in Tables 2 and 3 as a function of the correlator window size. For unsuccessful registration attempts, these are attempts with too few accepted offset estimates and attempts with standard deviations much larger than 1 pixel, no value was plotted for the corresponding correlator window size.

The registration between the different optical channels (Table 3) was successful with very low standard deviations even for the smallest window sizes considered. This is not astonishing as the three optical images are just different channels of the same acquisition. Consequently, this result is not representative for a registration between multi-temporal optical data.

The registration between the Landsat TM channels (Table 2) and the SAR backscattering image was only successful with larger correlator windows. This is not surprising when considering how different the images look. The standard deviations achieved were of the order or 1 pixel in Easting direction and below 0.5 pixel in Northing direction. Our explanation for the higher variability in the Easting direction are related to the special SAR imaging effects at the near and far range forest boundaries (corresponding closely to the eastern and western boundaries). The registration between the Landsat TM channels and the Tandem coherence map was found to be more robust than the registration with the backscatter image. The explanation of this is the quite high contrast between forest and non-forest in the coherence image. The standard deviations were similar as for the backscattering images but based on more accepted estimates. The standard deviation was again higher in Easting direction, for the same reason.

In addition, to satellite imagery the modified landuse inventory data set was considered. Both the optical and the SAR data registered successfully but more robustly with larger estimator windows, as can be expected for this data set based on 100m resolution data. This type of registration can be used to automate the refinement of the geocoding of SAR or optical imagery with map type data sets used as reference.

5 Conclusions

Automated and precise image registration is feasible in many cases. Two specific algorithms were presented. The coherence optimization method is predominantly for application in the context of SAR interferometry. The registration accuracy achieved clearly meets the "registration error < 0.2 pixel" requirement of SAR interferometry. The intensity cross-correlation method results in similarly accurate results but it is more robust and has a very wide applicability. Application examples include multi-temporal SAR imagery, multi-sensor SAR imagery, optical imagery, as well as registration between SAR and optical data, between satellite imagery and spatial data sets as a landuse inventory, and between SAR imagery and a simulated data set calculated based on a DEM.

6 Acknowledgment

ESA is acknowledged for providing ERS data under ERS AO3-175 project (PI U. Wegmüller) and the Swiss Federal Statistical Institute for the digital landuse inventory (Areal Statistik).

7 References

1. Werner C., U. Wegmüller, T. Strozzi, and A. Wiesmann, "*Gamma* SAR and Interferometric Processing Software", Proceedings of ERS-ENVISAT Symposium, Gothenburg, Sweden, 16-20 Oct. 2000.
2. Werner C., T. Strozzi, A. Wiesmann, U. Wegmüller, T. Murry, H. Pritchard, and A. Luckman, "Complementary measurement of geophysical deformation using repeat-pass SAR", Proceedings of IGARSS 2001, Seattle, Australia, 9-13 July 2001.
3. Strozzi T., A. Luckman, T. Murray, U. Wegmüller, and C. Werner, Glacier motion estimation using SAR offset tracking procedures, submitted to IEEE TGRS in 2001.
4. Wiesmann A., T. Strozzi, and U. Wegmüller, JERS SAR and Interferometric Processing for the boreal forest mapping project SIBERIA, Proceedings of IGARSS'99, Hamburg, 28 June - 2 July 1999 (DD01_03).
5. Wegmüller U., Automated terrain corrected SAR geocoding, Proceedings of IGARSS'99, Hamburg, 28 June - 2 July 1999 (CC03_02).

ADAPTIVE REDUCTION OF REGISTRATION-NOISE EFFECTS IN UNSUPERVISED CHANGE DETECTION

LORENZO BRUZZONE AND ROBERTO COSSU

Dept.of Information and Communication Technologies, University of Trento,

Via Sommarive 14, 38050, Povo, Trento, Italy

E-mail: lorenzo.bruzzone@ing.unitn.it

An approach aimed at reducing the effects of registration noise in unsupervised change detection is proposed. This approach is composed of two main phases. In the first phase, the distribution of the registration noise is adaptively estimated on the specific scene of interest. In the second phase, the estimated distribution is used to define a local decision strategy for carrying out change detection by reducing the effects of misregistration. Experimental results obtained on both a synthetic and a real multitemporal data sets confirm the effectiveness of the proposed method.

1 Introduction

In the past few years, there has been a growing interest in the use of multitemporal remote-sensing images for change-detection purposes [1]-[2]. This interest stems from the wide range of applications in which change detection methods can be used (e.g. environmental monitoring, urban studies, forest monitoring). Usually, change-detection techniques involve a pixel-by-pixel comparison of two multispectral remote-sensing images acquired on the same area at different times. For this reason, in order to perform change detection accurately, it is mandatory to apply a preliminary co-registration process to multitemporal images, so that pixels with the same position on the images correspond to the same area on the ground. However, in remote-sensing applications, it is not possible to obtain a perfect alignment between images acquired at different times, even by applying advanced co-registration procedures [3]-[4]. The residual misalignment results in the so-called "registration noise", which significantly affects the accuracy of the change-detection process [3]-[4]. Generally, it is difficult to reduce the effects of the registration noise, because it depends on several complex factors (e.g. the structure of the considered scene, the spatial resolution of the sensor, the stability of the acquisition platform). These factors prevent from formulating an analytical model of the aforementioned noise and, consequently, from defining effective processing algorithms.

In this paper, an approach aimed at reducing the effects of registration noise in unsupervised change detection is proposed. The proposed approach adaptively estimates the distribution of the registration noise in the specific scene of interest. Then, the estimated distribution is used to define a local decision strategy for carrying out change detection by reducing the effects of misregistration. The experimental results obtained on both an artificially generated data set and a real data set confirm the effectiveness of the presented method.

2 Methodology

The proposed approach is developed in the framework of the well-known and widely used change vector analysis (CVA) technique [1]. The CVA technique performs change detection by differencing the feature vectors of pixels with the same spatial coordinates in the two considered images. The change-detection map is obtained by applying, pixel by pixel, a decision threshold ν_{th} to the module of the difference vector. In some applications, the phase of the difference vector is also exploited to detect different kinds of changes.

For simplicity, in the presentation of the proposed method, let us assume that only two spectral channels are considered for each multispectral image in the application of the CVA technique. This means that difference vectors have only two components and, consequently, they can be represented in a 2-dimensional space described by their modules and phases (M-P space). If more spectral channels are considered, more phase contributions should be taken into account. However, it is straightforward to extend the proposed method to the N-dimensional case.

The proposed approach aims at estimating the distribution of the registration noise in the aforementioned M-P space, given a specific scene of interest. Then, the estimated distribution is used to define a local adaptive decision strategy for carrying out change detection by taking into account registration noise. The approach is composed of two main steps: i) adaptive estimation of the distribution of the registration noise in the M-P space; ii) definition of a decision strategy for producing the change-detection map. These steps are described in greater detail in the following.

2.1 *Adaptive estimation of the distribution of the registration noise in the M-P space*

As described before, the distribution of the registration noise in the M-P space depends on several complex factors like the structure of the considered scene, the properties of the specific sensor adopted for image acquisition, and so on. We propose to use a single-date image of the considered multitemporal data set misregistered in several different ways versus itself for extracting this dependence. In particular, a set of L misregistered images obtained for different values (x_p, y_q) of misregistration in the horizontal and vertical directions, respectively, is generated. Let $M = \{(x_i, y_j)\}$ (with x_i ranging from $-x_{max}$ to $+x_{max}$ and with y_j ranging from $-y_{max}$ to $+y_{max}$, both with a predefined step) be the set of misregistration values considered to misregister the single-image considered. Then, the CVA technique is applied to the original and misregistered images, and the distributions in the M-P space of the resulting difference vectors are analyzed. The analysis is carried out by quantizing the M-P space in $N \times N$ cells. Let $S = \{s_1, \ldots\ldots, s_{N \times N}\}$ be the set of cells obtained by quantization and let $g_i(x_p, y_q)$ be the number of pixels that occur in the cell s_i when the difference vectors are generated for a misregistration of $(x_p, y_q) \in M$ pixels. In this situation, the distributions of difference vectors in the M-P space only depend on the registration noise. On the basis of the value $g_i(x_p, y_q)$, it is easy to estimate the probability $P_i(n/x_p, y_q)$ to have a pixel associated with registration noise in the cell s_i, given a misregistration of (x_p, y_q) pixels. These probabilities can be estimated for all $N \times N$ quantization cells considered and for all misregistration values $(x_p, y_q) \in M$ selected. Then, from such probabilities, different strategies can be adopted to approximate the global probability $P_i(n)$ to have pixels affected by registration noise in s_i. In the following, we propose two different strategies: i) the *maximum* strategy; ii) the *correlation* strategy. It is worth noting that, from the $P_i(n), i = 1, .., N \times N$, it is possible to approximate the average global density $p(n)$ of the registration noise in the M-P space.

2.1.1 Maximum strategy

The maximum strategy models the local registration noise distribution in the M-P space on the basis of a global analysis of $P_i(n/x_p, y_q)$ obtained for all misregistration values $(x_p, y_q) \in M$ considered. In particular, the rationale of this strategy consists in modeling the registration noise in each quantization cell with the maximum possible noise resulting for the pairs of misregistation values considered in the set *M*. Consequently, it estimates the registration noise of each cell s_i on the basis of the maximum value of the noise that occur in s_i for the *L* different pairs of misregistration values considered, i.e.

$$P_i(n) = \max_{(x_p, y_q) \in M} \{P_i(n/x_p, y_q)\} . \tag{1}$$

2.1.2 Correlation strategy

The correlation strategy aims at estimating the pair (x_p^*, y_q^*) of misregistration values that characterize the considered multitemporal images and then carries out the estimation of the registration noise distribution on the basis of such values. A possible approach for estimating misregistration values consists in evaluating the residual misregistration computed on ground control points after the registration phase. However, it is well known that this estimation is usually inaccurate, and does not represent the real situation in the entire image. For this reason, we propose an alternative approach. Such an approach consists in comparing (in the M-P space) the difference vector distribution of the considered multitemporal images with the distributions of the difference vectors obtained for different values of misregistration. To this end, the CVA technique is applied to the considered multitemporal images and the number of difference vectors that occur in the cell s_i is used to estimate the probability $P_i(x)$ that a pixel of the difference image is associated with s_i. Then, the correlation $R_{x,n}(x_p, y_q)$ between the global distribution of difference vectors $p(x)$ and each registration noise density distribution $p(n/x_p, y_q)$ is computed as follows:

$$R_{x,n}(x_p, y_q) = \frac{1}{N \times N} \frac{\sum_{i=1}^{N \times N} \left[P_i(x) - \mu_x\right]\left[P_i(n/x_p, y_q) - \mu_{n,x_p,y_q}\right]}{\sqrt{\sigma_x^2 \sigma_{n,x_p,y_q}^2}} \tag{2}$$

where μ_x, μ_{n,x_p,y_q}, σ_x^2 and σ_{n,x_p,y_q}^2 are the mean values and the variances of the distributions $p(x)$ and $p(n/x_p, y_q)$, respectively, over the $N \times N$ quantization cells. The pair of misregistration values $(\hat{x}_p, \hat{y}_q)$ that maximize the correlation function $R_{x,n}(x_p, y_q)$ is considered as an estimation of the true (x_p^*, y_q^*). Consequently, we can write as follows:

$$P_i(n) = P_i(n/\hat{x}_p, \hat{y}_q). \tag{3}$$

It is worth noting that misregistration values often depend on the specific portion of the considered image. In order to take into account this dependence, the above-described strategy can be applied separately to different image areas, after partitioning the original image into subimages of pre-defined size.

2.2 *Definition of a detection strategy for producing the change-detection map*

At this point, the distribution in the M-P space of the difference vectors obtained by applying the CVA technique to the multitemporal images is analyzed with respect to the estimated registration-noise distribution. The analysis is carried out separately for each quantization cell. From the knowledge of $P_i(n)$, it is possible to define an adaptive decision strategy that allows one to detect changed areas by taking into account registration noise. Let us consider a generic pixel $x_d(i,j)$ in the difference image, whose difference vector lies in the cell s_i. This pixel is associated with changed or unchanged areas according to the following decision rule:

$$x_d(i,j) \in \begin{cases} change & if \ \{|x_d(i,j)| \geq v_{th} \ and \ [P_i(x) \geq P_i(n) + \varepsilon]\} \\ no\ change & if \ \{|x_d(i,j)| < v_{th} \ or \ [P_i(x) < P_i(n) + \varepsilon]\} \end{cases} \quad (4)$$

where $\varepsilon \in [0,1]$ is a parameter that tunes the decision threshold that takes into account the registration noise.

A critical aspect in the proposed technique concerns the relative position of pixels in the quantized M-P space for the multitemporal images and for the single image misregistered versus itself. If significant differences in light conditions affect the two considered multitemporal images, the distributions $p(n)$ and $p(x)$ may be shifted in the M-P space with respect to each other. As a consequence, the above-described strategy may result ineffective, since corresponding quantization cells are not associated with the same information on the multitemporal and reference M-P spaces. In order to overcome this problem, we propose to use an iterative procedure for the "registration" of the two M-P spaces. In particular, such a procedure shifts the quantization cells of the reference space so that $P_i(n)$ and $P_i(x)$can be matched. The shift is carried out in a predefined range of values that depend on the difference between light conditions in the multitemporal images (these differences can be evaluated on the basis of the mean value of the difference image associated with the multitemporal images). In order to define the direction and value of the shift that provide the best distribution matching, a correlation measure between $P_i(x)$ and the different shifted versions of the probability $P_i(n)$ is used.

3 Experimental Results

3.1 Synthetic Data Set

In order to quantitatively assess the effectiveness of the proposed method, an artificially generated data set was considered. The data set was obtained by using the procedure described in the following. An image acquired by an Airborne Thematic Mapper (ATM) multispectral scanner was used as a reference image. In particular, a section (230×250 pixels) of a scene acquired in an agricultural area near the village of Feltwell (U.K.) was selected (bands 2, 3 and 5 were considered) (see Figure 1) [2]. This image was assumed to be the t_1 image of the multitemporal data set (i.e. the image acquired at time t_1). The t_2 image was obtained by artificially inserting some differences in the t_1 image in order to simulate land-cover changes. Then, the mean value of the t_2 image was shifted to simulate differences in light conditions, and this image was misregistred with respect to the t_1 image.

Figure 1. Band 5 of the ATM image considered for the evaluation of the proposed algorithm.

In the first experiment, a misregistration of 1 pixel per direction was considered. The proposed technique was applied to detect changes occurred between the t_1 and t_2 images. Different pairs (x_p, y_q) of misregistration values of t_1 versus itself were considered in the subset M in order to estimate the misregistration noise distribution. In particular, we selected $x_{max}=y_{max}=2$ and considered all possible pairs of misregistration values adopting a step equal to 0.25. The three-dimensional M-P space (it is worth noting that with three spectral bands the difference vectors are described by the module and two phase components) was split into $50 \times 50 \times 50$ quantization cells. The threshold v_{th} was set equal to 100 (this value represents the minimum error threshold derived by a manual trial-and-error procedure).

The change-detection errors obtained with the proposed technique (using the *maximum* and the *correlation* strategies for estimating the $P_i(n)$) are given in Table 1. For a comparison, the results obtained by applying the traditional CVA technique to the difference image and to the difference image filtered with a 3×3 median filter are also reported. As one can see, both the *maximum* and the *correlation* strategies exhibit sharply higher overall accuracies than the standard CVA. This confirms the effectiveness of the proposed approach. It is worth noting the different behaviors that characterize the proposed strategies. As expected, the *maximum* strategy decreases the number of false alarms with respect to the *correlation* strategy, whereas the *correlation* strategy decreases the number of missed alarms with respect to the *maximum* strategy. Consequently, the choice of the best estimation strategy should be carried out by taking into account the importance of false and missed alarms in the application considered.

A qualitative analysis of the obtained results can be performed by comparing the map of the position of simulated changes (figure 2a) with the change-detection maps provided by the proposed method (figure 2b) and the CVA technique (figure 2c). As one can see, the changes erroneously detected by CVA technique (i.e. the

Table 1. Change-detection errors obtained by the proposed and the standard techniques on the simulated data set (misregistration of 1 pixel in both the horizontal and vertical direction)

Method	Missed alarms	False alarms	Total error
Proposed (*maximum* strategy)	38	177	215
Proposed (*correlation* strategy)	9	203	212
Standard CVA	22	1484	1506
Standard CVA (3x3 median filter)	9	868	877

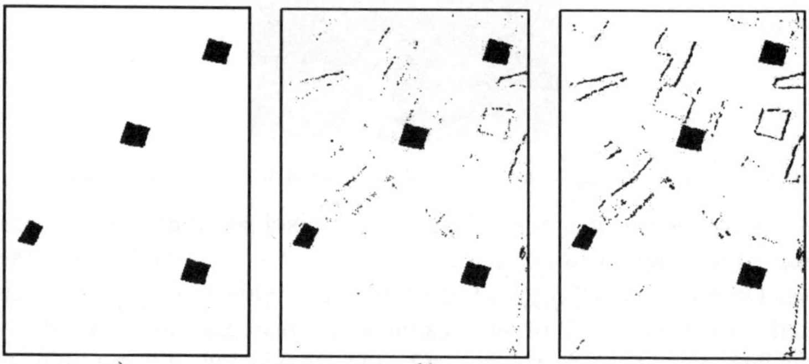

a) b) c)

Figure 2. a) Position of simulated changes; b) changes detected by the proposed technique (correlation strategy); c) changes detected by the CVA technique.

artifacts in correspondence of field boundaries) are significantly reduced in the map obtained by the proposed technique.

In the second experiment, the effectiveness of the proposed method by varying the entity of the misregistration values was analyzed. In particular, the considered misregistration values ranged from 0.25 to 2 pixels per direction with a step of 0.25 pixels. In table 2, the mean values of the results obtained with the proposed technique (maximum strategy) and the traditional CVA are reported. An analysis of the results reveals that the proposed method exhibits a higher accuracy as compared with the standard CVA technique.

3.2 *Real Data Set*

In this subsection the results obtained with a real multitemporal data set composed of two multispectral images acquired by the Thematic Mapper sensor of the Landsat 5 satellite are described. In particular, the images were acquired in April 1991 and May 1997 on the area of Parco of Ticino, Milan, North Italy (figure 3). Several changes affected the considered area due to the works for the construction of Malpensa2000 international airport.

Table 2. Results obtained by the proposed and the standard techniques on the simulated data set by varying the misregistration values from 0.25 to 2 pixels (by steps equal to 0.25)

	Method		
	Proposed (*maximum* strategy)	Standard CVA	Standard CVA (3x3 median filter)
False alarms	128	581	221
Missed Alarms	8	9	23
Total Error	136	590	245

Ten ground control points were used in order to co-registrate the images. The proposed technique was applied to detect changes occurred between the t_1 and t_2 acquisitions. The obtained change-detection map (figure 4a) was compared with the one yielded by applying the CVA technique (figure 4b). Since no ground truth was available, only a qualitative analysis of the obtained results was performed. An analysis of the two images points out: a) changes involved by the construction of the new airport; b) seasonal changes in the riverbed; c) registration noise. In greater detail, the change detection map obtained by the proposed technique exhibits a significant reduction of the false alarms due to the registration noise, thus confirming the effectiveness of the proposed method.

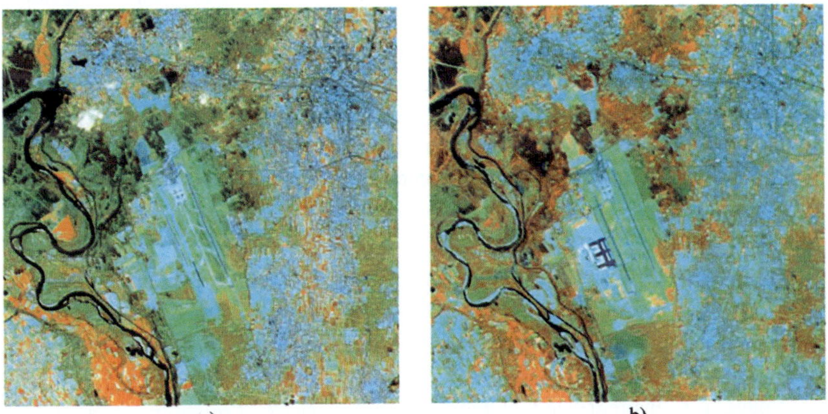

a) b)

Figure 3. False color composition (bands 4, 5, and 7) of the images acquired by the TM sensor of the Landsat 5 satellite on the on area of Parco of Ticino, Milan, North Italy. a) Image acquired on April 1991; b) image acquired on May 1997.

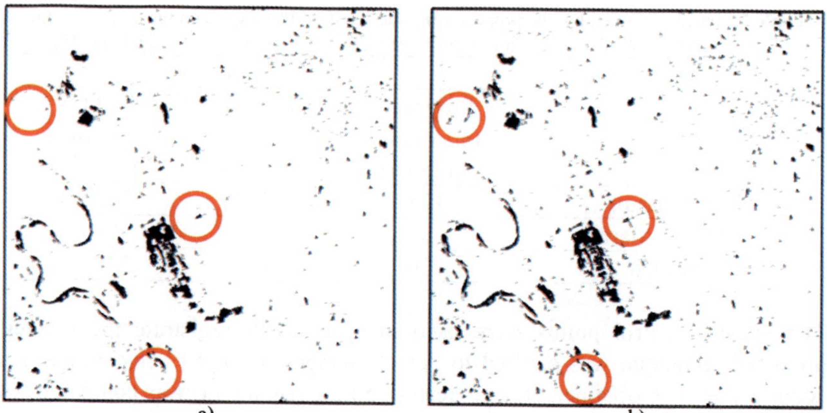

a) b)

Figure 4. Change-detection maps obtained on the real data set by a) the proposed technique (correlation strategy), b) the CVA technique. In red circles some examples of the artifacts introduced by misregistration noise and reduced by the proposed technique are pointed out.

4 Conclusions

An approach aimed at reducing the effects of registration noise in unsupervised change detection has been presented. Several experiments were carried out by applying the proposed approach to both a synthetic and a real multitemporal data set. The obtained results point out the effectiveness of the presented method.
As a future development of this work, we are considering the possibility to jointly exploit the spatial contextual information and the distribution of pixels in the M-P space to further increase the change detection accuracy.

References

1. Singh A., Digital change detection techniques using remotely-sensed data, *International Journal of Remote Sensing*, **10** (1989), pp. 989-1003.
2. Bruzzone L. and Fernández Prieto D., Automatic analysis of the difference image for unsupervised change detection, *IEEE Transactions on Geoscience and Remote Sensing*, **38** (2000), pp. 1171-1182.
3. Townshend J. R. G., Justice C. O. and Gurney C., The impact of misregistration on change detection, *IEEE Transactions on Geoscience and Remote Sensing*, **30** (1992), pp. 1054-1060.
4. Dai X. and Khorram S., The effects of image misregistration on the accuracy of remotely sensed change detection, *IEEE Transactions on Geoscience and Remote Sensing*, **36** (1998), pp. 1566-1577.

DETECTION OF PIXEL-LEVEL LAND-COVER CHANGES WITH MULTI-TEMPORAL IMAGERY: THEORY AND EXAMPLES WITH IMAGERY OF 1 METER AND 1 KILOMETER SPATIAL RESOLUTIONS

DOUGLAS STOW, DONGMEI CHEN, AND LLOYD COULTER

Department of Geography, San Diego State University, San Diego, California 92182-4493 USA E-mail: stow@mail.sdsu.edu

Detecting land-cover change features that are similar in size to the ground sampling distance of a multitemporal image data set is greatly dependent on the accuracy of spatial registration, or compensating for registration errors. Two case studies are presented from monitoring studies of habitat reserves and urbanization in southern California. Each case study represents a different approach to achieving pixel-level change detection by mitigating misregistration effects. For tests with 1 m and 1 km resolution imagery, the most promising aspect of the techniques that have been developed to minimize misregistration effects is the reduction of artifacts associated with false detection of change.

1 Introduction

The most stringent test of image change procedures is the detection of land-cover changes that occur at spatial scales close to that of the ground sampling distance (GSD) of the associated multi-temporal image data. The potential for noise sources obscuring land-cover change signals and creating false alarms is particularly great when attempting to detect land-cover changes at this scale. Primary sources of noise are spatial misregistration and differences in radiometric properties (e.g., from variations in sensor, illumination, and atmospheric properties) between multi-temporal images [2].

The objectives of this paper are to: (1) present theoretical requirements and implications for achieving pixel-scale change detection from multi-temporal image data sets captured by remote sensing systems; and (2) demonstrate two approaches to minimizing misregistration on land-cover change detection for two image data sets, each having very different spatial resolutions. Objective 2 is met by presenting two case studies of land-cover change applications in southern California. The first is a study on monitoring of shrubland habitat reserves, which incorporates multi-temporal imagery with 1 m spatial resolution and auto-registration techniques. The second is an exploration of the sensitivity of multi-temporal AVHRR data (1 km resolution) in detecting land cover changes associated with urbanization. A model that compensates for misregistration is applied to determine if land-use/land-cover signals are more detectable.

2 Theoretical Background

The components of image brightness (B) that influence the temporal change between corresponding pixels of a multi-temporal image data set are represented by the following equation [3]:

$$\frac{\Delta B}{\Delta t} = -\frac{\Delta B}{\Delta x} D_x - \frac{\Delta B}{\Delta y} D_y + R + S$$

where: ***B*** is the relative amount of reflected and/or emitted EMR received by the remote sensor or some continuous-value derivative from such remotely sensed radiation (e.g., spectral ratio);

$\frac{\Delta B}{\Delta t}$ is the discrete temporal change in image brightness (ΔB) over time interval Δt, *i.e., conventional image differencing;*

$\frac{\Delta B}{\Delta x}$ and $\frac{\Delta B}{\Delta y}$ are the spatial gradients of image brightness in the x and y directions (west-east and north-south directions for images that are geographically referenced), respectively;

D_x and D_y are misregistration magnitudes in x and y, respectively;

R is the temporal difference in B due to radiometric effects, e.g., sensor noise, sensor calibration drift, and illumination and atmospheric differences; and,

S is the temporal change in B resulting from actual land-cover changes, i.e., signals.

By rearranging equation, the change signal term (S) can be isolated. Thus, land-cover change signals can be quantified more reliably from image differencing if either misregistration and the spatial gradients of brightness are negligible or both are known, and if radiometric differences between dates are normalized.

Of the two noise sources, misregistration is the more critical to minimize as its effects are more localized and spatially variable [6]. Stow and Chen [4] define the total (h) noise equivalent temporal difference in brightness due to misregistration (MNEΔB) as:

$$\text{MNE}\Delta\text{B} = (\text{D}_\text{h} * \frac{\Delta B}{\Delta h})$$

where: $\text{D}_\text{h} = (\text{D}_\text{x}^2 + \text{D}_\text{y}^2)^{1/2}$ and

$$\frac{\Delta B}{\Delta h} = ((\frac{\Delta B}{\Delta x})^2 + (\frac{\Delta B}{\Delta y})^2)^{1/2}$$

Requirements for registration precision are generally assumed to be less than half of a pixel [1,5]. This is achieved routinely for moderate- and coarse-resolution satellite data (e.g, TM and AVHRR), but is extremely difficult to achieve for very high resolution image data (e.g., airborne digital cameras and IKONOS). The impact of spatial brightness gradients (D_h) will depend mostly on landscape heterogeneity at the scale of the ground sampling distance (GSD). Misregistration effects will tend to be more substantial for landscapes with locally heterogeneous land-cover compositions.

Radiometric offsets between dates are normally less critical, unless differences of absolute quantities (e.g., albedo or surface temperature) are important to derive. Radiometric effects such as drifts in radiometric calibration or satellite overpass times, atmospheric path radiance and transmittance, etc. are generally more regionalized and apply to an entire scene or image subset. These effects tend to create a radiometric bias in image differences, which can be compensated for by simple normalization procedures, or by adjustments in the process of selecting threshold difference values of land-cover change/no-change. Some radiometric effects can be localized and therefore, are more critical to compensate for. These include spatially varying atmospheric effects such as smoke and haze, and substantial differences in local solar illumination geometry caused by differences in time of year and/or day for image acquisition, and irregular terrain. These more localized effects can be minimized by careful replication of image acquisition specifications for time sequential images and/or with complex atmospheric-terrain correction models.

3 Case Study #1

We are examining the potential of resolving fine-scale changes in land cover resulting from anthropogenic disturbances in shrubland habitat reserves in southern California with the aid of multi-temporal digital multi-spectral data having 1 m spatial resolution. With this very high resolution imagery and the irregular terrain of the study area, root mean square registration errors have varied between 1.5 to 6 m (pixels), depending on the geometric processing and mosaicking approaches that were applied. This degree of offset has precluded reliable detection of pixel-level changes.

To minimize false detection of habitat change caused by misregistration and radiometric differences, we developed a coordinated image acquisition and processing scheme to ensure that digital image frames are captured as close to the same observation points in both time and space, for all pairs of multi-temporal images. Our Airborne Data Acquisition and Registration (ADAR) system (a commercial multispectral digital camera system manufactured by Positive Systems, Inc.) was flown at the same time of day on two near anniversary dates, one year apart. Frame centers of images on the second date were matched as close as possible to those of the first date with the aid of on-board global positioning data. Thus, illumination and view geometry for each pair of frames were similar for both image dates, which minimizes relief displacement and radiometric differences.

With the aid of the Digital Images Made Easy (DIME) software package from Positive Systems, Inc., a frame-by-frame image correlation matching process yielded image pairs with sub-pixel registration errors. Resultant change detection products depict small change features such as newly formed trails less than 0.5 m in width and shrubs as small as 1 to 2 m that had been removed (Figure 1). Large change features such as burn scar (upper right edge) and a utility access road (top center) are readily apparent as bright objects. A majority of the change features in the habitat reserve are detected adjacent to residential structures. Habitat reserve managers refer to these features as "urban edge effects." Also apparent as small dark circular objects are many small shrubs (that had differing amounts of green leaves, i.e., different leaf area indices) on the two acquisition dates. This resulted from differences in precipitation and antecedent moisture in the preceding winter rainfall seasons. If signals of land-cover change are defined to be only manifestations of anthropogenic disturbance effects, e.g., from the perspective of a habitat reserve manager, these transient changes in vegetation could be considered a third type of noise source.

4 Case Study #2

For the second case study of land-cover change we: (1) investigated the sensitivity of multitemporal image data from the Advanced Very High Resolution Radiometer (AVHRR) sensor on the National Oceanic Atmospheric Administration (NOAA) satellite for detecting land-use/land-cover changes primarily associated with urbanization; and (2) tested a model which compensates for misregistration. Two near-anniversary, single-date NOAA AVHHR images of a rapidly urbanizing landscape in southern and Baja California were analyzed. Analyses were facilitated by reference data from detailed GIS data layers of land-use/land-cover types for the two years corresponding to image acquisition dates (1990 and 1995). Detailed descriptions of procedures and results can be found in Stow and Chen [4].

Almost all AVHRR pixels containing land-use/land-cover changes had mixtures of no change areas, even when the extent of contiguous change features (i.e., polygons) was greater than the nominal 1 km^2 ground sampling area. Correlation between image difference magnitudes and percentage land-cover change within a pixel were only significant when at least 70% of a pixel had changed (Figure 2). The strongest signals of image brightness change occurred for NDVI and Channel 4 surface temperature products. 'Undeveloped to urban' and 'undeveloped to water' were the land-use/land-cover transition sequences with the most definitive AVHRR change signals.

Misregistration fields were derived by surfacing (via interpolation) the residual errors of a set of ground control points for each image date, and then differencing the error surfaces. Mean magnitudes of misregistration errors were estimated to be around 0.2 pixel units in x and y directions, while the mean values for misregistration noise equivalent in brightness change (MEΔB) were 0.02, 0.02, and 1.96K for image differences of Channel 1 reflectance, NDVI, and Channel 4 surface temperature, respectively. The misregistration compensation model appeared to reduce false detection of change, as indicated by a reduction of local variance for difference images generated with the model (Figure 3). However, improvements in detection of land-use/land-cover changes were not conclusive, as can be gleaned from Figure 2.

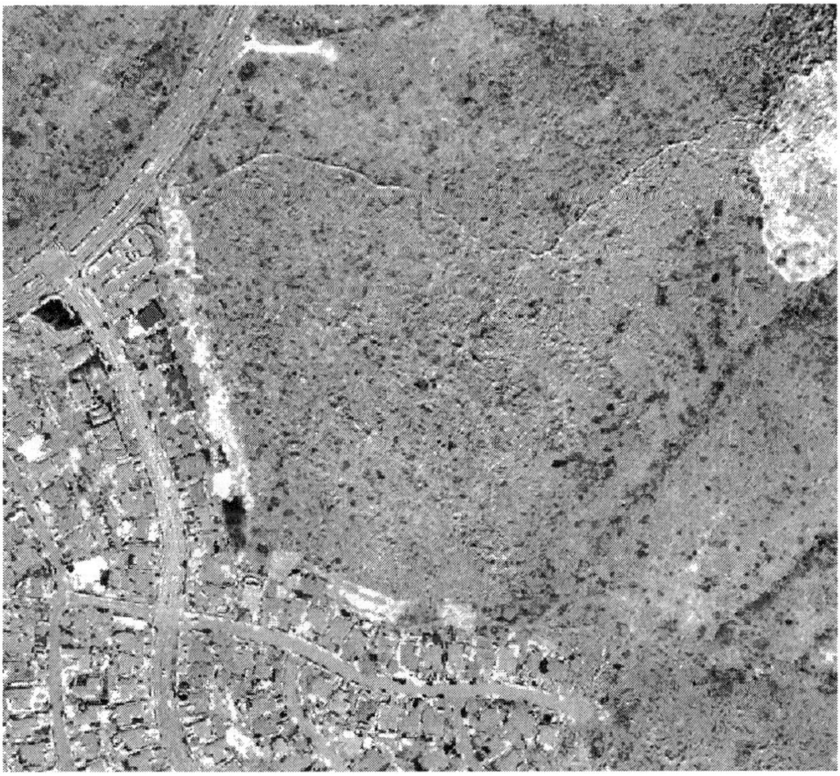

Figure 1. Red waveband difference image from multitemporal (June 27, 1998 and May 26, 1999) ADAR frames captured from 2,300 m at similar camera station points and registered with a semi-automatic image matching and stitching routine. Increases in reflectance are exhibited in brighter tones and decreases in darker tones. The scene covers approximately 2.0 x 1.5 km of the Mission Trails Regional Park and fringing suburban residential neighborhoods in San Diego, California, USA. Over 80 meters of vertical relief occurs within this scene.

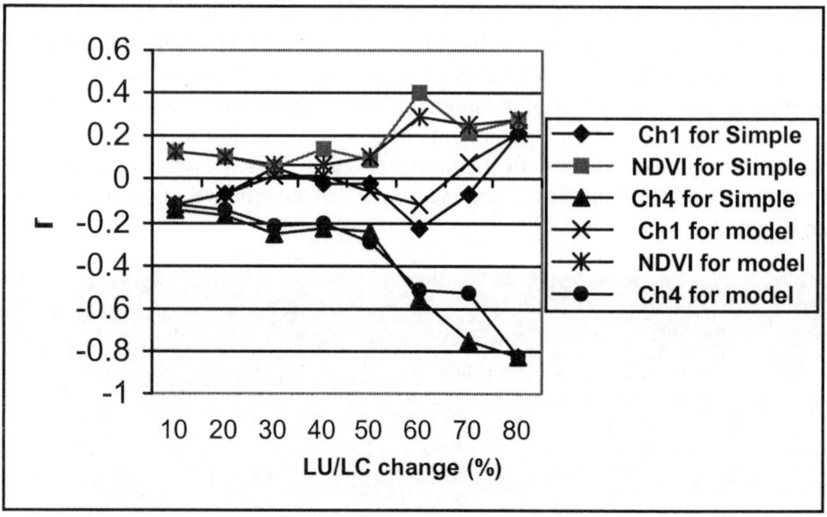

Figure 2. Correlation coefficients (r) for image difference magnitude vs. percentage of land-use/land-cover change for Channel 1 (visible) reflectance, NDVI, and Channel 4 (TIR) radiometric temperature images. Simple = standard image differencing and model = misregistration compensation model.

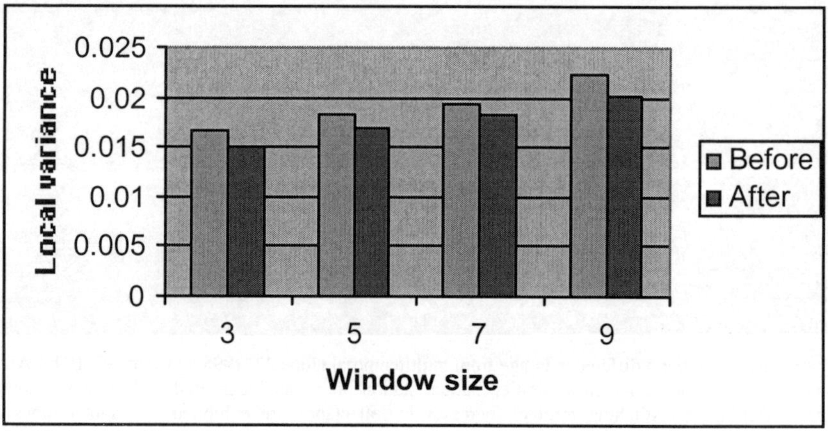

Figure 3. Graph of local variance statistics at various window sizes comparing results from standard image differencing (Before) and misregistration compensation model (After). Lower local variance for the compensated image difference suggests that artifacts of misregistration have been reduced.

5 Conclusions

The success of detecting land-cover change features that are similar in size to the GSD of a multitemporal image data set is greatly dependent on the accuracy of spatial registration, or the ability to compensate for registration errors. Less critical are radiometric differences, which tend not to be as localized. Two case studies were presented, each representing a different approach to achieving pixel-level change detection by mitigating misregistration effects.

Airborne multispectral image data with very high spatial resolution are required to monitor detailed changes in shrubland habitat reserves within southern California. Because of characteristically steep and variable terrain, image data sets with 1 m spatial resolution are very difficult to register accurately and registration errors are difficult to estimate precisely. By capturing digital image frames near the same station points of previous imaging campaigns and by utilizing image correlation matching procedures, sub-pixel registration accuracy was achieved. Duplication of observation perspectives and time of day also helped to minimize radiometric differences. While most land-cover changes of interest to reserve managers occur at characteristic scales greater than 1 m, imagery with meter to sub-meter resolution is generally required to identify the types of transitions associated with human impacts on habitat reserves. The major benefit of the combined image acquisition and semi-automatic matching approach is to substantially minimize false detection of change that result from misregistration.

In the context of the monitoring urbanization within the southern California metropolis, results from empirical exploration of the sensitivity of coarse (1 km) spatial resolution, mulitemporal imagery from the NOAA AVHRR satellite for detecting land-use/land-cover changes suggests that few AVHRR pixels represented 100% change. In other words, land-use/land-cover change features tended to be smaller than GSD of AVHRR and many pixels contained more than one feature with differing types of land-use/land-cover transitions. The sensitivity of multitemporal AVHRR difference images was limited until at least 70% of a pixel consisted of land-use/land-cover change. Given that most change features are smaller than the size of the AVHRR GSD, registration requirements are particularly stringent. Readily available information about residual errors of ground control points used in standard AVHRR processing routines, enabled misregistration fields to be estimated and a model that compensates for misregistration to be tested. Improvements in detecting sub-pixel changes were not conclusive, however, false changes appeared to be reduced. Thus, for tests with 1 m and 1 km resolution imagery, the most promising aspect of the techniques that have been developed to minimize misregistration effects has been the reduction of false change artifacts (i.e., errors of commission).

6 Acknowledgments

Funding for research elements reported in Case Study 1 was provided by the NASA Food and Fiber Applications of Remote Sensing program and for Case Study 2 from San Diego State University. Julie Raine of Positive Systems, Inc. assisted with the testing of ADAR frame-to-frame matching routines.

References

1. Dai, X., and Khorram, S., Quantification of the impact of misregistration on the accuracy of remotely sensed change detection. *Proc. IEEE Geosciences and Remote Sensing Symp.* (Texas, IEEE) (1997) pp.1763-1765.
2. Gong, P., LeDrew, E., and Miller, J., Registration-noise reduction in difference images for change detection. *Intl. J. Remote Sensing* **13** (1992) pp. 773-779.
3. Stow, D., Reducing misregistration effects for pixel-level analysis of land-cover change. *Intl. J. Remote Sensing* **20** (1999) pp. 2477-2483.
4. Stow, D., and Chen, D., Sensitivity of Multitemporal NOAA-AVHRR Data to Land-use/Land-cover Changes and Misregistration in an Urbanizing Region, *Remote sensing of Environment* in review.
5. Townshend, J., Justice, C., Gurney, C., and McManus, J., The impact of misregistration on change detection. *IEEE Trans. Geoscience and Remote Sensing* **30** (1992) pp. 1054-1060.
6. Verbyla, D., and Boles, S., Bias in land cover change estimates due to misregistration. *Intl. J. Remote Sensing* **21** (2000) pp. 3553-3560.

IMAGE THRESHOLDING FOR LANDSLIDE DETECTION BY GENETIC PROGRAMMING

Paul L. Rosin
Department of Computer Science
Cardiff University
UK
E-mail: `Paul.Rosin@cs.cf.ac.uk`

Javier Hervás
Institute for Systems, Informatics, and Safety
EC Joint Research Centre
Ispra, VA, Italy
E-mail: `javier.hervas@jrc.it`

This paper describes an approach to image thresholding that combines various multiscale and morphological features, including texture, shape and edge filtering, by using genetic programming, to detect the presence of landslides and their active sectors in change detected multitemporal aerial images.

1 Introduction

Detecting landslides and monitoring their activity is of great relevance for natural hazard assessment and disaster prevention in hilly areas. Very high resolution optical satellite data is now becoming available, and to evaluate the potential application of these new images to detect ground surface changes as a result of landsliding, in previous work we developed and applied change detection and thresholding methods on digital aerial photographs over the Tessina landslide in Veneto, Italy[1]. This is a nearly 3-km long complex landslide consisting of rotational slides in its head and a mudflow in the remaining, largest part of the body. The landslide has developed in Eocene flysch materials partly covered by colluvial and glacial till sediments. The landslide was first triggered in 1960 and has since undergone a number of reactivations. The physical setting of the area and landslide dynamics are further described in[2,1].

The landslide currently affects an area including a number of small villages and mixed woodland-grassland landcover. It generally appears as a distinct bright feature on visible-wavelength panchromatic images because of outcropping soil under disrupted vegetation (figure 1a&b).

This paper focuses on applying genetic programming techniques to detect and monitor landslides from optical remote sensing data. To this end, we have used radiometrically mormalised and orthorectified multitemporal aerial photographs at 1m ground resolution over the Tessina landslide area[1].

2 Genetic Programming

Genetic programming (GP) is an optimisation technique based on the concepts of Darwinian evolution [3]. A population of individuals is created, each representing a potential solution to the problem. The solution offered by each individual is assigned a fitness value which indicates how well that solution performs. Over time the fitter individuals live and reproduce while the others tend to die without reproducing, simulating the process of survival of the fittest. With each generation the overall population fitness improves, thus performing optimisation.

Whereas genetic algorithms usually code solutions by fixed-length strings which results in their solution space being constrained, GP uses a tree structure. The leaves of the tree (called terminals) represent input variables or numerical constants. Internal nodes (called functions) represent arithmetic, mathematical, boolean, etc. functions. Individuals are evaluated by traversing through the tree, applying function nodes to their children, and so can be considered to represent programs. The search space of GP is the set of all the possible compositions of the functions in the union of the terminal and function sets.

The initial population is formed by constructing random trees. Thereafter, individuals are chosen from the population according to the fitness of their solutions. In a similar manner to genetic algorithms, genetic operators are applied to these trees. Many variations of the operators exist, but the two main ones are as follows. In *mutation* an individual receives a random alteration. Either the value or variable at a terminal is replaced, or an entire subtree is replaced. In *crossover* two random nodes are selected from within each of two parent individuals and then the resultant subtrees are swapped, generating two new child individuals which typically replace less fit members of the population.

3 Automatic Thresholding

Our previous work on change detection considered a variety of algorithms, based on quite diverse principles, for determining appropriate thresholds for difference images [4]. Although some approaches were better than others, none were entirely satisfactory on their own as they tended to pick up spurious detail arising from residual effects of the radiometric normalisation. In an attempt to overcome this, a post-processing step was included to reduce the noise using shape properties of the thresholded regions. Area, width, perimeter, and rectangularity filters were applied and were shown to provide significant improvements.

In this paper an alternative approach is taken. Rather than manually design thresholding algorithms, one will be learnt using genetic programming. In this case the desired program is a binary classifier (although GP is more general purpose, and can also be applied to generate many other types of programs). That is, we wish to distinguish the "change" and "no-change" classes in the scene due to landsliding. Various schemes are possible. For instance, previously we have used GPs to generate classifiers made up from individual compositions of functions, one tree per class. These are then run in parallel and a winner takes all strategy used to select the class[5]. Here however, since only two classes are present a simpler strategy is sufficient and was found to work better. Programs are generated that take for each pixel a set of property values and compute a single real value. Pixels are assigned to the appropriate class according to whether their evaluation by the GP is a positive or negative value.

The function set was made up of a small set of standard operators: the arithmetic ones: *, +, -, /; as well as *absolute*, *sigmoid*, *minimum*, and *maximum*. The terminal set is more crucial and needs to be chosen with some care. In addition to random ephemeral constants and the difference image further input is required to enable a good classifier to be constructed. Whereas our previous work applied post-processing to the thresholded image it is desirable to retain the intensity information as long as possible before binarising. This is analogous to fuzzy reasoning where the data is only "crispened" at the end of processing chain. As already noted, much of the spurious detail was at a large scale, which suggests that providing coarser scale information could be helpful. To this end morphological opening and closing were applied over a range of scales to the difference image to remove the bright and dark details respectively. (See Stringa[6] for the use of mathematical morphology in change detection). Another technique for incorporating scale is to simply smooth the image, which again was applied over a range to provide a multiscale image representation. The spurious detail might show a distinct texture, and so some texture maps were included to enable their identification. Laws'[7] texture energy approach involving the convolution of masks for detecting spots, edges, lines, V's, etc. was used.

Finally, a modified distance transform was applied to the inverted difference image. The standard distance transform is applied to a binary image and efficiently computes the distance at each pixel to the nearest feature pixel[8]. While straightforward for binary data, it is more difficult for grayscale data which generally needs to be thresholded, bringing us back to our thresholding problem. One approach has been suggested for incorporating intensity information for edge-like features[9]. An alternative was developed here for arbitrary grayscale images in which the image is thresholded at all possible levels. The distance transform is applied to this image set and the results combined by adding them together and rescaling. The effect is to incorporate some coarser spatial information while retaining graylevel information. This is done without arbitrarily thresholding the image, and also without distorting the image by blurring. Low values are produced in the distance transformed image as a function both of the intensity of the input image (i.e. the darker the inverted difference image, the darker the transformed image) and also of the proximity to dark pixels (i.e. the closer they are, the darker the transformed image).

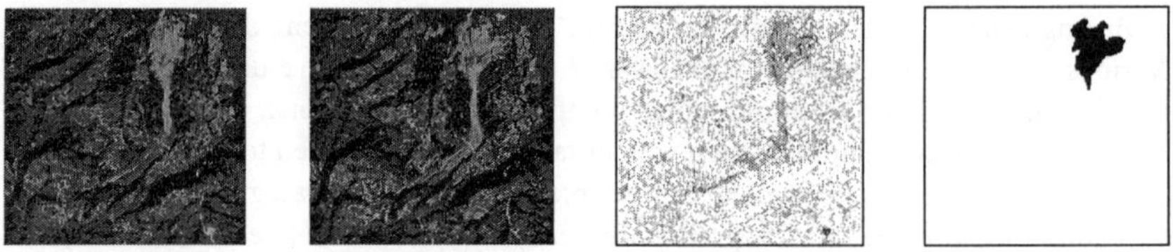

(a) 1988 (b) 1994 (c) difference image (d) ground truth

Figure 1: Images of Tessina before and after the landslide reactivation.

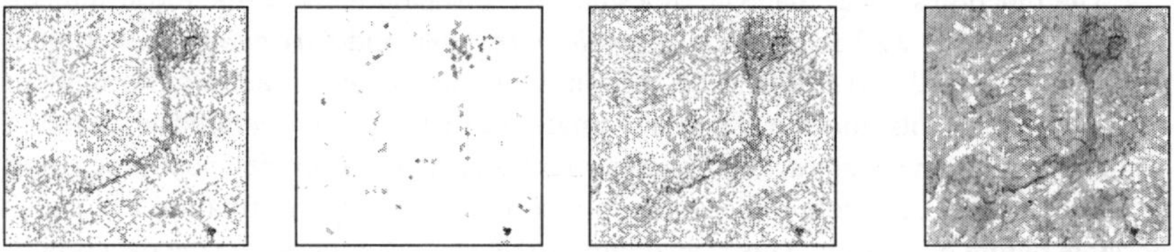

(a) 3 × 3 opening (b) 31 × 31 opening (c) 3 × 3 closing (d) 31 × 31 closing

Figure 2: Mathematical morphology operations applied to difference image

4 Results

Figure 1a&b shows orthorectified and radiometrically normalised images of the Tessina area covering the period before and after the most recent major landslide movement in 1992. The landslide runs from near the top right corner to the centre bottom left of the images; the thinner sector represents the mudflow. After some cropping their size is 2703 × 2590. The difference image (figure 1c; black indicates high change, white low change) is calculated by taking the absolute difference of the pixel intensity values over the two dates.

Various categories of change can be specified. We have taken the head scarp and upper accumulation area of the landslide as ground truth (figure 1d) as it gives a reasonably substantial set of training data for both classes (change and no-change), but future work will look at the smaller reactivation area as defined in[2,1]. The selected area of change makes up about 3% of the image, while the reactivation area is only 0.3%. Figures 2–5 show the various features extracted from the difference image. For reference, the result that we previously obtained using the corner thresholding method[4] is given in figure 6.

As only one image is available it has to be used both for training and testing.

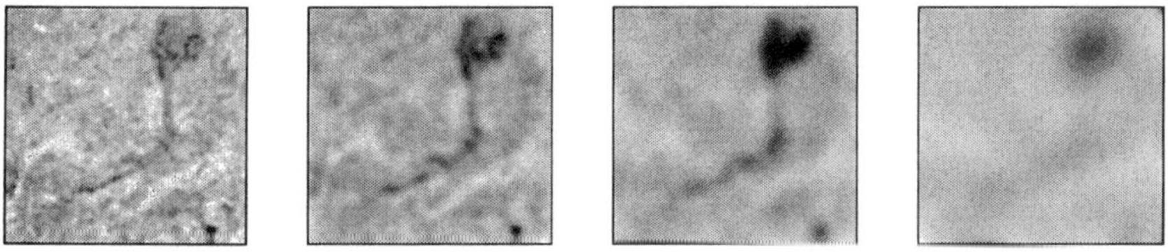

(a) $\sigma = 5$ (b) $\sigma = 10$ (c) $\sigma = 20$ (d) $\sigma = 40$

Figure 3: Smoothed version of the difference image

(a) horiz. edge (b) vert. edge (c) horiz. line (d) vert. line (e) spot

Figure 4: Texture energies of the difference image

About 2700 training pixels were sampled from the image (i.e. about 4% of the data) for training the classifier which is then tested by applying it to the full image. For some classifiers, applying an even sampling of the image such that the number of class samples is proportional to the size of the class builds into the classifier expectations of the class likelihoods. This is the case for the classifier in this paper as the fitness function is the percentage of correct classification. While it is desirable to use these expectations it was found to sometimes lead to problems due both to the larger no-change class dominating the optimisation criterion, and also because of the small number of change class training pixels.

The GP was run with a population of 20,000 trees which were allowed to grow to a maximum depth of 7. For each experiment 10 runs with different random initialisations were made, with each population undergoing about 200-300 generations. Substantial computing power is required for such development as each run takes about 4 days to complete on a Sun Ultra-5 workstation. The following figures (7-12) show the results using either representational sampling (left image) or quasi-equal sampling (right image).

The problem with a large variation in class sizes is demonstrated when the classifier is trained with just the difference image (figure 7). When representational sampling is used it is found that the best classification is obtained by simply blanking the image. Although the change class is almost completely wrongly classified the

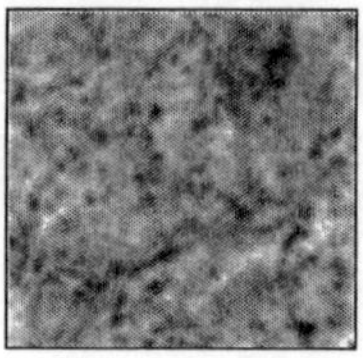

Figure 5: Summed distance transform over all thresholds

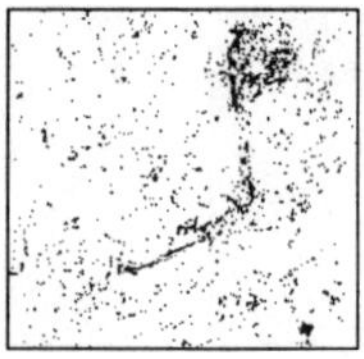

Figure 6: Thresholding the difference image using the corner method

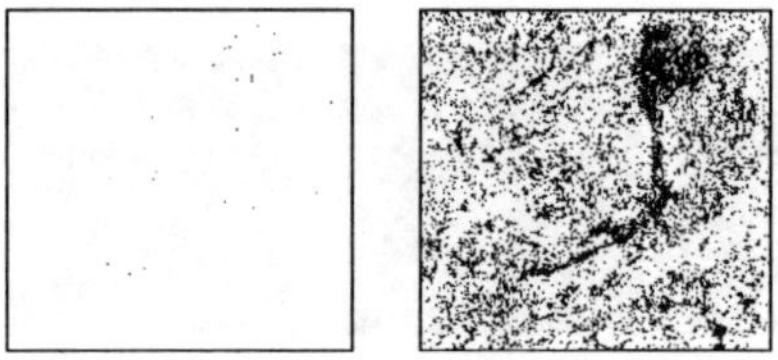

Figure 7: Training using the difference image alone

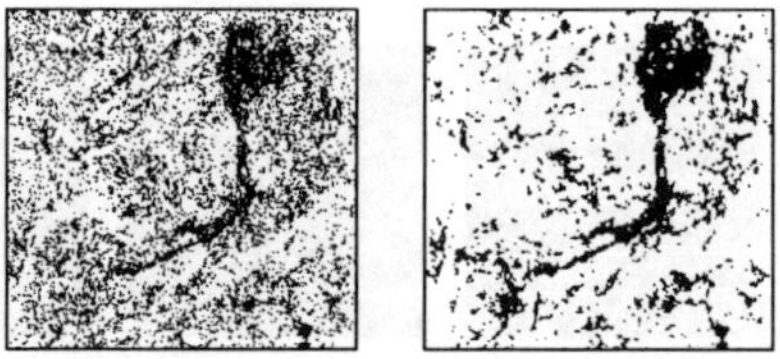

Figure 8: Training using the difference image plus opening and closing

remaining 97% of the image is correct, and so the GP solution obtained a high fitness score. Taking equal (or at least roughly equal) numbers of samples avoids this trivial solution, although there is not enough information available to the GP system to generate a solution better than a standard thresholding method.

Incorporating the data from the filtered images produces a noticeable improvement. Combining the difference image with its opened and closed versions enables the coarse structure of the landslide to be identified even though much residual noise is still picked up (figure 8). Adding textural information to the difference image gave extremely poor results (figure 9), and its addition to the mathematical morphology filtering also provides no benefit (figure 10).

The smoothed images provide much coarser scale spatial information than the opened and closed images (figure 11). This is evident when they augment the difference image as only a few large-scale blobs are extracted, and the majority of noise has been eliminated. Including opening and closing and the distance transformed image provides some further improvement so that the result in figure 12 is getting close in size and shape to the ground truth.

The fittest individual found with representational sampling that was executed to produce figure 12a is shown below (written as a prefix expression) where the terminals `s5`, etc. are the difference image smoothed by $\sigma = 5$ etc., `o15` is the image opened with a 15×15 mask, and `dt` is the distance transformed image.

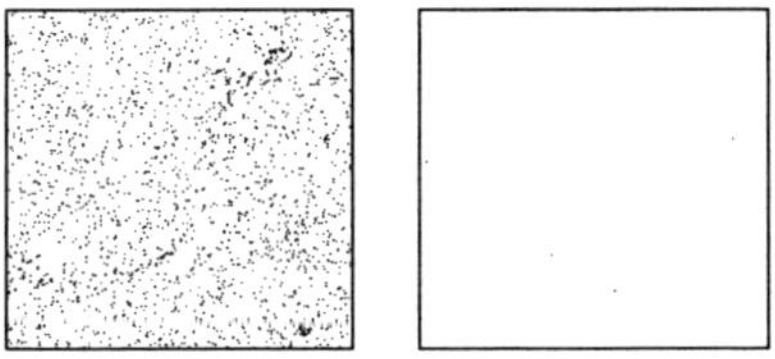

Figure 9: Training using the difference image plus texture

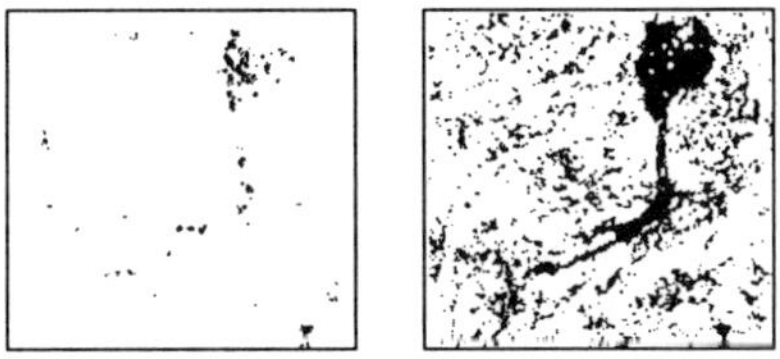

Figure 10: Training using the difference image plus opening, closing, and texture

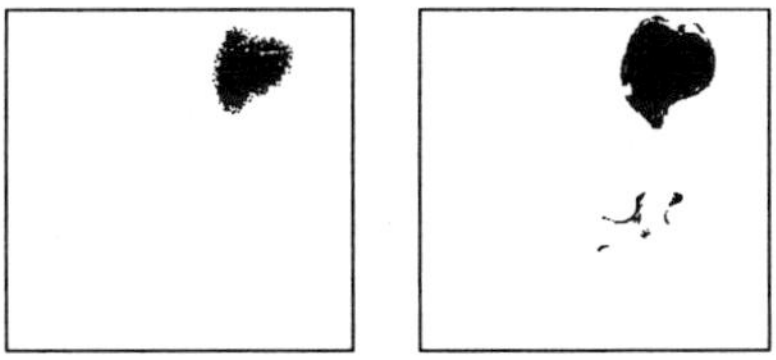

Figure 11: Training using the difference image plus smoothing

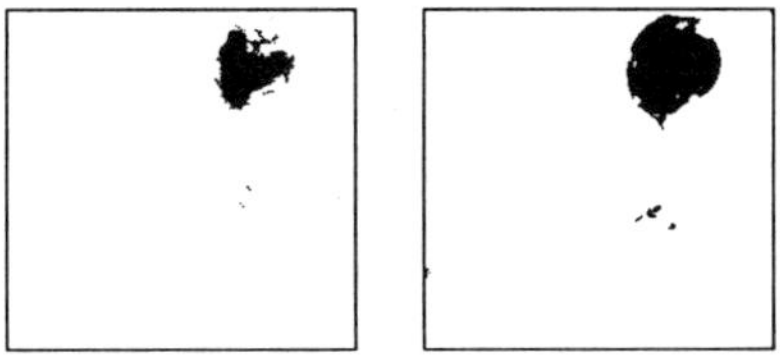

Figure 12: Training using the difference image plus opening, closing, and smoothing, and distance transform

```
(+ (minimum (minimum (- (- s10
                        (* difference
                           (+ s20 s20))) dt)
               (maximum (maximum dt
                           (- s40 dt))
                    (* (- s10 dt) s40)))
          (minimum (* (- o15 s10)
                   (sigmoid (* difference
                               (+ s20 s20))))
               (maximum (minimum (+ -0.74079 s20)
                           (- (absolute difference)
                              (* s40 s10)))
                    (- (* s10 o15) s20))))
   (- (* (+ s20 s10)
         (* s40 s10))
      (maximum (- (* (absolute difference)
                  (* (+ s20 s10)
                     (* s40 s10)))
                (maximum (* (* s10 o15)
                            (- s40 dt))
                         (- s40 s10)))
            (maximum (* (- s10 dt) s40)
                 (maximum (- (absolute difference)
                             (* s40 s10))
                          (- s40 dt))))))
```

5 Conclusions

Our experiments showed that simple per pixel thresholding was not entirely satisfactory on its own in providing a good segmentation of a difference image (obtained from time-separated acquisitions) for detecting landslide occurrence and/or reactivation. Utilising coarser scale information as well as texture, shape, and edge filtering provided a set of additional features. These were combined using genetic programming, and formed a new classifier that provided better identification of active sectors within a landslide body.

6 Acknowledgements

Program development was done using the lil-gp genetic programming system developed at Michigan University[10].

References

1. J. Hervás, J.I. Barredo, P.L. Rosin, A. Pasuto, F. Mantovani, and S. Silvano. Monitoring landslides from optical remotely sensed imagery: the case history of Tessina landslide, Italy. *Geomorphology*, in press.
2. F. Mantovani, A. Pasuto, S. Silvano, and A. Zannoni. Collecting data to define future hazard scenarios of the tessina landslide. *Int. J. of Applied Earth Observation and Geoinformation*, 2(1):33–40, 2000.
3. J.R. Koza. *Genetic Programming: On the Programming of Computers by Means of Natural Selection*. MIT Press, 1992.
4. P.L. Rosin, J. Hervás, and J.I. Barredo. Remote sensing image thresholding for landslide motion detection. In *1st Int. Workshop on Pattern Recognition Techniques in Remote Sensing*, pages 10–17, 2000.
5. P.L. Rosin and H.O. Nyongesa. Combining evolutionary, connectionist, and fuzzy classification algorithms for shape analysis. In S. Cagnoni *et al.*, editor, *Real-World Applications of Evolutionary Computing*, pages 87–96. Springer, 2000.
6. E. Stringa. Morphological change detection algorithms for surveillance applications. In *British Machine Vision Conf.*, pages 402–411, 2000.
7. K.I. Laws. Rapid texture identification. In *Proc. SPIE Image Processing for Missile Guidance*, pages 376–380, 1980.
8. G. Borgefors. Distance transformations in digital images. *Computer Vision, Graphics and Image Processing*, 34(3):344–371, 1986.
9. P.L. Rosin and G.A.W. West. Salience distance transforms. *CVGIP: Graphical Models and Image Processing*, 57:483–521, 1995.
10. D. Zongker and B. Punch. *http://isl.cps.msu.edu/GA/software/lil-gp*. 1996.

A MULTITEMPORAL CHANGE-DETECTION ALGORITHM FOR THE MONITORING OF BURNT AREAS WITH SPOT-VEGETATION DATA

L. BOSCHETTI, S. FLASSE, S. TRIGG AND A. JACQUES DE DIXMUDE

Natural Resources Institute, University of Greenwich, Central Avenue, Chatham Maritime, ME4 4TB, Kent, United Kingdom,

tel: 44(0)16343945, fax: 44(0)16343232, e-mail: l.boschetti@gre.ac.uk

A change detection algorithm has been developed, focusing on Southern Africa, in the framework of the Global Burned Area initiative (GBA2000), whose aim is the production of monthly burned area maps for the year 2000 using SPOT-VEGETATION data. The burned areas are detected by analysing the changes induced by the fire in the red and near-infrared bands between the *current* image, and a *reference* image (updated regularly). To minimise the effect of observation and illumination geometry, the algorithm uses 5-days-apart images. The validation of the results, undertaken with high-resolution Landsat-TM data acquired over three test sites, shows the robustness of the algorithm and the good global accuracy of the resulting maps.

1 Introduction and framework

The year 2000 is a reference year for environmental assessment in relation to various activities, and in particular the United Nation's Ecosystem-related International Conventions. With a network of partners around the world, the Global Land Cover 2000 project (GLC 2000) aims at providing a harmonised land cover database over the whole globe. In this framework areas burned by fire have been recognised as playing an important role in environmental processes.

Consequently the Global Burned Area initiative 2000 (GBA 2000), co-ordinated by the Global Vegetation Monitoring Unit (GVM) of the Joint Research Centre of the European Commission, has been initiated to provide GLC 2000 with monthly burned area maps at global scale for the year 2000, using SPOT-VEGETATION data. GBA 2000 comprises a network of partners each responsible for the study of different regions of the world. The team of the Natural Resources Institute focuses on Southern Africa.

2 Material

2.1 SPOT VGT data

The VEGETATION Instrument (VGT), onboard SPOT-4, has four spectral bands: blue (**b0**, 0.440-0.470 μm), red (**b2**; 0.610-0.690 μm), Near InfraRed (**b3**; 0.7660 – 0.890 μm) and Short Wave InfraRed (**MIr**; 1.590 – 1.690 μm). The spatial resolution is 1 km at nadir. All the data used in this study are provided in S-1 format daily synthesis, *i.e.* calibrated, precisely geo-located and atmospherically corrected.

The cycle of the SPOT satellite is 26 days, but, due to the wide angle of view of VGT, data are mostly available daily. The viewing conditions of every geographic point vary for each of the 26 days of the cycle, but every five days the point is observed under similar angle geometry [1].

VGT data is usually characterised by good coverage and consistent image quality and consequently it is particularly suitable for investigations using change detection techniques [2]. Six months of daily images, covering the fire season between May and November were used in this study.

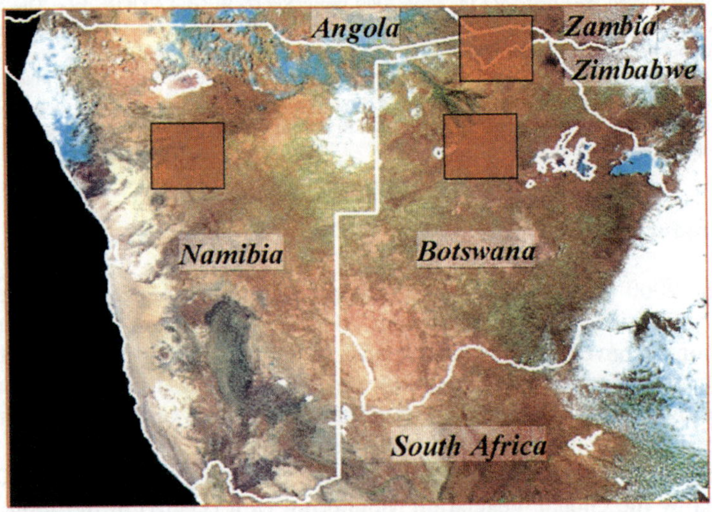

Figure 1. Sample SPOT-VGT image of the study area (Red: mir, Green: b3, Blue:b2); the red boxes mark the test sites where TM data have been used for the validation

2.2 Study area

The work has been conducted over a study area in Southern Africa covering Botswana and Namibia (17° to 28°S and 21° to 29°E, see Fig. 1),. The fire season usually starts in this area in May and ends in December or January, but we observed that the year 2000 fire season started later, in June, and by the end of November most part of the study area was substantially covered by thick clouds almost every day. As a consequence the detection of the areas burned in the last months of the fire season was nearly impossible. The three test sites used for the validation are also indicated in Fig. 1.

3 Method

3.1 Burnt area detection approach

A vegetation fire completely changes the characteristics of its local environment. The change is drastic and sudden. In order to build a robust algorithm applicable to various ecosystems as well as their seasonal variations (*i.e.* not to have to adjust thresholds for each case), the approach adopted here is to exploit this drastic change and therefore detect burned areas by looking at the changes of vegetation status and characteristics through time.
The main features of the approach are as follows:

- the basic condition to apply a change detection algorithm is that the changes in the radiance due to the land cover change induced by a fire are significantly greater than the changes due to other factors [3]. **b3** and **Mir** bands are used to detect the changes resulting from a fire event, following Trigg and Flasse [4, 5] who recently indicated that the **b3** and **Mir** bands provide the best combination of the SPOT-VGT spectral region to detect the transition between vegetation and burnt area.
- the change detection is based on the comparison between the 'new' (or 'current') image and a 'reference' image. The latter is built originally: it is initiated at the start of the processing (beginning of the season) and updated at every step (each time a new image is available) with those pixels that are 'clean', i.e. pixels not affected by clouds, shadows or other perturbing factors.

The next sections briefly cover the way the algorithm has been built to take into account issues affecting the detection, such as:

- variation of the viewing conditions
- clouds and smoke and other perturbing factors

3.1.1 Influence of the viewing geometry

It is widely recognised that the variation of the observation angles heavily influences the radiance received at the sensor [6,7]. As an example of the way this effect interferes with the burned area detection, Fig. 2 shows the reflectance plot of two adjacent pixels during one month of observations (4 - 30 August 2000). One of the pixels (pixel B) burns during the period (on August 21st), while the other remain unburned. Looking at the observations before the fire, when the two pixels have similar reflectance values, a periodical variation of each pixel reflectance is noticeable, mostly due to the variation of the geometry of illumination and observation. The amplitude of this variation (about 5% of reflectance) is in the same range of variation than the difference of reflectance values observed between pixel A and pixel B after the fire. Fig. 3 illustrates how, by only analysing an image every five days, variations due to the viewing conditions are minimised, and consequently the change variation induced by the fire is enhanced.

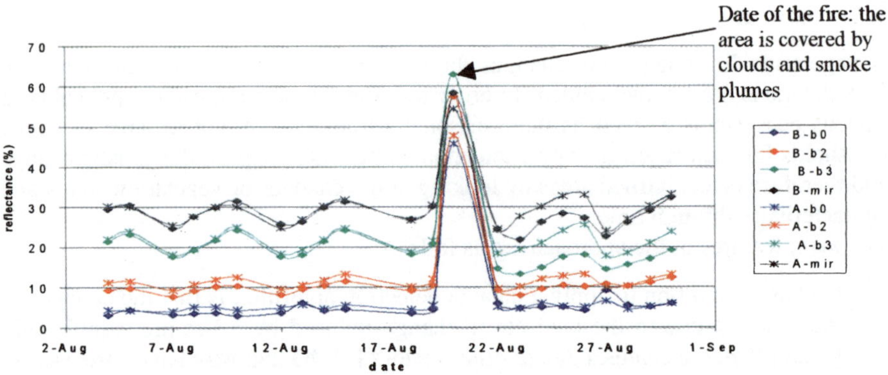

Figure 2. Reflectance of two adjacent pixels during 1 month of observations. Pixel A corresponds to unburned vegetation, pixel B burns during the period (August 21).

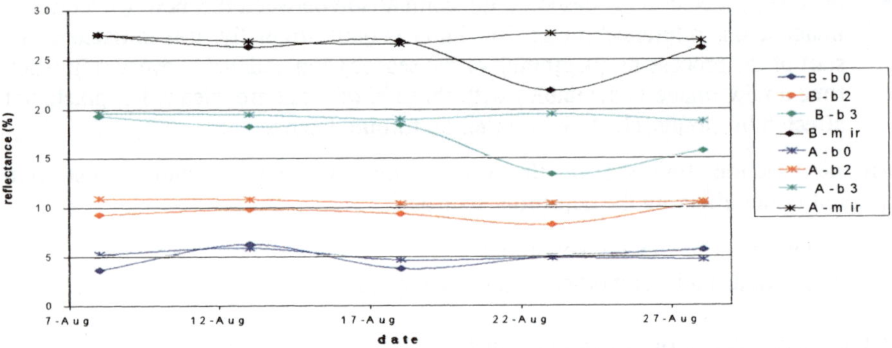

Figure 3. Plot obtained from Fig.2 with a 5 day sampling.

3.1.2 Clouds and smoke

As in most remote sensing studies, cloud and smoke masking is a critical issue. Pixels covered completely or partially by clouds usually show higher reflectance than a vegetated surface. If a cloudy pixel is not recognised by the cloud masking algorithm, the first following cloud-free observation for the same pixel will show a strong decrease in reflectance, even if the pixel is still 'unburned'.

The cloud mask adopted by GBA 2000 is based on multiple thresholds on reflectance values, and the cloud shadow mask is using the position of the Sun to identify those pixels that should be affect by cloud shadow [8]. A further test was introduced in this study, based on a multi-temporal analysis, to identify strong increases in the reflectance between two consecutive images, in order to eliminate those increases that can not be explained in terms of change of the surface conditions but that probably correspond to extra pixels partially affected by clouds, smoke or other perturbing factors.

3.2 Scheme of the algorithm

As discussed above, the algorithm compares an image every five-day in order to minimise the influence of the geometry of illumination and observation [NB: since the swath of the satellite is wide, it covers different time zones, i.e. different position of the Sun too, which also influences the reflectance]. This means that the whole data set is split into five separate sets each containing images with similar viewing conditions. These data sets, are analysed independently.
When a new image is available the algorithm processes as follows:

- The image is assigned to its corresponding set (based on the acquisition date).
- The difference between the new image and the reference image is computed pixel by pixel.
- If the difference is positive (i.e. the reflectance increases) and greater than a fixed threshold, then the pixel is marked as contaminated, as described in 3.1.2, and it is excluded from further processing.
- If the difference in the **b3** and **mir** bands is negative (*i.e.* the reflectance decreases) and greater than a threshold, then the pixel is marked as burned and the date of the detection is recorded in the burnt area map.
- The reference image is updated, taking from the new image all the pixels which are not marked as cloud/cloud shadow in the pre-processing, and not marked as cloud or burned area in the above steps of the processing. This means that the reference image is at any time the best one among the latest observations of each pixel.

4 Results

The algorithm produces daily burned areas, *i.e.* maps showing the pixels affected by burned areas and the date of detection, assuming that a pixel burns only once in the fire season. As each of the five series runs independently, five burned area maps are obtained. When, in the post processing phase, these maps are combined, two characteristics can be extracted:

- The date of first detection among all the series.
- The number of times the pixel has been detected as burned.

It is possible to use the latter as confidence level indicator, assuming that the more times the pixel has been detected, the more reliable the detection is. The map in Fig. 4 has been obtained considering only the pixels detected at least twice.

Figure 4. Burned area map obtained processing the whole data set for the year 2000-fire season; the test sites for the validation are outlined in red.

5 Algorithms evaluation

The algorithm performance has been evaluated using a multi-criteria methodology [9]; besides the traditional accuracy indexes, such as omission and commission errors, a specific analysis of the errors induced by the lower spatial resolution of the SPOT VGT compared with the high resolution reference data was undertaken.

The validation, based on Landsat7 ETM data, involved three test sites in different regions of the study area (Fig. 1). Although only a small fraction of the whole window is covered, the three sites represent the main ecosystems of the area; moreover the TM data have been acquired in different moments of the fire season, therefore also covering variations due to the season. In each site the reference map derived from high resolution TM data is compared with all the burned areas detected in the VGT data set from the beginning of the processing (end of May) to the acquisition date of the TM.

Results summary for the three sites:

- **Namibia**: reference data acquired on 31 October, at the end of the fire season. The burned areas are very small, with a very low level of fragmentation when observed at 1km resolution. The omission error is about 20% and the commission error, about 40%. The latter, is mainly due to pixels that have been classified as burnt in the VGT data but are actually only partially covered by burned areas.

- **Botswana**: reference data acquired on 10 September, at the beginning of the fire season. In the window there are only few burned areas (less than 2% of the TM image). The low number of false detection (few hundreds of pixels for over 40.000), indicates that the algorithm is quite robust to run unsupervised on several months of data without gross errors.
- **Caprivi strip** (Namibia-Botswana): reference data acquired on 4 October, when the fire season in the area is almost finished. In this window a strong omission error (more than 50%), can be due to the very low change induced by the fire in the reflectance of some areas, because of very dry vegetation before the fire (which has already a very low reflectance in the infrared bands) or to the size of the fires being too small to induce a significant signal change at 1km resolution.

Additional validation approaches and results discussion will soon be available in another publication.

6 Conclusions

This work was conducted in the framework of the GBA 2000 initiative, whose objective is the production of monthly burned area maps of the whole globe for the year 2000, from SPOT VEGETATION data, providing daily images of consistent quality. Consequently the use of change detection technique was possible. In this work an algorithm was developed to detect burned areas by looking at the strong decrease in the infrared bands **b3** and **mir** induced by the fire. Instead of introducing a specific correction for the influence of the variation of the geometry of illumination and observing, this effect was minimised by making use of the 5-day cycle of the satellite and consequently comparing only 5-day-apart images.
The validation was conducted comparing the VGT derived burned area maps with high resolution TM data, acquired over three different sites. This permitted to observe the behaviour of the proposed algorithm over different environments and in different moments of the fire season. The accuracy assessment proved that the algorithm is robust enough to process several months of data without any supervision, while avoiding gross errors. Using a unique set of thresholds (relative to the changes in radiance) for the whole area and period, the results are generally good.

7 Acknowledgments

The first author is supported by a grant from Politecnico di Milano.

References

1. Spot-Vegetation Technical Information, available at http://vegetation.cnes.fr
2. Singh, A., 1989, Digital change detection using remotely sensed data. *International Journal of Remote Sensing*, 10:989-1003
3. Ingram, K., Knapp, E., and Robinson, J.W., 1981, Change detection procedure development for improved urbanized area delineation, technical memorandum CSC/TM-81/6087, Computer Sciences Corporation, Silver Springs, Maryland, U.S.A.
4. Trigg, S and S. Flasse (2000) 'Characterizing the spectral-temporal response of burned savannah using in situ spectroradiometry and infrared thermometry' *International Journal of Remote Sensing*, **21**, 3161-3168.
5. Trigg, S. and S. Flasse (2001) 'An evaluation of different bi-spectral spaces for discriminating burned shrub-savannah' *International Journal of Remote Sensing*, in print
6. Flasse, S., (1993) Extracting quanatitative information from satellite data: empirical and physical approaches. PhD Thesis, Faculté des Sciences Agronomiques, Université Catholique de Louvain, Louvain-La-Neuve, Belgium, and Joint Research Centre, European commission, EUR 15409 EN, p.199
7. Gobron, N, Pinty, B., Verstraete, M. and Widlowsky, J-L, (2000) Advanced Vegetation Indices Optimised for Up-Coming Sensors: Design, Performance and Applications. *IEEE Transactions on Geoscience and Remote Sensing* 38(6):2489-2505
8. Kempeneers, P., Lissens G., Fierens, F., Van Rensbergen, J, 2000, Detection of Clouds and Cloud-Shadows for VEGETATION images, *Proceedings of the Vegetation 2000 Conference* (Belgirate, Italy, 3-6 April 2000), in print
9. Boschetti, L., Flasse, S., Trigg., S, Brivio, P.A. and Maggi, M., 2001, A methodology for the validation of low resolution remotely sensed data products, accepted for *Proceedings of the 4th ASITA Conference* (Rimini, Italy, 9-12 October 2001)

SHAPE CHANGE DETECTION BY FUZZY MEASURES

Paul C. Smits
Joint Research Centre, TP262, 21020 Ispra (VA), Italy. Tel. +39-0332-785279. Fax +39-0332-789803. E-mail: paul.smits@ieee.org

Silvana G. Dellepiane
University of Genoa, Via all'Opera Pia 11A, 16145 Genoa, Italy. E-mail: silvana@dibe.unige.it

Most methods that support users in the analysis of multi-temporal remote sensing images rely on changes in the spectral signatures of pixels. This is useful when dealing with images that are collected with identical or similar sensors, but inappropriate when a scene is acquired with two sensors that measure different physical parameters.
An alternative to using the spectral information of multi-temporal and multi-sensor images is to exploit the pictorial information in the images, obtained by suited segmentation approaches rather than the spectral information. Segmentation approaches need to be optimised or tuned for the images under analysis and for the phenomena of interest, which implicitly includes the nature of the observed scene.
The approach proposed in this paper follows the direction of using segmentation algorithms. However, rather than using the crisp regions computed from the multi-temporal data, the difference in pictorial information between the two dates is determined by the fuzzy segmentation's membership maps computed for the two data sets. The fuzzy membership maps can be considered an "intermediate product" of a fuzzy segmentation algorithm.
The results show that this approach, which in fact produces a measure for change of shape, could be of interest to remote sensing applications, but care should be taken to filter the input data such that the identified changes are representative for the phenomena of interest, and not for some undesired noise component.

1 Introduction and problem definition

There are many possible applications of remote sensing image change detection that rely on data originating from different sensors. Often these sensors are similar in terms of spectral response. An example is the combined use of Landsat and SPOT multispectral images in agricultural monitoring, where the Landsat bands are chosen such that they match the available spectral bands of the SPOT data.
However, it frequently occurs that the sensors are fundamentally different in nature, which is the case when a scene is acquired first by an optical sensor, and later by a Synthetic Aperture Radar (SAR) instrument, but also when a newly acquired image is combined with previously computed for manually collected information stored in a geographic information system (GIS).

In these cases, the range of suited automatic or semi-automatic techniques that can assist in the analysis becomes rather limited. Pixel-wise techniques are not always able to provide a useful basis for tools to identify changes. And if they do in theory, they might run into practical problems, like the requirement of a large amount of training samples that may not always be available. An example of a pixel-wise method is the Post Classification Comparison (PCC), which is in fact known to be problematic in terms of requirements for training samples and regarding its accuracy. Another example regards the use of a stacked feature vector which combines the bands of the two multi-temporal data sets. The latter technique offers the possibility to either use the stacked vector directly in what can be called a multi-temporal classification, or the vector can be use in a principal component transformation (PCT) [6].

An alternative to pixel-wise analysis is the comparison of pictorial information (e.g., regions or edges) present in the imagery collected at different times. Under certain conditions, the changes of shapes between the regions found in the first image and those found in the second image are an indication for the changes that have occurred in an area [7].

One way of overcoming problems related to differences in sensor characteristics is by using segmentation algorithms, optimised for the sensors, the contents of the scene, and the phenomena of interest, followed by a comparison of the shapes of the regions thus obtained from the multi-temporal data sets.

The approach proposed in this paper is based on the concept of segmentation algorithms. If classification can be regarded a form of clustering in the feature space, segmentation algorithms can be seen as a clustering in the image space, grouping pixels into regions which are homogeneous in some respect [8].

However, rather than using the crisp regions determined from the multi-temporal data, the difference in pictorial information between the two dates is determined by the fuzzy segmentation's membership maps computed for the two data sets. The fuzzy membership maps can be considered an "intermediate product" of fuzzy logic-based segmentation algorithms. In particular, this paper follows the concept of comparing shapes derived from the different data sets, using a technique based on the fuzzy intensity connectedness measure proposed and described in [1], [2], and [3].

The organization of the rest of this paper is as follows. Section 2 provides an overview and describes the concept of fuzzy intensity connectedness, which is used to express the pictorial content. Section 3 shows results obtained with synthetic and real data, and section 4 discusses and concludes the paper.

2 Proposed method

2.1 Overview

Given are an image data set acquired at t_1 and an image data set acquired at t_2, represented by two feature vectors $X_i^{t_1}$ and $X_i^{t_2}$, respectively, defined on single lattice S of points, and $i \in S$ denotes a generic pixel. Let I_i denote some measure computed form X_i at location *i*. It is assumed that the data sets are co-registered, and that *i* represents the same location in both data sets.

The objective of the proposed method is to create a new image ψ defined on S in which each pixel is assigned a value that is a measure for change of the shape of the local region to which the pixel partakes. This makes ψ_i a function of $R_i^{t_1} = \phi(X_i^{t_1})$ and $R_i^{t_2} = \varphi(X_i^{t_2})$, where R indicates the region to which pixel i belongs, and ϕ and φ are suited segmentation operations (not necessarily identical), optimized for the scene under analysis and the phenomena of interest. If the regions resulting from a segmentation are identical in the images of both dates, there is no change of shape. If, on the other hand, the segmentation results differ in a region of the image, we want to know to what extent the shapes differ.

It is possible to work directly with the segmentation results. However, methods exist that create a probabilistic or fuzzy membership value before a threshold actually creates the crisp regions. Here it is felt that rather than using measures computed from the crisp segmentation result, it is worthwhile investigating the possibility of using a "soft" segmentation result, or, in other words, the half-product of a fuzzy segmentation: the membership map or similar .

Of particular interest here are region growing methods that exploit some monotonic decreasing fuzzy membership maps. Based on the exploitation of a membership function that has its highest value at a location of interest a, these approaches generate a local membership map that assigns high membership values to pixels with a similar spectral signature as a, and that are also close to a in terms of distance in the image space; lower values are assigned to pixels that have different spectral signatures. Groups of pixels with membership values higher than a user provided threshold correspond to what are regions in normal segmentation results.

This paper investigates if and how measures can be computed from the local differences in the fuzzy membership maps computed for the two dates.

Based on those premises, the proposed shape change detection technique consists of the following steps:

1. Compute for pixel a a local fuzzy χ-connectedness field from both data set $X_a^{t_1}$ and data set $X_a^{t_2}$. The fields are denoted by $\Gamma_\chi^a(t_1)$ and $\Gamma_\chi^a(t_2)$, respectively.
2. Compute the distance ψ_a between the two fields.
3. Repeat steps 1 and 2 until all $a \in S$ are processed.

The following paragraphs will explain the concept of fuzzy χ-connectedness (step 1) and outline the distance measure mentioned in step 2.

2.2 *Fuzzy χ-connectedness*

A fuzzy image H is a fuzzy subset

$$H = \{(p, \eta(p)) \mid p \in S\}, \tag{1}$$

with η in the range [0,1]. It can be derived from the fuzzification of an image I by finding a rational value v such that the relationship $\eta(p) = v \cdot I(p)$ gives rise to values for η in the range [0,1], p begin a pixel location. For 8-bit gray scale images, a typical value of v would be 1/255.

The fuzzy χ-connectedness (intensity connectedness) is defined as in [3]:

$$c_{\chi^a}(p) = \max_{P(a,p)}\left[\min_{z \in P(a,p)} \chi^a(z)\right], \tag{2}$$

which includes the transformation of the field H into the field χ^a defined as

$$\chi^a = 1 - |\eta(p) - \eta(a)|. \tag{3}$$

The reference point a is called the *seed point*. A fuzzy χ-connected component associated with the seed point a, is the fuzzy field

$$\Gamma_\chi^a = \{(p, c_{\chi^a}(p)) \mid p \in F\}, \tag{4}$$

where F is a subset of S. Figure 1a shows an example of a fuzzified image. Figure 1b shows the χ-connected component where r is taken as the seed point. Although in Fig. 1b pixel r and s are spatially spoken close, they are distant in terms of membership value: the fuzzy connectedness path that connects the two is relatively long. The path follows as much as possible pixels with similar connectedness values.

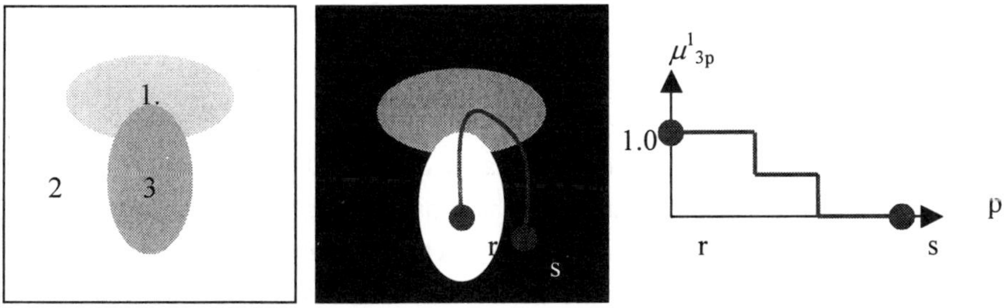

a. b. c.

Fig. 1. Schematic example of how the long-range interaction works. a) Fuzzified synthetic image. b) Fuzzy χ-connected component for seed point r. White corresponds to 1.0, black to 0.0.

The above description has been for grey level images, but it is possible to extend the Fuzzy χ-connectedness to multispectral images. A straightforward way is to rewrite (3) as a multispectral fuzzy χ-connectedness:

$$\chi^a = \sum_{b=1}^{B} 1 - \left|\eta^b(p) - \eta^b(a)\right|, \tag{5}$$

where b is an index for the multispectral band, and B is the total number of bands considered.

To sum up, starting from a generic seed point a, a fuzzy connectedness map is grown with is a soft representation of the object to which the seed point partakes.

2.3 Distance between the fuzzy fields

The distance between two local fuzzy fields Γ_χ^{a,t_1} and Γ_χ^{a,t_2} needs to be defined. This distance should be chosen such that it can be considered a measure for the change of shape. In this article, the distance is the average distance a fuzzy distance and computed as follows:

$$\psi(a) = \frac{1}{|F|} \sum_{i \in F} \left|\Gamma_\chi^{a,t_1} - \Gamma_\chi^{a,t_2}\right| \tag{6}$$

In words, (6) says that the distance between the two fuzzy pictorial representations of the two dates is equal to the sum of the pixel-wise difference between those representations. The sum is computed over a specified area *F*.. Here, *F* is the local extent of the fuzzy field and it determined by

$$F = \{(p, c_{\chi^a}^{t_1}(p)) \mid c_{\chi^a}^{t_1}(p) > \gamma\}, \tag{7}$$

γ being a suited threshold that determines the extent of the fuzzy field *F*. Note that γ does not turn the fuzzy fields into a hard partitioning, but determines only the local (spatial) limit of the subimage from which the distance measure is computed. The range of γ is $]0,1]$.An alternative to (7) is to fix *F* such that the area is sufficiently large for the phenomena of interest.

3 Results

3.1 Synthetic data set

Figure 2 shows the results of the method applied to a synthetic image data set. The differences between t1 and t2 are only minor, reflected by the result in Fig. 2c.

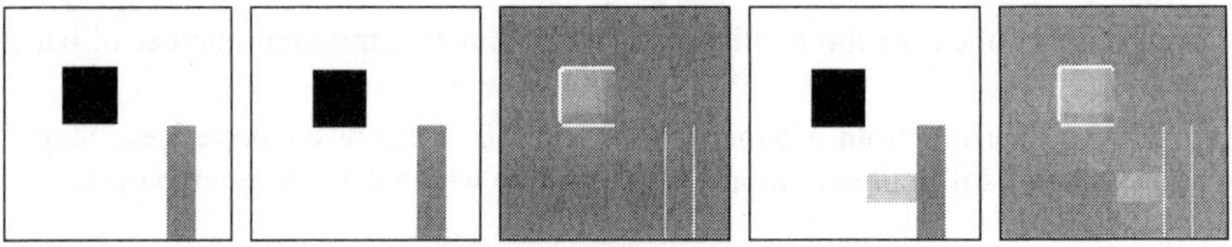

a. b. c. d. e.

Fig. 2. Results with the synthetic image data set. a) Image t1. b) Image t2. c) Detected changes between t1 and t2 with the proposed approach, enhanced for visualization. d) Image t3. e) Detected changes between t1 and t3, enhanced.

3.2 Real data set

The data set used for the experiments is a portion of GRSS_DFC_0004 [4]. It is a 1024 x 750 pixel AIRSAR data set of Flevoland, Netherlands, with C, L, and P bands in their HH, HV and VV polarizations. The AIRSAR image is acquired on August 16, 1989. The second image, acquired on 18 October 1999 over the same area is a 1025 x 750 Landsat 7 TM image. The original image has a resolution of 30 x 30 m for the multispectral bands. The ETM image is resampled (nearest neighbourhood) and co-registered to the AIRSAR image to 12.1 x 6.6 m resolution. For the experiments, a 300x300 pixel portion is used, with the L-hh band for 1989 and ETM band 4 for 1999 (Fig. 3). The L-hh band is filtered with the Lee filter [5], with the Multiplicative Noise Mean set to 0.75, and the Noise Variance to 0.75.

a. b.

Fig. 3. 300x300 pixel portion of the Flevoland data set [4]. a) L-hh intensity image, 1989. b) Landsat 7 TM image, 1999 (band 4)

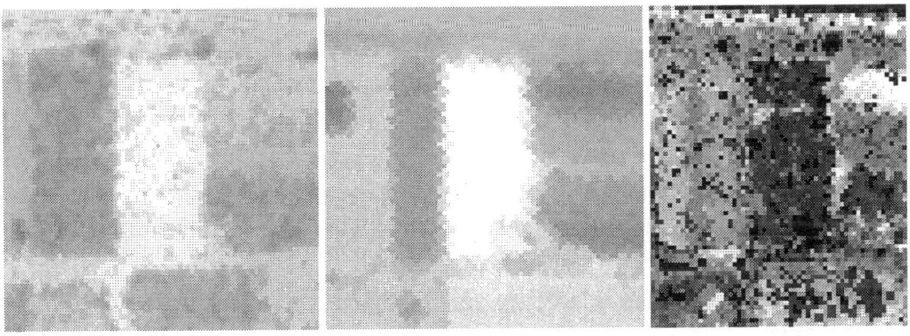

a. b. c.

Fig. 4. Illustration of the proposed approac. a) Fuzzy connectedness map computed for seed pixel (81,95) from the filtered L-hh image. b) Computed from the ETM image from the same location. c) The local change image using the measure described in subsection 2.3, histogram equalised to emphasize the changes.

Figures 4a and 4b show the χ-connectedness maps produced for location (92,92) of the images in Fig. 3 (i.e., the seed point *a*). Figure 4c shows a detail of the change map produced by using the distance measure defined in subsection 2.3, computed from χ-connectedness maps as in Figs. 4a and 4b, but computed for each pixel in the image.

4 Discussion and conclusions

In this paper a new approach to shape change detection has been proposed. The method relies on fuzzy measures computed from images acquired by different sensors at two different times. For each location, the fuzzy objects belonging to that location are extracted from the two images and compared. The power of the method is that is can combine sensors with different characteristics. However, a relevant correlation should exist between the images acquired by the two sensors; appropriate filtering may be needed to enhance the features of interest. The drawback of the method is that it is rather computation-intensive. The full paper will contain more results and will further detail the theory behind the approach. The results show that this approach, which in fact produces a measure for change of shape, could be of interest to remote sensing applications, but care should be taken to filter the input data such that the changes are representative for the phenomena of interest, and not for some undesired noise component.

5 References

1. S.G. Dellepiane and F. Fontana, "Extraction of Intensity-Connectedness for Image Processing," Pattern Recognition Letters, vol. 16, no. 3, p.313-324, March 1995.

2 M. Mari and S. Dellepiane, "A non-linear image processing approach through fuzzy measures", Pattern Recognition Letters, vol. 18, p. 1109–1115, November 1997.

3 Dellepiane, S., Fontana, F., Vernazza, G., 1996. Non linear image labelling for multivalued segmentation. IEEE Trans. Image Process. 5 3 , 429–446.

4. IEEE Geoscience and Remote Sensing Society, Data Fusion Committee benchmark images. Data set GRSS_DFC_0004. Online. http://www.dfc-grss.org/.

5. Lee, Jong-Sen, "Digital Image Enhancement and Noise Filtering by Use of Local Statistics," IEEE Transactions on Pattern Analysis and Machine Intelligence, Vol PAMI-2, No.2, March 1980, pp. 165-168.

6. R.A. Schowengerdt, *Remote Sensing – models and methods for image processing*, Academic Press, 1997.

7. P.C. Smits and A. Annoni, "Towards operational knowledge-based remote-sensing image analysis," *Pattern Recognition Letters*, vol. 20, no. 11-13, November, 1999, pp. 1415-1422.

8. Haralick, R.M., and Shapiro, L.G., 1993, *Computer and robot vision* (New York: Addison-Wesley).

PROPOSAL OF DIFFERENT APPROACHES TO SPATIO-TEMPORAL CONTEXTUAL CLASSIFICATION OF REMOTE-SENSING IMAGES

FARID MELGANI AND SEBASTIANO B. SERPICO

Dept. of Biophysical and Electronic Eng., Univ. of Genoa
Via Opera Pia, 11a, I-16145 Genova, Italy.
E-mail: vulcano@dibe.unige.it

In addition to multisensor data fusion, the integration of spatial and/or temporal contextual information in the classification scheme may represent an effective way to improve the classification accuracy with respect to the single-time non-contextual classification. The supervised spatio-temporal contextual classification problem can be viewed as a problem of fusion of multisource (spectral, spatial and temporal) information in order to assign each pixel to the most likely class. In this paper, we propose and compare fusion models based on different approaches, namely, the statistical, neural networks and Markov Random Fields (MRF) approaches. Experimental results on a multitemporal data set consisting of two multisensor (Landsat TM and ERS-1 SAR) images are presented. The performances of the proposed fusion approaches are compared with a reference classifier based on Markov Random Fields in terms of classification accuracy.

1 Introduction

The fusion of spectral information with spatial and temporal contextual information can represent a valuable way to improve the accuracy of the classification of multitemporal remote-sensing images. Contextual information allows to eliminate possible ambiguities and to correct errors in a classification problem. Contextual classification has been deeply investigated by the pattern recognition community in particular, with reference to the spatial context [8]. By contrast, the investigation concerning the use of the temporal context for classification has not been so extensive. Focusing on the classification of remote sensing images based on multitemporal information, we recall the "cascade" classifier introduced by Swain [12], who generalized the Bayes optimal strategy to the case of multiple observations by assuming a class-conditional independence of feature vectors derived from different temporal data sets and by allowing class changes over time. The stochastic model-based approach proposed by Kalayeh and Landgrebe [6] is based on a stochastic system representing land-cover types through a non-stationary Gaussian process, as input, and the temporal spectral behavior, as output. The probability of transition from a class at one date to a different class at another date is used as intuitive information to express the correlation of two or more images separated by time. An iterative technique, deduced from a derivation of the Bayes rule for minimum error, is proposed by Bruzzone and Serpico [1] to estimate the matrix of the probabilities of transitions (TPM). In [2], the TPM is replaced with the Joint Probability Matrix (JPM), computed by applying a specific formulation of the Expectation-Maximization algorithm. Two techniques, derived differently from the notion of the minimum expected cost and dealing with the multitemporal aspect at the decision level, were presented by Jeon and Landgrebe [4]. Finally, only few papers have been published dealing with both the spatial and temporal contexts for remote-sensing image classification. An interesting (but quite expensive from the computational viewpoint) autocorrelation model was proposed by Khazenie and Crawford [7]; it results from the superposition of a class-dependent process and an autocorrelated (spatial and temporal) noise process. Markov Random Fields (MRFs) are characterized by a remarkable ability to easily integrate the spatial and temporal contexts in a scene model by using the concept of energy functions. Jeon and Landgrebe [5] proposed a cascade spatio-temporal contextual classifier that involves MRFs to model spatial correlation. MRF capabilities are fully exploited by a general model that permits one to easily merge not only different data sources, including ancillary data (e.g. derived by a GIS), but also the spatio-temporal aspect by simple energy functions [11].

In this paper, we propose and compare fusion models based on different approaches, namely, the statistical, neural networks and Markov Random Fields (MRF) approaches. Experimental results on a multitemporal data set consisting of two multisensor (Landsat TM and ERS-1 SAR) images are presented. The performances of the proposed fusion approaches are compared in terms of classification accuracy with a reference classifier based on Markov Random Fields.

2 Problem Formulation

Let us consider a multitemporal data set consisting of two images, I_1 and I_2, acquired at times t_1 and t_2, respectively. I_1 and I_2 are assumed to be registered and, consequently, to refer to the same ground area; each image may, in general, be of the multispectral and/or multisensor type. We regard a (multitemporal) pixel as a pair of elements of the images I_1 and I_2 corresponding to the same ground point and characterized by two feature vectors, x_1 and x_2, extracted from the images I_1 and I_2, respectively. $C=\{c_1,c_2,\ldots,c_{N1}\}$ and $D=\{d_1,d_2,\ldots,d_{N2}\}$ are the two sets of possible class labels to which each pixel may be assigned at times t_1 and t_2, respectively. According to the Bayes Rule for Minimum Error (BRME), the optimal joint classifications C and D of all the pixels of the two images at times t_1 and t_2, respectively, are defined by the following maximization problem:

$$\underset{C,D}{Max}\ \{P(C,D|X_1,X_2)\} \quad (1)$$

where X_1 and X_2 represent the sets of feature vectors for the entire image at times t_1 and t_2, respectively. This global joint classification scheme provides the optimal solution in terms of the BRME. However, the huge computational load required by the direct computation and maximization of (1) makes the problem hard to deal with. Instead, a practical, though suboptimal, solution can be obtained by applying some transformations and adopting some assumptions [10]. In such a way, one can reduce the complex problem of finding the optimal classification of the whole of the image to the problem of an iterative optimal classification of each single pixel taking into account a local spatio-temporal context where most of the correlation is contained. Accordingly, for each pixel of the scene, the two labels c_m^k and d_n^k, i.e., the solutions at the current k-th iteration for the images at times t_1 and t_2 respectively, can be searched for independently of each other and are such that:

$$P(c_m^k \mid x_1, C_N^{k-1}, D_{N*}^{k-1}) = \underset{c_i \in C}{Max}\{P(c_i \mid x_1, C_N^{k-1}, D_{N*}^{k-1})\} \quad (2)$$

$$P(d_n^k \mid x_2, D_N^{k-1}, C_{N*}^{k-1}) = \underset{d_j \in D}{Max}\{P(d_j \mid x_2, D_N^{k-1}, C_{N*}^{k-1})\} \quad (3)$$

x_1 and x_2 represent the feature vectors of the pixel of the images at times t_1 and t_2, respectively. If we consider a 3×3 window for both spatial and temporal neighborhoods, C_N and D_N are sets of 8 class labels, whereas C_{N*} and D_{N*} are sets of 9 class labels (as they include also the label of the pixel itself, at the other date). By means of the implementation of the classification through an iterative scheme for each temporal image, the spatio-temporal neighboring labels computed at the previous iteration are used to generate the pixel labels at the current iteration.

The problem of estimation of the multisource class posterior probabilities expressed in Eqs. (2) and (3) can be approached as a problem of multisource (spectral, spatial and temporal) fusion for which different solutions can be considered.

3 Statistical Fusion Model

In this Section and the following ones, as the same reasoning applies to the multisource class posterior probabilities expressed in Eqs. (2) and (3), we shall focus only on the case related to image I_1. If we assume that these multiple sources of information are independent, the following relation can be easily proved [9]:

$$P(c_i \mid x_1, C_N^{k-1}, D_{N*}^{k-1}) = \frac{P(c_i \mid x_1).P(c_i \mid C_N^{k-1}).P(c_i \mid D_{N*}^{k-1})}{P^2(c_i)} \quad (4)$$

where $P(c_i)$ is the prior probability of the i-th class in the image I_1. This probability can be estimated on the basis of the relative frequency of each class in the training set. $P(c_i|x_1)$ represents the single-time posterior probability conditioned by the information carried by the feature vector x_1 (or in other words, it represents spectral information). In our approach, this probability is estimated by a Multilayer Perceptron neural network (MLP) trained by the error back-propagation algorithm, according to the minimum Mean Square Error (MSE) criterion. The probability estimates also allow to generate (by applying the Bayes rule for minimum error) the pixelwise classification of the image I_1, which (together with the pixelwise classification of I_2 provided by another MLP) is required for the initialization of the iterative classification algorithm.

As an estimate of spatial contextual information expressed by the probability $P(c_i \mid C_N^{k-1})$ that the current pixel may belong to the class c_i at time t_1, given the spatial context C_N^{k-1}, we propose to utilize the relative frequency of the class c_i in C_N^{k-1}, on the basis of the classification map obtained at the previous iteration:

$$P(c_i \mid C_N^{k-1}) = \frac{1}{8} \sum_{c_h \in C_N^{k-1}} I(c_i, c_h), \text{ where } I(c_i, c_h) = \begin{cases} 1, \text{if } c_i = c_h \\ 0, \text{otherwise} \end{cases} \quad (5)$$

A possible way to model the temporal link between images acquired at two dates, for classification purposes, lies in using the Transition Probability Matrix (TPM). In the present case, the elements of the TPM provide the probabilities of transition $P(c_i|d_h)$ from any class d_h in $D=\{d_1,d_2,\ldots,d_{N2}\}$ to any class c_i in $C=\{c_1,c_2,\ldots,c_{N1}\}$. To compute the TPM, we propose to consider the expert's knowledge of the possible transitions between classes at two different times, expressed in terms of allowed ($\lambda(c_i|d_h)=1$) or forbidden ($\lambda(c_i|d_h)=0$) transitions. Then we weigh this information with the prior probability of each class c_i and normalize it as follows:

$$P(c_i \mid d_h) = \frac{\lambda(c_i \mid d_h).P(c_i)}{\sum_{c_j \in \Phi} [\lambda(c_j \mid d_h).P(c_j)]} \quad (6)$$

The TPM elements given by (6) are finally used to compute the posterior probability term based on the temporal context (i.e., the temporal context derived from the classification map of the image I_2 at the previous iteration) as the average transition probability from all the labels in D_{N*}^{k-1} to the class c_i at time t_1:

$$P(c_i \mid D_{N*}^{k-1}) = \frac{1}{9} \sum_{d_h \in D_{N*}^{k-1}} P(c_i \mid d_h) \quad (7)$$

4 Fusion Model based on Neural Networks

The assumption about the independence of the three sources of information introduced in the statistical fusion model in general is not verified and involves an evident loss of information. Another possible way to estimate the multisource class posterior probabilities of Eq. (2) is to express them as a non-linear combination of the three above information sources. Multilayer Perceptron (MLP) neural networks represent a good candidate to learn complex relationships that may characterize different information sources and, consequently, to deduce implicitly and directly from examples the source-specific reliability weights in the combination. As the network configuration can influence the classification performances, three MLP-based configurations are considered. For all of them, the classification decision is carried out using the classical "winner-takes-all" rule. The first network configuration is the classical Fully Connected Architecture (FCA) in which the multisource class posterior probabilities are estimated by a single MLP. The second network configuration, called the Class-Sensitive Architecture (CSA), implements a "one-net-one-class" classification scheme where the multisource class posterior probability estimations are decoupled by using for each of them a separate MLP subnet. All subnets share the same input vector of multisource information but are trained independently for best performance. As an extension and simplification of the CSA, the third network configuration, called Independent Subnets Architecture (ISA), performs a local aggregation of the multisource information about the same class, thus disregarding the multisource information from the remaining set of classes.

5 Fusion Model based on Markov Random Fields

Markov Random Fields (MRF) have been found to provide a useful and theoretically well-established tool to integrate spatial and temporal contextual

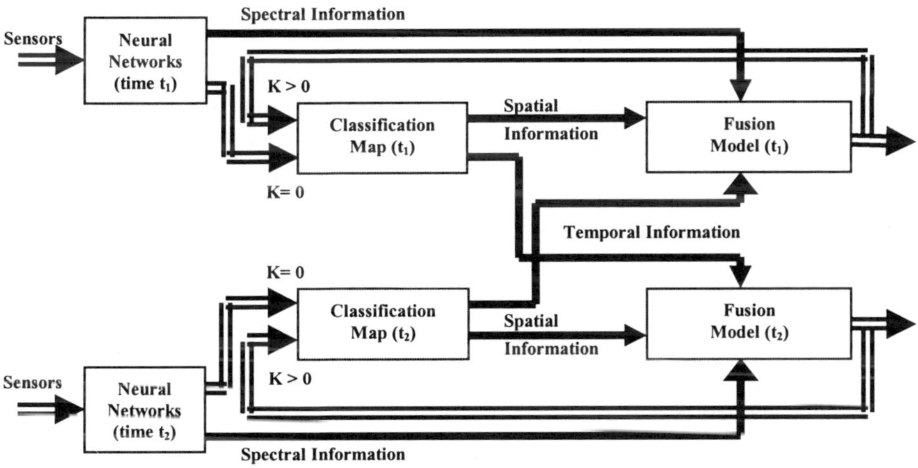

Figure. Block diagram of the proposed spatio-temporal contextual classifiers

information in the classification process. As an alternative to the "cascade" approach [11] adopted to deal with a temporal sequence of images, a "mutual" approach is proposed that is based on a coupling of the MRF models defined for each single-time image. Each MRF model is similar to the one described in [11], with the difference that the sensor-specific class-conditional energy function is expressed using the single-time class posterior probabilities provided by the MLPs, instead of the multivariate normal distribution. The "cascade" approach [12] consists to analyze the temporal sequence, image by image, in the order in which they were obtained. For instance, considering two images acquired at times t_1 and t_2, respectively, it means that the image at time t_1 will be processed basing only on spatial information. Once the image at time t_1 is processed, the resulting classification map is used as temporal information for the image at time t_2. In consequence, the image at time t_2 takes advantage of temporal information in addition to spatial information. Obviously, this approach simplifies much the computational requirements compared to processing both images jointly (simultaneously). The proposed "mutual" approach is at the midway between the "cascade" and "joint" processing approaches in the sense that it couples the MRF models associated to each image in the energy minimization process carried out by the recursive Iterative Conditional Modes (ICM) algorithm, practically similar to our recursive label estimation algorithm schematized in the above figure. It is expected that a better exploitation is achieved thanks to the iterative bi-directional exchange of temporal information between the coupled MRF models.

6 Experimental Results

Situated in the Basin of Po River in northern Italy, the study area is represented by two temporal registered images (April and May 1994) taken by the Landsat TM and ERS-1 SAR sensors. For each date, eleven features were extracted, namely: 6 intensity features from Landsat TM, 1 intensity feature from ERS-1 SAR (C band, VV polarization) and 4 texture features (entropy, difference entropy, correlation and sum variance) computed from the ERS-1 SAR image using the gray-level co-occurrence matrix. In our experiments, two training sets were necessary. The first one was used to train the non-contextual neural networks classifier while the training of the neural networks devoted to source fusion (the FCA, CSA and ISA fusion models) and the empirical estimation of the parameters of the MRF models was done with the second training set. In addition, an independent test set was used to assess the accuracy of the classifiers. The set of information classes considered in the April and May images are provided in Tables 1 and 2, together with the number of training and test pixels.

Table 1. Numbers of training and test pixels for April.

CLASS	TRAINING 1	TRAINING 2	TEST
WET	606	1130	743
BARE	2315	1614	1702
CEREALS	248	190	190
WOOD	3131	2000	1463
TOTAL	6300	4934	4098

Table 2. Numbers of training and test pixels for May.

CLASS	TRAINING 1	TRAINING 2	TEST
CORN	466	344	428
WET	418	338	338
DRY	2037	2062	1679
CEREALS	248	190	190
WOOD	3131	2000	1463
TOTAL	6300	4934	4098

For each date, a first stage of two-layer perceptron neural network (MLP) with 11 input nodes, 10 hidden nodes and a number of output nodes depending on the number of classes (4 for April and 5 for May), was trained using the back-propagation algorithm. The two MLPs made it possible to produce the initial classification map for each date, in addition to the single-time posterior probabilities required by the different spatio-temporal contextual classifiers that are the Statistical Spatio-Temporal Contextual classifier (SSTC), the neural networks classifiers (FCA, CSA and ISA), the "Mutual" MRF classifier (MMRF) and the reference "Cascade" (CMRF) as described in [11] (with the difference that the spectral energy function is expressed using the single-time class posterior probability provided by the first stage of MLPs). The classification results are reported in Tables 3 and 4. The results obtained by the MLPs are already satisfactory both in terms of Overall Accuracy (OA) and Average Accuracy (AA). However, the main difficulty faced by the MLPs was the discrimination of the class "cereals" in April (accuracy of 26.84%). This is explained by the early state of such kind of plant in this period of the year.

A preprocessing step was required by the CMRF and MMRF in order to find empirically a good set of MRF parameters while the SSTC was exempted from this step. The SSTC, CMRF and MMRF required 6 iterations to reach the convergence. The weights of the FCA, CSA and ISA classifiers were updated at each iteration so to take into account the evolution of the typology of the spatio-temporal contexts during iterations. As expected, due to its reduced architecture, the ISA reached the convergence faster and, consequently, required less learning effort with respect to the CSA and FCA to reproduce the desired multisource-information / class-information mapping.

The integration of spatio-temporal contextual information showed its benefits by a significant improvement of accuracy for the class "cereals" but also for the other classes for both dates April and May. In particular for the April image, it was obtained a gain in the Overall Accuracy (OA) of 3.66% (CSA), 3.47% (MMRF), 2.96% (CMRF), 2.39% (SSTC), 2.37% (FCA) and 2.05% (ISA). The gain in the Average Accuracy (AA) for the April image is clearly underlined with 17.47% (CSA), 16.25% (FCA), 15.48% (MMRF), 14.34% (ISA), 10.82% (CMRF) and 10.40% (SSTC). This improvement was expected from the recovery of the class "cereals" (from 26.84% (MLP) to 66.84% (SSTC) and 95.79% (CSA)), thanks to the temporal contextual information. For the May image, the MMRF and CSA made the difference with respect to the non-contextual classification with a significant gain in OA and AA of 2.2% and 4.69%, respectively, for the MMRF and a gain in AA of 6.22% for the CSA.

Table 3. Percent accuracy obtained for the April image

Method	OA	AA	WET	BARE	CER	WOOD
MLPC	94.75	80.22	97.71	97.70	26.84	98.63
CMRF	97.71	91.04	95.96	99.76	68.42	100
SSTC	97.14	90.62	97.98	98.06	66.84	99.59
MMRF	98.22	95.70	97.04	98.47	87.37	99.93
FCA	97.12	96.47	96.50	98.53	94.74	96.10
CSA	98.41	97.69	97.44	99.24	95.79	98.29
ISA	96.80	94.56	96.64	98.12	86.84	96.65

Table 4. Percent accuracy obtained for the May image

Method	OA	AA	CORN	WET	DRY	CER	WOOD
MLP	92.75	83.31	81.30	69.82	97.97	68.94	98.49
CMRF	93.27	82.03	79.91	66.57	99.52	64.21	99.93
SSTC	94.17	84.58	80.84	76.04	99.64	66.84	99.52
MMRF	94.95	88.00	81.54	69.82	99.82	88.95	99.86
FCA	92.53	89.10	73.13	86.69	95.18	94.21	96.31
CSA	93.36	89.53	69.39	88.17	95.65	95.79	98.63
ISA	92.68	86.29	72.90	76.33	97.68	87.37	97.20

7 Conclusions

Experimental results confirm that the integration of the spatio-temporal contextual information is able to improve the classification accuracy. Promising results were obtained by the proposed "mutual" MRF-based classifier. Thanks to a better exploitation of spatio-temporal contextual information with respect to the reference "cascade" classifier based on MRF, the “mutual” approach provided more accurate results. The usual “cascade” approach may suffer from the propagation of the classification error from an image at a date to another image at another date. The effectiveness of the "mutual" approach is also showed from the results obtained by the statistical fusion model which makes use in some sense of this approach. For this fusion model, slightly better results were achieved with respect to the reference MRF-based classifier, with the advantage of the absence of parameter to be optimized. Concerning the fusion model based on neural networks, the relevant classification results obtained by the "one-net-one-class" scheme of the CSA confirms the power of neural networks when applied to the spatio-temporal contextual classification problem. However, the computational load required by such a fusion model is so that it does not represent an attractive practical solution to the problem. With respect to the CSA, the FCA and ISA represent simplified network architectures involving a potentially reduced functional approximation capability, as it was confirmed by the results.

References

1. Bruzzone L. and Serpico S. B., An iterative technique for the detection of land-cover transitions in multitemporal remote-sensing images, *IEEE Trans. Geosci. Remote Sensing* **35** (1997) pp. 858-867.
2. Bruzzone L., Prieto D. F. and Serpico, S. B., A neural-statistical approach to multitemporal and multisource remote-sensing image classification, *IEEE Trans. Geosci. Remote Sensing* **37** (1999) pp. 1350-1359.
3. Dubes R. C. and Jain A. K., Random field models in image analysis, *J. Appl. Statist.* **16** (1989) pp. 131-163.
4. Jeon B. and Landgrebe D. A., Decision fusion approach for multitemporal classification, *IEEE Trans. Geosci. Remote Sensing* **37** (1999) pp. 663-672.
5. Jeon B. and Landgrebe D. A., Classification with spatio-temporal interpixel class dependency contexts, *IEEE Trans. Geosci. Remote Sensing* **30** (1992) pp. 663-672.
6. Kalayeh H. M. and Landgrebe D. A., Utilizing multitemporal data by a stochastic model, *IEEE Trans. Geosci. Remote Sensing* **24** (1986) pp. 792-795.
7. Khazenie N. and Crawford M. M., Spatio-temporal autocorrelated model for contextual classification, *IEEE Trans. Geosci. Remote Sensing* **28** (1990) pp. 529-539.
8. Kittler J. and Föglein J., Contextual classification of multispectral pixel data, *Image and Vision Computing* **2** (1984) pp. 13-29.
9. Lee T., Richards J. A. and Swain P. H., Probabilistic and evidential approaches for multisource data analysis, *IEEE Trans. Geosci. Remote Sensing* **25** (1987) pp. 283-293.
10. Melgani F. and Serpico S. B., A statistical approach to the fusion of the spectral and spatio-temporal contextual information for the classification of remote sensing images, in Proc. of the 1st *Patt Recog, for Remote Sensing Workshop, Andorra (2000)* pp. 56-61.
11. Solberg A. H. S., Taxt T. and Jain A. K., A Markov Random Field model for classification of multisource satellite imagery, *IEEE Trans. Geosci. Remote Sensing* **43** (1996) pp. 100-113.
12. Swain P. H., Bayesian classification in a time-varying environment, *IEEE Trans. Geosci. Remote Sensing* **8** (1978) pp. 880-883.

FUZZY, NEURAL AND NEURO-FUZZY CLASSIFICATION OF PRE- AND POST-EVENT SAR IMAGES FOR FLOOD MONITORING AND DISASTER MITIGATION

G. AMICI, F. DELL ACQUA, P. GAMBA, G. PULINA

Department of Electronics, University of Pavia, Via Ferrata, 1, 27100 Pavia Italy

E-mail: gamba@ele.unipv.it

In this paper we discuss some approaches to change detection for flood monitoring using SAR images. In particular, the aim of this work is to have a comparative assessment of neural, fuzzy, and neuro-fuzzy classifiers applied to satellite images of the 2000 flood in Pavia, Northern Italy. We address both data fusion with multitemporal fuzzy or neural classification and decision-level fusion by combining classifications or fuzzy membership matrixes for the pre- and post-event images. We found that the neuro-fuzzy approach is the most suitable to extract changes and to provide the required information for flood monitoring and disaster mitigation.

1 Introduction

The need of multitemporal information in disaster management is urgent, because of the large effects that are caused, especially in urbanized areas, by natural as well as man-induced extreme phenomena. Floods, fires, earthquakes and droughts are only examples of how far we can go by using suitably realized and tested tools for the multitemporal analysis of satellite data.

In this area, a particular attention has been paid to the use of satellite images for flood monitoring. In particular, in technical literature many papers have been presented on SAR data analysis with respect to flood events, or on SAR and multispectral data combination for a better evaluation of the inundated areas [1-4]. When only SAR images are available, many interesting works have considered the possible joint use of amplitude and interferometric coherence SAR maps to delineate the flood extent. The most recent ones [5,6] show that the use of the latter data is extremely helpful and allows extracting flood maps suitable for Geographic Information System applications.

In all these works, the stress has been on the characterization of the best data set to be considered, and this could be considered now a well-assessed result for the remote sensing research community. The open point is now the definition of suitable multitemporal classifiers and their comparison for a given data set. Therefore, this work intends to give a contribution to this task by comparing different classifiers (or classification strategies) on a common background, the SAR data available for the flood events occurred in the last decade in the area around Pavia, Northern Italy. Moreover, since the last event is extremely recent (October

2000), a detailed ground truth made by means of aerial surveys and ground data collecting is available, and could be used to quantitatively evaluate the considered approaches.

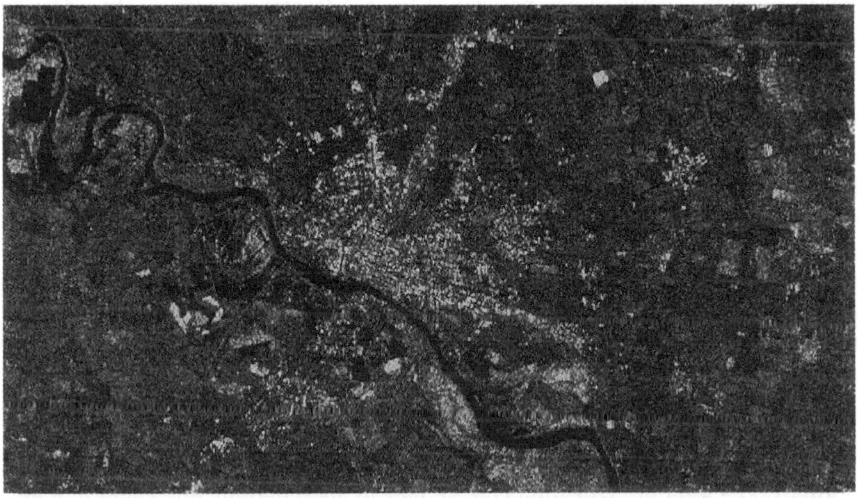

(a)

(b)

Figure 1. (a) RADARSAT image of the area around the town of Pavia, Northern Italy, on October, 20th, 2000; (b) ERS-2 image of the same area on October, 28th, 2000.

2 The multitemporal classification chain

As already outlined, there are many methodologies applied to flood monitoring, and some of them are based on the use of fuzzy or neural classifiers. In this work, we want to provide a detailed comparison of three different procedures, based on the Fuzzy C-Means (FCM) or the Adaptive Resonance Theory (ART) concepts. In particular, we want to investigate if it is better to estimate the flooded areas by means of a multitemporal contextual classification or by separate classifications, with fuzzy rules to fuse the outputs. The limits of this approach are due to the inherently spectral classification that we are performing, where no spatial information is introduced. However, as shown in [7], such characterization can be introduced as a further step, a so-called *kernel-based classification*, performed by means of the same classifiers.

The steps of the processing chain are therefore as follows: the input data are raw SAR images, suitably co-registered to the regional 1:2000 thematic map of the studied area. These data comprise pre- and post-event images of a flood event. Each couple of pre- and post-event images may be analyzed by means of a joint classification or by classifying each image and then fusing in different manners the classification results.

Using a fuzzy approach, the classification is provided by means of the FCM algorithm. The joint classification corresponds to applying this procedure to a multi (actually, bi-) dimensional data set. Note that for flood analysis we are interested only to the class corresponding to areas were the radar response has changed, due to the presence of water. So, flooded areas could be recognized and investigated.

However, we can extract the same information by separately classifying the pre- and post-event image and comparing the fuzzy partition matrices. This could be obtained by looking for areas that had a very small membership value to the water class in the first image and acquired a large membership value after the event (i.e. in the second image). More generally, each pixel changing so deeply its membership from one class to another is part of an area where some changes have happened.

If hard and not fuzzy logic is followed, this processing step corresponds to make a logical XOR between the classification maps. Indeed, if a pixel continues belonging/not belonging to a class (membership = 1/0 for both the analyzed images) it does not belong to the changed class (its membership value should be set to 0).

In fuzzy logic, we implemented three fuzzy operators:

$$XOR_1 = \max(a\,,\,b)\; - \;\min(a\,,\,b) = |a\; - \;b| \qquad (1)$$

$$XOR_2 = a + b - 2 \cdot a \cdot b \qquad (2)$$

$$XOR_3 = \min(1\;,\;a + b) - \max(0\;,\;a + b - 1) \qquad (3)$$

These three formulas have been developed starting from the fuzzy logic operators summarized in [8] and correspond to using the triangular norm to generalize the

AND operators and the triangular co-norm to generalize **the OR operator**. We note that, by this choice, the implemented fuzzy XOR does not provide either a conjunctive or a disjunctive behavior.
As we will see in the next section, the use of these operators and the hard XOR give very interesting results, while the analysis by means of a fuzzy joint classification does not. So, to refine this approach we implemented a different classification tool, based on ART neural networks. In particular, we refer here to the ART-2 and fuzzy ARTMAP networks.
The ART-2 is a self-organizing network that requires real vectors as inputs and looks for similarities in their patterns. It stores the *memory* of previously considered data into vectors called *categories* and performs a comparison (a *resonance*) between the input and each of the memories. If a sufficiently high resonance (similarity) is found, the memory is updated with new informations, otherwise a new category is introduced. This network can be used to perform both a spectral and a spatial unsupervised classification of multitemporal data, by clustering the original data into significant classes with respect to their temporal pattern i.e. what we are looking for.
This operation could be described, extremely simplifying, by the sequence of the following steps: first, the input vector is normalized (in the input layer F_0). The, it is compared it in the F_1 layer with the prototypes stored in the memory layer F_2. If the comparison (essentially, a scalar product) provides a value higher than a fixed threshold, then a resonance has been found and the memory vector is updated taking into account the new input. If no resonance occurs, a new memory vector is added to the F_2 layer. The fuzzy ARTMAP neural network is defined by means of a structure as in Fig. 2, where there are highlighted the two ART blocks, called "ART_a", "ART_b" and a "Map Field".

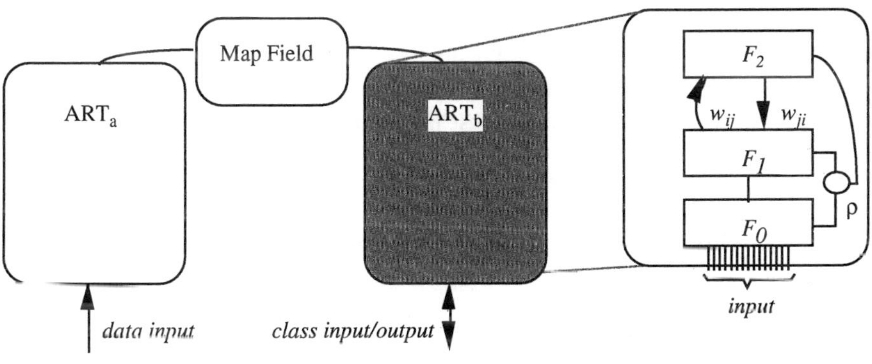

Figure 2. The basic structure of the fuzzy ARTMAP classifier.

The ART-2 network is an *unsupervised* system, and the number of memory vector (i.e. of classes) that it provides at the end of the classification is unknown when the procedure starts. Instead, a different implementation of the ART concepts is the so called fuzzy ARTMAP neural networks, that combines two ART-2 networks in a *supervised* system. The difference is that in this version two ART modules incorporate *fuzzy* rules and are connected via a "Map Field" (see fig. 2). We do not give here a description of the algorithm, well presented in [9,10], we just recall the main steps to understand which are the parameters ruling the procedure and their role.

In the training phase, after a new input is presented, we look for the maximum (with respect to j) of the *choice function*

$$T_J = \max_j \left|\boldsymbol{I} \wedge \boldsymbol{w}_j\right| / \left(\alpha + \left|\boldsymbol{w}_j\right|\right) \tag{4}$$

where α is the *signal rule parameter* and $\boldsymbol{w}_j$ represent the ART internal connection weights (that is, its *memories*). The corresponding *match function* $\left|\boldsymbol{I} \wedge \boldsymbol{w}_J\right| / \left|\boldsymbol{I}\right|$ is compared with a threshold, called the *vigilance parameter* ρ: only if a resonance is found (i.e. T_J exceeds the threshold), $\boldsymbol{w}_J$ is updated as follows:

$$\boldsymbol{w}_J^{NEW} = \beta\left(\boldsymbol{I} \wedge \boldsymbol{w}_J^{OLD}\right) + (1-\beta)\boldsymbol{w}_J^{OLD} \tag{5}$$

In the testing phase, the resonance assigns each new input to one memory, and the corresponding output class is offered as a result.

We note that the behavior of the fuzzy ARTMAP is ruled by the three parameters α, β, and ρ. The usual choice for α is very small, if not equal to 0, as suggested in literature. Instead, β can be chosen adaptively in (0,1), although usually close to one, to speed up the training phase. Finally, the value of ρ rules directly the number of memory elements stored in the ART networks, since it is the reference threshold for the resonance test. So, it is the most important parameter when evaluating the performance of the system as a classifier.

Indeed, it is intuitive that the representation that the network builds during training affects the results of the testing phase. We found that it is most useful to allow the internal representation to be redundant, with many memory elements, to allow a better capacity of the system to react to new inputs not considered yet.

Finally, we should note that, for classification problems there is no need for the complete ART_b (see again fig.2), and a simplified structure was proposed in [9], and investigated also in [10]. This simplified fuzzy ARTMAP exploits only one fuzzy ART network in a *supervised* system, and has been considered in this paper as the working tools for fuzzy ARTMAP classifications.

Table 1. Classification accuracy for the multitemporal analyses compared in this paper (flooded areas only).

Classifier	Classification accuracy (15/08/92 ^ 20/10/00)	Classification accuracy (15/08/92 ^ 28/10/00)
FCM with *hard* comparison	65.5 %	60.3 %
FCM with *fuzzy XOR* comparison	30.4 %	29.0 %
Multitemporal ART-2	= %	38.3 %
Multitemporal Fuzzy ARTMAP	83.1 %	56.2 %

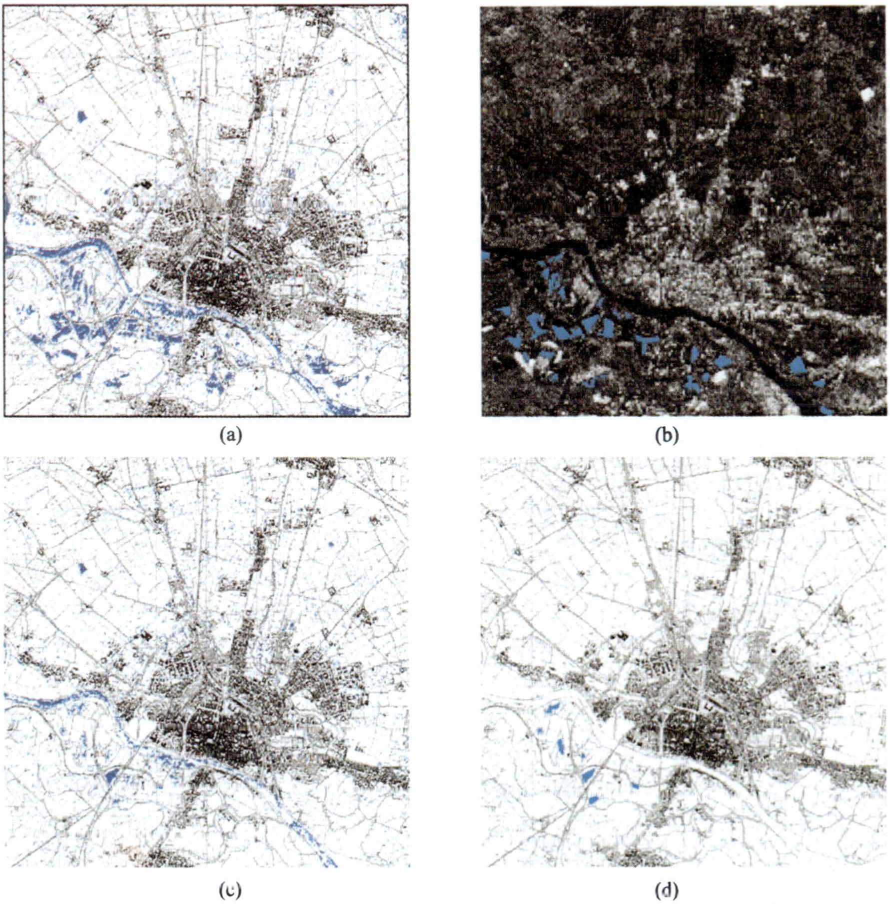

Figure 3. Fuzzy multitemporal classification: (a) the FCM with hard comparison output, superimposed to a map of the town of Pavia (1994 ERS image — 2000 RADARSAT image); (b) the ground truth at the same date; (c) the FCM with hard comparison output, superimposed to a map of the town of Pavia (1994 ERS image — 2000 ERS image); (d) the ground truth at the same date.

3 Experimental results

We base our investigation on a complete data set of ERS data covering the area and the dates of the major flooding events in the last decade in the Province of Pavia, Northern Italy. They were acquired in the framework of an internal project of the University of Pavia aiming at providing spatial information about the floods of the rivers Ticino and Po, and at extracting significant inputs for hydrological models of the basin.
In this paper we will concentrate only on the last flood event, happened in October 2000, a temporal period in which we also acquired a fine beam RADARSAT image to provide a comparison between the two sensors. A subset of the data sets, corresponding to the area just around the town is depicted in fig. 1. The best results obtained after applying to these images the approaches outlined in the previous section are shown in fig. 3, while the overall accuracy values for the most significant evaluations are proposed in Table I.
Even this very preliminary results show that there are interesting hints to be extracted from the investigation. In particular, we observe that the hard comparison is always better than the fuzzy combination of partition matrices. Moreover, the ART-2 neural classifier sometimes is not able to recognize the change class, while the fuzzy ARTMAP procedure provides the best achievable results, considering both dates. In particular, we have excellent detection of flooded areas in the image with better SNR, i.e. the ERS-2 data on October, 20th.

4 Acknowledgements

This work was partially funded by ASI. We thank also the University of Pavia and the Internal Project Monitoring catastrophic events at different spatio-temporal scales by means of aerial/satellite images and Geographic Information Systems . Finally, many thanks to Dr. F. Zucca for the help in recording the ground truth and the useful discussion about the classification results.

References

1. D.J. Weydhal, Flood monitoring in Norway using ERS-1 SAR images , Proc. of IGARSS '96 Vol. 1, pp. 151 —153.
2. S. Takeuchi, T. Konishi and Y. Suga, Comparative study for flood detection using JERS-1 SAR and Landsat TM data , Proc. of IGARSS '99, Vol. 2, pp. 873 —875.
3. W. Hagg, M. Sties, Monitoring the Oder/Germany flood with ERS, RADARSAT and optical data , Proc. of IGARSS '98, Vol. 3, pp. 1307 —1309.

4. D. Geudtner, R. Winter and P.W. Vachon, Fllod monitoring using ERS-! SAR interferometry coherence maps , Proc. of IGARSS '96, Vol. 2, pp. 966 —968.
5. S. Dellepiane, G. Bo, S. Monni and C. Buck, SAR images and intereferometric coherence for flood monitoring , Proc. of IGARSS 00, Vol. 6, pp. 2608 —2610.
6. G. Nico, M. Pappalepore, G: Pasquariello, A: Refice and S. Samarelli, Comparison of SAR amplitude vs. coherence flood detection methods — a GIS application , *Int. J. of Remote Sensing*, vol. 21, no. 8, pp. 1619-1631, 2000.
7. P. Gamba, B. Houshmand: An efficient neural classification chain for optical and SAR urban images , *Int. J. of Remote Sensing,* in press.
8. I. Bloch, Fuzzy classification for multi-modality image fusion , Proc. of ICIP-94, Vol. 1, pp. 628 —632.
9. Mannan, B., Roy, J., and Ray, K.A., 1998, Fuzzy ARTMAP supervised classification of multi-spectral remotely-sensed images. *Intl. J. Remote Sensing*, **19** (4), 767-774.
10. Carpenter, G.A., Gjaja, M.N., Gopal, S., and Woodcock, C.E., N., 1997, ART neural networks for remote sensing: vegetation classification from Landsat TM and terrain data. *IEEE Trans. Geoscience and Remote Sensing*, **35**, 308-325.

USING LINEAR REGRESSION FOR THE AUTOMATION OF SUPERVISED CLASSIFICATION IN MULTITEMPORAL IMAGES

[1]Feitosa, R.Q., [2]Meirelles, M.S.P., [3]Blois, P. A.

[1,2]Dept. of Electrical Engineering ;
Catholic University of Rio de Janeiro
Rua Marquês de São Vicente, 225
22453-900, Rio de Janeiro, Brazil
Tel.: +55 21 5299438

[1,3] Dept. of Computer Engineering
State University of Rio de Janeiro
Rua São Francisco Xavier, 524
20550-013, Rio de Janeiro, Brazil
Tel.: +55 21 587 7442

e-mail: raul@ele.puc-rio.br, maggie@eng.uerj.br, prblois@dog.com

A method for the automatic estimation of spectral signatures in multitemporal remotely sensed images is proposed. The method resembles relative atmospheric corrections techniques where a linear relationship between image bands across time is assumed. Before computing the regression coefficients an outlier filtering procedure is applied resulting in an improved accuracy in the estimated spectral signatures. The work compares also band-to-band regression and multiple regression. Moreover the use of single model for all classes and one model per class is investigated. Scenes of two areas undergoing a rapid environmental degradation process were used as database to evaluate the proposed approach. The combination of outlier filtering with one regression model per class produced the best results for moderate to high percentage of change pixels in the multitemporal images.

1 Introduction

Supervised classification of remotely sensed images involves three basic steps: *a*) the photo interpreter selects some representative training data for each classe in the image, *b*) an automatic tool estimates the spectral signatures based on the training data and generates the first version of a thematic map using some classification model, and *c*) the photo interpreter refines the initial thematic map by incorporating some a priori knowledge not represented in the classifier used in step *b* .

A great effort has been done worldwide to build image interpretation systems able to carry out all these tasks automatically [1] [2]. This work presents an approach to generate the training data (step *a*) in an automatic fashion.

Although no explicit radiometric correction is performed, there is a close analogy between the methods considered here and relative atmospheric correction techniques [3] [4] [5]. In fact simple intuitive extensions to the procedure discussed in this work would suffice to convert this approach into a relative correction technique. The proposed methods, as well as relative atmospheric correction, are based on the assumption of a linear relationship between image bands across time [6]. Pixels that change from one class in the earlier image to a different class in the later image (*change* pixels) behave like outliers and affect the accuracy of these techniques.

This work proposes and investigates methods to filter the training data from outliers before estimating the coefficients of the linear relationship between image bands. The procedure discriminates between *change* and *stable* pixels by evaluating how well the linear model fits to them.

The next section describes the outlier filtering procedures considered in this study. The experiments performed to evaluate the performance associated to this method are explained in the section 3. Section 4 discusses the experiment results. The paper ends summarizing the main conclusions of this work.

Figure 1: Structure of the Automatic Classification System

2 Filtering Outliers

2.1 *Problem Formulation*

The problem considered in this work is illustrated in Figure 1. An automatic classification system has three inputs: i) the multispectral image I_A of some area acquired at date t_A, ii) a reliable thematic map TM_A relative to the image I_A, and iii) a second multispectral image I_B of the same area acquired at later date t_B. The task of the system consists of generating a thematic map (TM_B) for image I_B without any user intervention.

This work focuses on the step of automatic selection of training data. So the task for which an automatic solution is required consists of *developing an automatic procedure to collect data examples from image I_B for each land cover type, on the basis of the three system inputs.*

2.2 *First Basic Assumption*

The basic assumption underlying the methods proposed in this work is that the differences in the land cover between I_A and I_B are moderate. This is in most cases a realistic assumption if the time elapsed between t_A and t_B is not long.

According to this assumption the thematic map TM_A is a good *initial guess* for the thematic map TM_B. Since most pixels are assumed to remain in the same class in both images (*stable* pixels), a representative training set can be built by picking up pixels from image I_B assuming the class assignment defined in TM_A.

The inaccuracy of the spectral signatures estimated on the basis of such data will increase with the proportion γ of *change* pixels in the training set.

A more accurate estimate for the spectral signatures could be obtained, if one were able to identify the *change* pixels and refine the *first guess* by eliminating them from the training set before estimating the signatures. The second assumption provides a way to identify potential *change* pixels.

2.3 Second Basic Assumption

Changes in the atmospheric conditions, illumination and view angles and sensor calibration, among other factors, influence the spectral signatures. The proposed method assumes that all these factors affect the spectral response of a single land cover type quite in the "same way", so that the intensities of a *stable* pixel in the later image (I_B) can be estimated with a good accuracy by a linear function of the intensities of the same pixel in the earlier image (I_A).

Such a function is expected to fit well to *stable* pixels. For the "small" proportion of *change* pixels the function will in most cases not fit well. Therefore the distinction between *stable* and *change* pixels can be made on the basis on how well the function fits to them.

2.4 The Methods

The methods studied in this work start with the initial guess and refine it by identifying the *change* pixels and excluding them from the initial training data.

Clearly the performance associated to this proposal will depend on the function used to map pixel intensities from one image to another. Inspired by many previous approaches for relative atmospheric calibration four linear regression methods are considered in this study:

a) Band-to-band regression, one model for the entire image,
b) Multiple regression from several bands, one model for the entire image,
c) Band-to-band regression, one model per class, and
d) Multiple regression from several bands, one model per class.

The band-to-band regression models (cases *a* and *c*) take the form:

$$y_{ij} = \beta_0 + \beta_1 x_{ij} + \varepsilon_{ij} \qquad 1$$

where y_{ij} denotes the value of pixel i for band j in an later image and x_{ij} in an earlier image, β_0 and β_1 are the regression coefficients, and ε_{ij} is the corresponding error term.

Equation 2 shows the model relative to cases *b* and *d* for images with *p* bands.

$$y_{ij} = \beta_0 + \beta_1 x_{i1} + \beta_2 x_{i2} + \cdots + \beta_p x_{ip} + \varepsilon_{ij} \qquad 2$$

In this equation y_{ij} denotes de value of pixel i for band j in an later image and x_{ij} in an earlier image, β_0 to β_p are the regression coefficients, and ε_{ij} is the error term for images with p spectral bands.

In the cases a and b the land cover information is not taken into account, and a single model for all land cover types is assumed. In the cases c and d the model is fitted to each land cover type individually.

Many statistical tests have been proposed for detecting and rejecting outliers in linear regression models. (see [7] for a discussion of some approaches). Some of these tests base on the residuals r_i given respectively by equations 3 (cases a and c) and equation 4 (cases b and d).

$$r_i = y_{ij} - (\beta_0 + \beta_1 x_{ij}) \qquad 3$$

$$r_i = y_{ij} - (\beta_0 + \beta_1 x_{i1} + \beta_2 x_{i2} + \cdots + \beta_p x_{ip}) \qquad 4$$

By assumption, these residuals have independent normal distributions with zero mean and a constant variance. It is a common practice to scale the residuals so that they have unitary variance. So the normalized residuals r_i^* will be given by

$$r_i^* = \frac{r_i}{\sqrt{\mathrm{var}(r_i)}} \qquad 5$$

where var(r_i) is the estimated error variance. In the equation above the normalized residual has a Student's t distribution with (n-p-1) degrees of freedom, where n is the total number of data points and p is the number of predictor variables. The confidence interval for the mean of each error will be given by

$$c_i = r_i\left[1 \pm t_{(n-p-1)}(1-\alpha/2)\sqrt{\mathrm{var}(r_i)}\right] \qquad 6$$

where $t_{(n-p-1)}(1-\alpha/2)$ is the upper $100(1-\alpha/2)$th percentile of a Stutendt's t distribution with (n-p-1) degrees of freedom. A confidence interval given by equation 6 not including 0 (zero), is a good evidence that the data point is an outlier. This work follows this reasoning to discriminate between *change* and *stable* pixels in the initial guess. Accordingly, the following sequence of actions is proposed to estimate the spectral signatures in the later image.

1. build a training set for the earlier image (TM_B) based on the later thematic map (TM_A) assuming that all pixels are *stable,*
2. apply the linear regression model,
3. eliminate from the training set pixels for which the error confidence interval does not contain the value 0 (zero),
4. compute the spectral signatures using remaining the pixels.

It is worth mentioning that for $\alpha = 0$ (zero) corresponds to a method in which no outlier filtering takes place.

3 Experiment Design

3.1 Study Areas

The imagery available for this investigation consisted of spectral bands 3 to 5 of two pairs of Landsat TM-5 scenes mapped to RGB channels. The first pair of scenes was obtained in 1996 and in 1999 both during the dry season. They cover part of the Taquari Watershed in western Brazil.

In the second pair of scenes part of the Atlantic Forest in southeastern Brazil is depicted. They were obtained in 1993 and 2000 also during the dry season. Only segments with 400 by 400 pixels cropped from the scenes were used in the experiments.

A well trained photo interpreter performed the registration and classification of all images according to 9 different land cover types with the help of automatic software tools, eventually exploring additional information available in a GIS and applying some a priori knowledge about the covered areas.

In the experiments reported in section 4 the earlier image (I_A), its thematic map (TM_A) produced by the photo interpreter, along with the later image (I_B) were used as input to the system shown in Figure 1: Structure of the Automatic Classification System. The thematic map corresponding to the later image produced by the photo interpreter (TM_{Bref}) was considered as the reference training data for performance evaluation.

3.2 Building the Dataset

The first basic assumption (section 2.2) requires a low percentage (γ) of *change* pixels in the initial guess. In order to have control over this parameter during the experiments not all *change* pixels were included in the data set. For each land cover type a number of *change* pixels were randomly selected for training. This number was given by the product between γ and the number of *stable* pixels of the corresponding land cover type in TM_A. Beyond the so selected *change* pixels all *stable* pixels were included in the training data.

3.3 Performance Metrics

The most representative training set of the land cover types for the image I_B consists of all pixels of I_B with the classes assigned in TM_{Bref}. The spectral signatures estimated on the basis of such a training set were used in the experiments as reference.

The performance measure will be given by the average percentage of agreement between the outputs of a maximum likelihood classifier trained with the reference training set and with the filtered training set according to the proposed methods, whereby all classes were assumed to have a multivariate normal distribution.

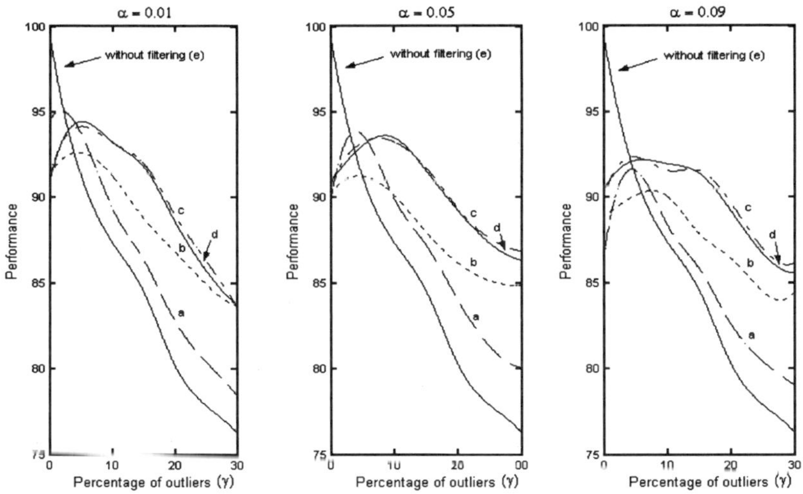

Figure 2: Average percentage of classification agreement between the reference and the proposed training set as a function of the percentage of *change* pixels (γ) for different confidence levels (α); images from the Taquari Watershed; curves correspond to *a)* band-to-band regression, one model for the entire image, *b*) multiple regression from several bands, one model for the entire image, *c*) band-to-band regression, one model per class, and *d*) multiple regression from several bands, one model per class, *e)* no outlier filtering.

4 Experiment Results

Figure 2 and Figure 3 show the performance results respectively from the experiments on the imagery from the Taquari Watershed and from the Atlantic Forest. Each figure presents three graphs corresponding to different levels of confidence (α= 0.01, 0.05 and 0.09). Each graph presents five curves corresponding to the four methods outlined in section 3 (curves *a* to *d*), and to a method involving no outlier filtering (curve *e* in solid line). This last curve represents the *initial guess* without any outlier filtering.

Each performance curve presents quite the same form across the graphs for both sites. As expected, the results are heavily dependent on the percentage of *change* pixels (γ) in the images. For γ below 3% to 5% outlier filtering brought no performance improvement.

For γ above 5% the use of one regression model per class (curves *c* and *d*) consistently brought a performance improvement in relation to a single model for all classes (curves *a* and *b*). As a matter of fact the use of outlier filtering with one single model for the entire data resulted in no relevant performance improvement relative to no filtering for the Atlantic Forest.

The graphs of Figure 2 and Figure 3 allow also to compare the performance associated to the band-to-band regression (equation 1) and to the multiple regression from several bands (equation 2) . When just one model is applied to all classes band-to-band regression (curve *a*) performs a little better than multiple regression (curve *b*) for γ below 5% to 10%. Above these values multiple regression presented a clear superiority only in the imagery from the Taquari Watershed.

For one model per class band-to-band regression (curve *c*) performed as well as multiple regression (curve *d*), being eventually better for some values of γ. In these cases the increased complexity of the multiple regression was not worth the trouble.

The last important observation is concerned with the level of confidence. Comparing the three graphs in each figure one notices that the best performance corresponds to the lowest value assigned to α in these experiments (=0.01). This happened in both sites for γ as large as 15%.

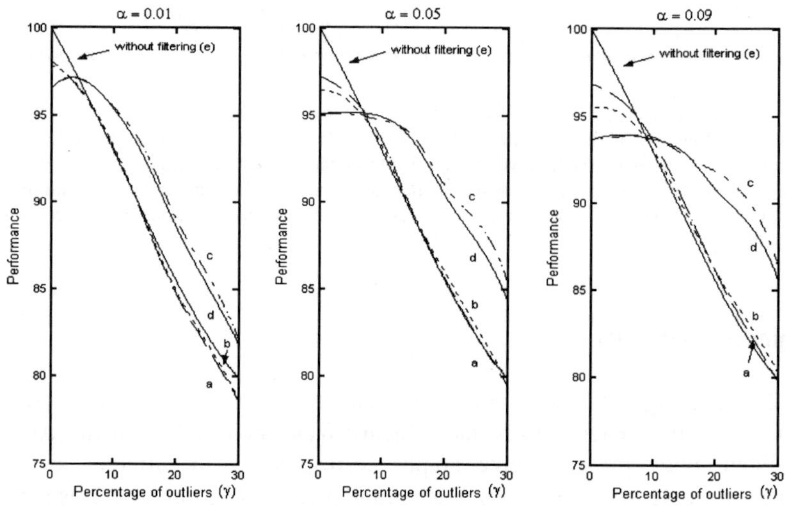

Figure 3: Average percentage of classification agreement between the reference and the proposed training set as a function of the percentage of *change* pixels (γ) for different confidence levels (α); images from the Atlantic Forest; curves correspond to *a)* band-to-band regression, one model for the entire image, *b*) multiple regression from several bands, one model for the entire image, *c*) band-to-band regression, one model per class, and *d*) multiple regression from several bands, one model per class *e)* no outlier filtering.

5 Conclusion

This work investigates automatic methods to generate training data for change detection in multitemporal images. The method assumes a linear relationship between image bands across time, and applies an outlier filtering procedure to increase accuracy. The work also compares the use band-to-band regression and the multiple regression. Moreover the use of single model for all classes and one model per class is investigated.

Experiments performed on scenes from two different sites indicated that outlier filtering combined with one regression model per class generally improves classification performance. Multiple regression did not shown any important superiority over the band-to-band alternative.

Acknowledgement

This work was supported by CNPq, an agency of the Brazilian government for scientific and technologic development.

References

1 Richards, J. A., Jia, X., "*Remote Sensing Digital Image Analysis - An Introduction*", 3rd Ed. Springer Verlag, 1999.
2 Mather, P., "*Computer Processing of Remotely-Sensed Images- An Introduction*", 2nd Ed. Wiley, 1999.
3 Chaves, P. S., Jr. Mackinson, D. J., "*Automatic detection of vegetation changes in the southwestern United States using remotely sensing images*", Photogramm. Eng. Remote Sens. 60(5), 1994, pp. 571-583.
4 Ouaidrari, H., Vermote, E. F., "*Operational Atmospheric Correction of Landsat TM Data* ", Remote Sensing and Environment, vol. 70, 1999, pp. 4-15.
5 Song, C., Woodcock, C. E., Seto, K. C., Lenney, M. P., Macomber, S. A., "*Classification and change Detection Using Landsat TM Data: When and How to Correct Atmospheric Effects?*", Remote Sensing and Environment, vol. 75, 2001, pp. 230-244.
6 Tokola, T., Löfman, S., Erkkilä, A., "*Relative Calibration of Multitemporal Landsat Data for Forest Cover Change Detection*", Remote Sensing and Environment, 68, 1999, pp. 1-11.
7 Douglas C. Montgomery, Elizabeth A. Peck, G. Geoffrey Vining, "*Introduction to Linear Regression Analysis*", 3rd Ed. John Wiley and Sons, 2001.

AN APPLICATION-ORIENTED CHANGE-DETECTION TECHNIQUE

D. FERNANDEZ-PRIETO

EO Applications Department, European Space Agency, ESA-ESRIN, Via Galileo Galilei, 00044 Fracasti (Rome), Italy

E-mail: diego.fernandez@esa.int

In many real world applications of change-detection techniques, the main operational objective consists in identifying a specific land-cover transition of interest for the end-users. To carry out this task, supervised or unsupervised approaches are usually applied. The former requires an exhaustive collection of training samples for all the land-cover classes present in the area under analysis at the two times considered, which is complex, expensive and time consuming. The latter does not provides the sufficient information about the nature of the different land-cover transitions occurred in the area under analysis between the two dates under consideration. In this paper, a partially supervised approach to change detection is proposed, which overcomes the above mentioned problems. It allows the identification of one or few land-cover transitions of interest by using exclusively training samples belonging to the land covers involved. Experiments carried out on a multitemporal data set demonstrate the validity of the proposed approach.

1 Introduction

The automatic detection of land-cover transitions in multitemporal Earth-observation data [1]-[3] represents one of the most interesting topics in remote sensing from both the methodological and the operational points of view. Several real-world applications benefit from this type of techniques ranging from forestry and agricultural surveys to urban studies and natural disaster management. In many real world applications, the objective of change detection consists in mapping only one (or only few) land-cover transition(s) of interest for the end-user (e.g., agricultural land to urban area, forest to burned area). Usually, to achieve this objective, supervised change-detection techniques [3] are applied, which require the availability of suitable ground truth information for each single-date image under consideration. This represents a serious drawback, as labeling training samples for all the land covers present in the area of interest, either by means of in-the-ground campaigns or by using photo-interpretation, requires an enormous effort and considerable time. In addition, in change-detection problems this task is particularly critical. In particular, training samples should be gathered at the two times under consideration, and, depending on the particular technique applied, samples should be related to the same points on the ground and should represent accurately the proportions of all the transitions in the whole images [3]. This represents a serious operational limitation.

In this paper, we propose an application oriented approach, based on a partially supervised change-detection technique [4], that overcomes the above-mentioned problem. In particular, the presented technique allows the identification of one (or few) land-cover transition(s) of interest for the end-user, by using training samples belonging exclusively to the land covers involved in the specific change(s) to be mapped. This represents a serious advantage, as avoids the need to label training samples for all the land covers present in the area of interest at the two considered times.

This novel technique is based on a partially supervised formulation of the compound classification algorithm, which makes use of a Markov random field approach [2] to model the spatio-temporal correlation between images. In addition, a Radial Basis Function (RBF) network [5] is used to model the probability density function of both the classes of interest (for which training samples are available) and the unknown classes present in the scenes under consideration (i.e., for which training samples are not available).

2 Methods

Let us consider two co-registered remote-sensing images, $\mathbf{X}_1$ and $\mathbf{X}_2$, acquired in the same geographical area at two different times, t_1 and t_2. Let $X_t(i,j)$ denote the feature vector associated with the pixel (i,j) of image $\mathbf{X}_t$. Let $\Omega_1 = \{\omega_1^1,...,\omega_{c1}^1\}$ and $\Omega_2 = \{\omega_1^2,...,\omega_{c2}^2\}$ be the sets of land-cover classes that characterize the geographical area under analysis at the times t_1 and t_2, respectively. For simplicity, let us assume that we wish to map one land-cover transition of interest, $(\omega_{\text{int}}^1,\omega_{\text{int}}^2)$, from those included in Ω_1 and Ω_2 (the extension of the method to the multi-class case is straightforward. Let us also assume that N_{int}^1 and N_{int}^2 training samples are available for the land-cover classes of interest ω_{int}^1 and ω_{int}^2, and that no prior knowledge is available concerning the unknown classes ω_{unk}^1 and ω_{unk}^2, which integrates the other land covers present in the area under consideration at the times t_1 and t_2, respectively. Let $\mathbf{C}_1$ and $\mathbf{C}_2$ denote the set of labels for the whole images $\mathbf{X}_1$ and $\mathbf{X}_2$, respectively, being $C_1(i,j) \in \{\omega_{\text{int}}^1,\omega_{unk}^1\}$ and $C_2(i,j) \in \{\omega_{\text{int}}^2,\omega_{unk}^2\}$ the corresponding labels for the single pixel (i,j).

The proposed approach is based on a partially supervised formulation of the compound classification [3], [4], which takes into account the spatio-temporal contextual information present in the images under analysis. In particular, we want to select the sets of labels, $\mathbf{C}_1$ and $\mathbf{C}_2$, that maximize the posterior probability

$p(C_1, C_2 / X_1, X_2)$. Simplifying this model by assuming time-conditional independence [3], the identification of the land-cover transition of interest $(\omega^1_{int}, \omega^2_{int})$ can be performed by maximizing the likelihood function

$$L(C_1, C_2 / X_1, X_2) = p(X_1 / C_1) p(X_2 / C_2) P(C_1, C_2), \tag{1}$$

where $p(X_i / C_i)$ represents the class-conditional probability at the time t_i given the set of labels C_i and $P(C_1, C_2)$ is the joint prior model for the class labels at both times.

We propose a MRF to model the spatio-temporal context in $P(C_1, C_2)$:

$$\begin{aligned} &P(C_1(i,j), C_2(i,j) / \{C_1(g,h), C_2(g,h);\ (g,h) \neq (i,j)\}) = \\ &P(C_1(i,j), C_2(i,j) / \{C_1(g,h), C_2(g,h);\ (g,h) \in \mathbf{N}(i,j)\}) = \\ &= \frac{1}{Z} \exp[-U_{context}(C_1(i,j), C_2(i,j) / \mathbf{N}(i,j))] \end{aligned} \tag{2}$$

where $\mathbf{N}(i,j)$ is the spatio-temporal neighborhood of pixel (i,j), Z is a normalizing factor and $U_{context}(\cdot)$ is the Gibbs energy function defined as [2]:

$$\begin{aligned} &U_{context}(C_1(i,j), C_2(i,j) / \mathbf{N}(i,j)) = \\ &\sum_{(g,h) \in \mathbf{N}(i,j)} \beta \delta_k [(C_1(i,j), C_2(i,j)), (C_1(g,h), C_2(g,h))], \end{aligned} \tag{3}$$

where δ_k is the Kronecker delta function and β is a constant that tunes the influence of the spatio-temporal context on the change-detection process.

In this context, the partially supervised change detection problem described by (2) is equivalent to minimize the energy function

$$U(X_1, X_2 / C_2, C_2) = \sum_{t=1}^{2} U_{data}(X_t) + U_{data}(C_1, C_2), \tag{4}$$

where $U_{context}(\cdot)$ is given by (4) and $U_{data}(\cdot)$ is the single-date class-conditional energy function

$$U_{data}(X_t, C_t(i,j)) = -Ln[p(X_t(i,j) / C_t(i,j))], \tag{5}$$

with $C_t(i,j) \in \{\omega^t_{int}, \omega^{t1}_{unk}\}$.

2.1 *Image statistical model*

The computation of (5) requires the estimation of the class-conditional probabilities of both the class of interest $p(X_t/\omega^t_{\text{int}})$ and the unknown class $p(X_t/\omega^t_{unk})$ for each time t. The former can be derived from the available training samples, whereas the latter cannot be estimated by using classical means. In this context, we propose an RBF network, which allows the iterative estimation of the probability density function of the unknown classes at each time by exploiting the spatio-temporal context.

It is based on the observation that the probability density function $p(X_t)$ of the image data $\mathbf{X}_t$ can be modelled as a mixture density of two classes, ω^t_{int} and ω^t_{unl}. This allows the class-conditional probability $p(X_t/\omega^t_{unk})$ to be described in terms of other statistical terms, which can be derived by exploiting the different information available: i.e. the training samples, the image data, and the spatial context,

$$p(X_t/\omega^t_{unk}) = \frac{p(X_t) - p(X_t/\omega^t_{\text{int}})P(\omega^t_{\text{int}})}{\left[1 - P(\omega^t_{\text{int}})\right]} \tag{6}$$

RBF models are modular, non-linear networks [5], which have proven a powerful tool to provide non-parametric estimates of complex data distributions. The architecture of the proposed RBF involves a common set of locally tuned kernel functions $\mathbf{\Phi} = \{\phi_{1,t}(\cdot),\ldots,\phi_{K,t}(\cdot)\}$, Gaussian functions characterised by a centre $\mu_{k,t}$ and a width $\sigma_{k,t}$, and a two set of weights $\mathbf{W}_t = \{w_{1,t},\ldots,w_{K,t}\}$ and $\mathbf{W}^t_{\text{int}} = \{w^{\text{int}}_{1,t},\ldots,w^{\text{int}}_{K,t}\}$, which regulates the contribution of the kernels to the final outputs. For each pixel $X_t(i,j)$, the two outputs of the network (associated with $p(X_t)$ and $p(X_t/\omega^t_{\text{int}})$, respectively) are given by

$$\hat{p}(X_t(i,j)) = \sum_{k=1}^{K} w_{k,t}\phi_{k,t}(X_t(i,j)), \tag{7}$$

and

$$\hat{p}(X_t(i,j)/\omega^t_{\text{int}}) = \sum_{k=1}^{K} w^{\text{int}}_{k,t}\phi_{k,t}(X_t(i,j)). \tag{8}$$

The training of the network involves the computation of the sets $\mathbf{\Phi}_t$, $\mathbf{W}_t$ and $\mathbf{W}^t_{\text{int}}$, such that equations (7) and (8) may be verified. To accomplish such a task, the EM algorithm will be used [6]. It is a general approach to maximum-likelihood parameter estimation and is an efficient tool for training RBF networks [7], [8].

2.1.1 Estimation of $p(X_t)$

This task involves the computation of $\mathbf{\Phi}_t$ and $\mathbf{W}_t$. For simplicity, kernel parameters are initialised using the classical *k*-means clustering algorithm [5]: i.e. the centres and widths of the kernels are initialised to the centres and variances of the resulting clusters; weights are initialised to $1/K$. The number of kernels K can be selected as a trade-off between computational load and final accuracy (it was observed that in a reasonable range, the selection of K is not a critical issue. The EM algorithm is then used to compute the final estimates. The equations to be used are [6]:

$$w_{k,t}^{+1} = \frac{\sum_{X_t(i,j)\in \mathbf{X}_{ti}} \frac{w_{k,t}^{+}\phi_{k,t}^{+}(X_i(i,j))}{p^{+}(X_t(i,j))}}{IJ}; \tag{9}$$

$$\mu_{k,t}^{+1} = \frac{\sum_{X_t(i,j)\in \mathbf{X}_t} \frac{w_{k,t}^{+}\phi_{k,t}^{+}(X_t(i,j))}{p^{+}(X_t(i,j))} X_t(i,j)}{\sum_{X_t(i,j)\in \mathbf{X}_t} \frac{w_{k,t}^{+}\phi_{k,t}^{+}(X_t(i,j))}{p^{+}(X_t(i,j))}}; \tag{10}$$

$$(\sigma_{k,t}^{2})^{+1} = \frac{\sum_{X_t(i,j)\in \mathbf{X}_t} \frac{w_{k,t}^{+}\phi_{k,t}^{+}(X_t(i,j))}{p^{t}(X_t(i,j))} \left[X_t(i,j) - \mu_{k,t}^{+1}\right]^2}{\sum_{X_t(i,j)\in \mathbf{X}_t} \frac{w_{k,t}^{+}\phi_{k,t}^{+}(X_t(i,j))}{p^{+}(X_t(i,j))}}, \tag{11}$$

where the superscripts + and *+1* denote the values of the parameters at the current and next iterations, respectively. At each iteration, the estimated parameters provide an increase in the log-likelihood function $L(\theta_t) = \ln p(\mathbf{X}_t | \theta_{ti})$, where $\theta_t = \{\mu_{1,t},..., \mu_{K,t}, \sigma_{1,t}^2,..., \sigma_{K,t}^2, w_{1,t},..., w_{K,t}\}$, until a local maximum is reached [6].

2.1.2 Estimation of $p(X_t / \omega_{\text{int}}^{t})$

Once the set $\mathbf{\Phi}_t$ was computed, $\mathbf{W}_{\text{int}}^{t}$ can be obtained using the following equation [6]:

$$\left[w_{k,t}^{\text{int}}\right]^{+1} = \frac{\sum_{X_t(i,j)\in \mathbf{T}_{\text{int}}} \frac{\left[w_{k,t}^{\text{int}}\right]^{+} \phi_{k,t}(X_t(i,j))}{p^{+}\left(X_t(i,j)/\omega_{\text{int}}^{t}\right)}}{N_{\text{int}}} . \tag{12}$$

The final weights are obtained by starting from the initial values (i.e. $1/K$) and by iterating the above equation until convergence.

2.2 *Iterative partially supervised change detection*

The minimisation of (4) is carried out using an iterative procedure based on the Iterated Conditional Modes (ICM) algorithm [9]:

Step 1. For each date t, after computing $p(X_t)$ and $p(X_t/\omega_{\text{int}}^{t})$ using the RBF network, initialise $P(\omega_{\text{int}}^{t})$ by applying a significant testing approach, as described in (Jeon and Landgrebe, 1999); On the basis of that first value, compute the initial estimate of $p(X_t/\omega_{\text{int}}^{t})$ using equation (6).
Step 2. Derive the first set of labels by assigning pixels $X_1(i,j)$ and $X_2(i,j)$ to the classes $C_1(i,j) = \{\omega_{\text{int}}^{1}, \omega_{unk}^{1}\}$ and $C_2(i,j) = \{\omega_{\text{int}}^{2}, \omega_{unk}^{2}\}$, respectively, that minimises the non-contextual energy function (5).
Step 3. Assign pixels $X_1(i,j)$ and $X_2(i,j)$ to the classes $C_1(i,j) = \{\omega_{\text{int}}^{1}, \omega_{unk}^{1}\}$ and $C_2(i,j) = \{\omega_{\text{int}}^{2}, \omega_{unk}^{2}\}$, respectively, that minimises (4).
Step 4. On the basis of the resulting set of labels, update $p(X_1/\omega_{\text{int}}^{1})$ and $p(X_2/\omega_{\text{int}}^{2})$ (by computing the new prior probabilities $P(\omega_{\text{int}}^{1})$ and $P(\omega_{\text{int}}^{2})$, and then using (6)).
Step 5. Iterate steps 3 and 4 until convergence is reached [9].

3 Experimental analysis

The considered data set is related to an agricultural area in the basin of the Po river (northern Italy). It consist of two Landsat-5 TM images (484×882) acquired in April and May 1994. Four land cover classes characterized the area of interest in April (i.e., wet-rice field, bare soil, cereals, wood), whereas, after some land-cover changes, five classes characterized the region in May (i.e., corn, dry-rice field, wet-rice field, cereals and wood). For the experiments, the land-cover transition (bare soil, dry-rice field) was considered as the change of interest. Only 3547 training samples belonging to the class "bare soil" were used for the April images, whereas

5955 samples were used for the class "dry-rice field" in May (training samples belonging to the other land-cover classes were assumed to not be available). A multitemporal test set, with 5695 pixels corresponding to the land-cover transition of interest and 37360 pixels belonging to the other land-cover transitions, was also derived from the ground truth available. In this context, different experiments were carried out in order to evaluate the accuracy of the proposed approach. To this end, the change-detection accuracy provided by the proposed technique was compared to the accuracy provided by the fully supervised *Post-Classification Comparison (PCC)* technique [3]. The *PCC* was applied by comparing, at a pixel level, two land-cover maps derived for each single date by using Maximum-likelihood classifiers trained with "complete" training sets (using samples belonging to all the classes present in the area under analysis at the two times).

Table I shows the change-detection accuracy provided by the proposed and the PCC techniques, respectively. As one can see, the results provided by the proposed approach are comparable to those derived from the use of the fully supervised PCC technique. This confirms the effectiveness of the presented technique. In particular, the accuracy provided by the proposed approach in the detection of the land-cover transition of interest (90.6%) is significantly higher than the one resulted from the PCC technique (75.7%). This may be a direct result of the use of the spatio-temporal correlation in the change-detection process.

Table 1. Change detection accuracy (%) provided by the partially supervised proposed technique and the fully supervised PCC

Transition	Proposed	Superv. PCC
(bare soil, dry-rice f.)	90.6	75.7
Unknown	87.5	89.0

4 Discussion and conclusion

A partially supervised technique for change detection in multitemporal remote-sensing images has been presented. The proposed technique allows the automatic identification of one (or few) land-cover transitions of interest for the end-user by using exclusively training samples belonging to the land-covers involved. This represents a significant advantage over traditional approaches, which requires an exhaustive collection of training samples for all the land-cover classes present in the area under analysis at the two times under consideration.

Experimental results obtained on a multitemporal data set confirm the effectiveness of the proposed approach.

5 Acknowledgements

This research was supported by the European Space Agency (ESA) (Project Ref. C1F.1033). The author wish to thank Prof. Lorenzo Bruzzone from University of Trento, Italy, for providing the remote-sensing images used in the experiments.

References

1. A. Singh, Digital change detection techniques using remotely-sensed data, *Int. J. Remote Sensing,* **10** (1989) pp. 989-1003.
2. L.Bruzzone and D. Fernàndez Prieto, Automatic analysis of the difference image for unsupervised change detection, *IEEE Trans. on Geoscience and Remote Sensing*, **38** (2000) pp. 1171-1182.
3. L. Bruzzone and S.B. Serpico, An iterative technique for the detection of land-cover transitions in multitemporal remote-sensing images, *IEEE Transactions on Geoscience and Remote Sensing*, **35** (1997) pp. 858-867.
4. B. Jeon and D.A. Landgrebe, Partially supervised classification using weighted unsupervised clustering, *IEEE Trans. on Geoscience and Remote Sensing*, vol. **37** (1999) pp. 1073-1079.
5. L. Bruzzone and D. Fernàndez Prieto, A Technique for the Selection of Kernel-Function Parameters in RBF Neural Networks for Classification of Remote-Sensing Images, *IEEE Transactions on Geoscience and Remote Sensing*, **37** (1999) pp. 1179-1184.
6. A. P. Dempster, N. M. Laird, and D. B. Rubin, "Maximum likelihood fro incomplete data via the EM algorithm," *Journal of Royal Statistic. Soc.* , **39** (1977) pp. 1-38.
7. Miller, J.D., and Uyar, H.S., 1998, Combined learning and use for a mixed model equivalent to the RBF classifier, *Neural Computation.* **10**, 281-293.
8. Tilsias, M., and Likas, A., 2000, A probabilistic RBF network for classification, *in Proc. of the IEEE International Joint Conference on Neural Networks,* **4**, 238-243.
9. Besag, J, 1986, On the statistical analysis of dirty pictures. *J. Royal Statist. Soc.*, **48**, 259-302.

MULTITEMPORAL CHANGE DETECTION FOR UPDATES OF TOPOGRAPHIC MAP DATA BASES

THOMAS KNUDSEN

National Survey & Cadastre—Denmark (KMS), 8 Rentemestervej, DK-2400 Copenhagen NV, Denmark, E-mail: thk@kms.dk

Improving the update procedure for a digital topographic map data base in an operational environment is a non-trivial task. Typical update procedures involve manual change detection on (bi)temporal sets of aerial photos and/or direct comparison of the latest map with a newer photo. Full automation of this process is not within immediate reach. In the short term, however, much can be done to reduce the burden of manual updating, for example by introducing automated change detection methods. This involves the comparison of the topographic data base (which can be thought of as a highly refined/generalized image) and a new, image, which must be segmented/generalized prior to the comparison. In an operational setup, this also involves long term storage of large anounts of image data, making lossy compression necessary. This compression must be used with great care in order to avoid negative influence on the spectrally based segmentation. Due to the complex nature of the mapping from topographic classes to spectral signatures, the segmentation must typically be carried out using some level of supervised classification. This is not necessarily desirable for operational tasks, so in this initial study, unsupervised clustering of a spectrally decorrelated colour photo into a very small number of classes is used, in an attempt of facilitating the transition to operational levels. This approach results in a quite meaningful set of clusters, but obviously not in something that can be used in a fine grained classification.

1 Introduction

TOP10DK, the Danish digital topographic data base, is the raw material behind all map production at the National Survey & Cadastre—Denmark (Kort & Matrikelstyrelsen, KMS). The development and update of TOP10DK is based on aerial photos: Each spring, one fifth of the Danish area is photographed. These photographs are used for the subsequent work with the updating of TOP10DK. Hence, currently a full update cycle is to be completed every 5 years.

The actual detection that something have changed is the first step in the process of updating a digital spatial database. Traditionally, labour intensive (= costly) methods have been used in this step: field inspections by topographers or direct comparison between the current map and a new orthophoto or stereo photo pair. In some cases, these methods will still be necessary for many years to come: we cannot expect to reach full automation of the map update process in the short term. In the meantime we may, however, significantly simplify the topographer's task by doing some kind of automated comparison between the data base objects, and the photos.

Detecting changes for mapping is, however, not a particularly simple task: Al-

Figure 1. Topographic map features are not necessarily simple to recognize, neither in the spatial, nor in the spectral domain: The residential area shown in the left panel does not have much in common with the other residential area shown in the right panel

though the intention is always to carry out the photo flights at approximately the same time of year, the diurnal, annual, and interannual variations of the vegetation coverage is of a magnitude that *dwarfs* the primarily human generated changes we want to detect. Additionally, as demonstrated in figure 1, the spectral and spatial characteristics of individual members of the same topographic base class may differ wildly. Hence, some kind of generalization of the surface elements is necessary in order to pinpoint the relevant changes.

2 Data

2.1 *Image data*

Until recently, KMS has only aquired black-and-white aerial photos. Now, however, all new photos are required to be in colour (RGB). While giving much better possibilities for detecting changes through variations in spectral response, RGB is not necessarily enough: In recent work, Petzold, Walter and Frisch[1,2,3,4,5] studied the possibilities for automated updates of the German topographic database ATKIS [6]. These studies indicate that information from the near infrared spectral band is extremely important for succesful map updates—primarily, of course, in order to discern between vegetation and man made objects. Also laser scanner height observations were found to be very useful, especially in order to discern between agricultural

Figure 2. Test area in *Kongens Lyngby*, near Copenhagen, Denmark: Plain RGB aerial photo from april 2000.

areas and forests, and between roads and buildings.

As a consequence of the German experiences with ATKIS, a number of test scenes have been photographed in plain RGB, as well as in colour-infrared (CIR). Laser scanner data are still not available over our test areas, but some height information may be extracted from stereo-processing of the aerial photos. Future plans include the use of radiometry and interferometry from airborne synthetic aperture radar (SAR) images, and satellite borne RGB+IR radiometric images. Currently, however, these data sources are still considered to have too coarse resolution (typically a few meters) for our purposes: The aerial photos, for comparison, are taken from an altitude of approximately 3800 m, resulting in a scale of 1:25000, and the scanning is carried out at a resolution of 21 μm (48 lines/mm) for a nominal ground resolution of 0.525 m, which is very close to the TOP10DK precision requirements.

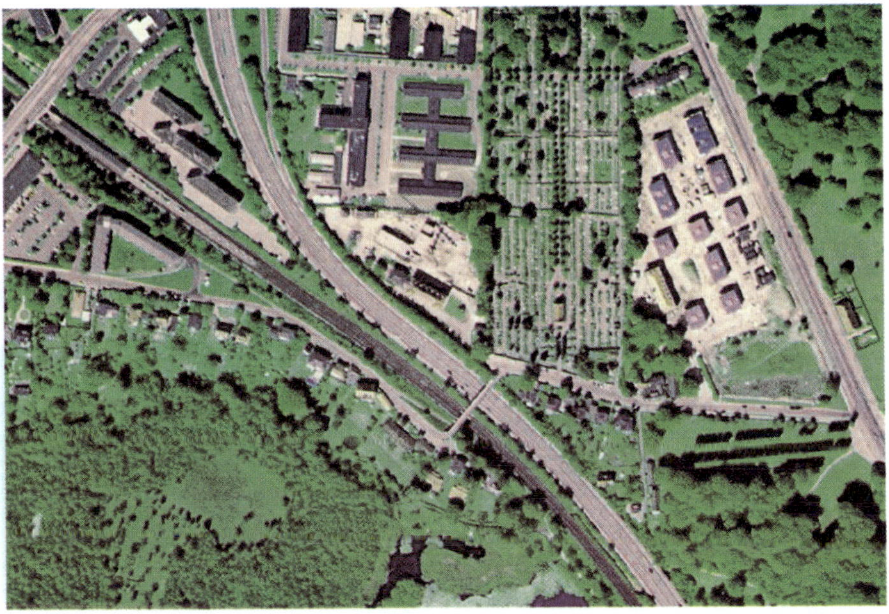

Figure 3. Same area as in figure 2, but using a colour-infrared (CIR) image from june 2000 (the first two channels have been swapped in order to facilitate comparison with the RGB version). The higher solar angle (a function of differences in both time-of-day and time-of-year) reduces the shadow prone areas, while the higher leaf area index (mostly a function of differences in time-of-year) makes it harder to detect ground features in the vicinity of trees

2.2 *TOP10DK*

The mean registration error of the TOP10DK data is better than 1 m, horizontally and vertically[7]. The registration is fully 3D and the data are fully topologic. The data base contains 43 different data classes of point, line, and area base type, which are complemented by a comparable number of labelling classes. Since the data include classes like administrative boundaries/borders, not all of the TOP10DK data base can be updated using aerial photos. In the present case, the emphasis is to be put on the areal data types *lake, building, and wetlands* and the line data types *road, railway, path, and river/creek.*

Figure 4. Same as figure 2, but overlaid by the building theme of the map data base. It is evident that a lot of building and demolition activities has been going on in this area since the original mapping.

2.3 Data storage and compression

Successful migration of automated update methods to the operational levels requires us to consider data handling aspects as well. A full coverage of 1:25000 aerial photos of Denmark consists of 5465 images. Each photo measures 230 mm by 230 mm, adding up to a total of approximately 115 MB per layer (345 MB per image) when scanned at 21 μm. This means that the storage requirements for a full coverage of Denmark is 5465×345 MB = 1.8 TB. Terabyte storage systems are not uncommon, but most ordinary storage systems are still an order of magnitude smaller than this. In a few years, this will probably change, but even in the longer term, we still need to consider image compression. Experiments have shown that lossless compression of aerial images will not result in compression levels higher than about 4:5 (i.e. a 20% reduction), so lossy compression is necessary to get a reasonable reduction.

JPEG is the most widespread, and hence most viable, compression method. As it is based on truncation of a discrete cosine transform, JPEG, tends to introduce artifacts into the reconstructed image. Studies of the photogrammetric effect of these

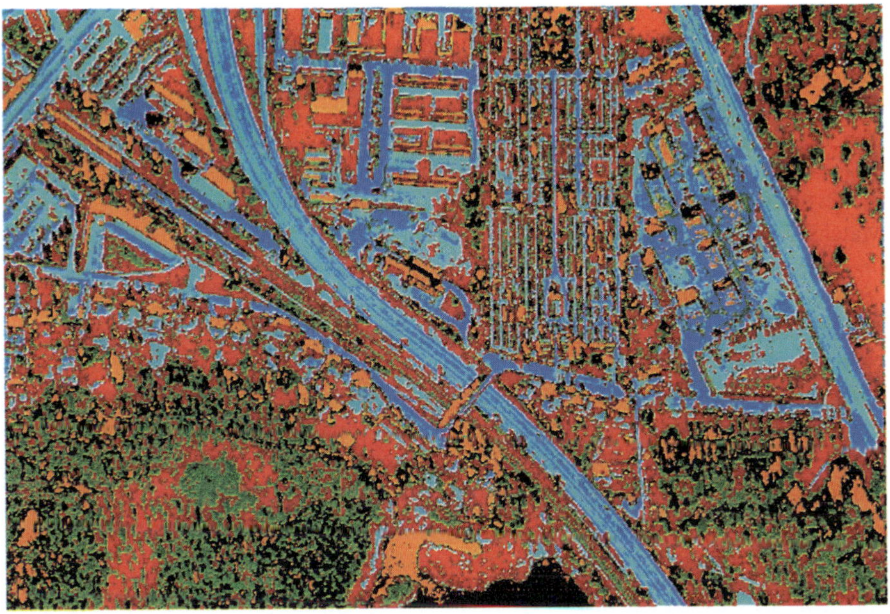

Figure 5. Simple unsupervised clustering into 6 classes of a decorrellated version of the image shown in figure 2: The *shadow* class is mapped rather robustly into the orange cluster group, although the highly eutrophic parts of the *water* class gets intermixed here as well. In general *trees* and *low vegetation* are split neatly into the green an red cluster groups, respectively, although low vegetation in very wet areas tend to become misclassified as trees. The two blue cluster groups catches all man made surfaces.

artifacts [8,9,10] have primarily focused on the degradation of orthorectification, introduced by the degradation of areal matching (this occurs when the matching routines mistakenly matches the compression artifacts, rather than the actual ground features). Based on greyscale photographs, Kiefner and Hahn[8], found that by keeping the compression at a maximum of 1:5 (i.e. a 80% reduction), acceptable results were still obtainable for orthorectification. In our case of colour photographs and spectral clustering, this level is used as a guideline, but to avoid too heavy degradation of the colour signal, the 3 channels must be treated independently, rather than subsampling the chromaticity information, as is common in JPEG processing. A compression of 1:5 will reduce the storage necessary from the 1.80 TB range to the 450 GB range. This is still a large amount of data, but not unreasonable, even with today's technology. Preliminary experiments have shown that the clustering process does not suffer from this level of compression.

3 Methods

Following the studies of ATKIS updates by Petzold and Walter, [1,2,3,4] the TOP10DK classes are clustered into 5 higher level super classes: *low vegetation, high vegetation, building, road, and water*. A sixth class, *shadow* is added during clustering of image data. This class is especially important, since it is a very good indicator of the level of confidence one should have to a specific matching of a data base item. Matching in shaded areas is highly error prone.

The clustering process used here is based on an iterative ISODATA-like procedure carried out on a spectrally decorrelated version of the RGB images. The spectral decorrelation is simply carried out by pixelwise subtraction of the green channel value from the red and the blue channel, which causes some reduction of the otherwise very high correlation between the channels.

Finally, the clustered photo is compared to the data base registrations, so basically we end up comparing a very highly generalized "photo" (= the data base objects) with a somewhat generalized photo (= the clustered version of the aerial photo), trying to detect each individual data base item in the clustered photo.

4 Results

The clustering of the decorrellated RGB image from figure 2 is shown in figure 5. Comparing this to the registrations from the TOP10DK building theme (figure 4, it is evident that most of the still existing buildings are caught by the "man-made" cluster group. Shadows are caught with high confidence, and in general high and low vegetation is discerned, despite problems in wet areas.

The main problem when using this clustered data set as input to the step of refinding the data base registrations will be in discerning the street and the building blocks. For older buildings, which are typically surrounded by vegetation, this is not as big a problem as for new buildings, where the vegetation is still sparse.

5 Discussion and Conclusion

Currently, at the initial stage of this work, the experiences from the clustering of the decorrelated RGB images alone seems rather encouraging. Additional input data is, however, desirable: Incorporating colour-infrared image data may stabilize especially the distinction between low and high vegetation, and a-priori knowledge derived from the topographic data base will be a valuable addition in general.

Algorithmically, the rather plain ISODATA-like clustering algorithm used here may be exchanged for never, more advanced algorithms (although the speed of ISODATA is an important virtue in an operational setup).

There is, however, still a long way to go before fully automated map updating becomes a reality.

Acknowledgement: Brian P. Olsen, M.Sc., carried out the orthorectification and the TOP10DK overlays for the images used here.

References

1. Volker Walter. Automatic change detection in GIS databases based on classification of multispectral data. *International Archives of Photogrammetry and Remote Sensing*, XXXIII(B4):1138–1145, 2000.
2. Bettina Petzold. Revision of topographic databases by satellite images—experiences and expectations. In Knudsen [11], pages 15–23.
3. Volker Walter. The design and implementation of an automated system for revision of topographic databases. In Knudsen [11], pages 7–14.
4. Bettina Petzold and Volker Walter. Revision of topographic databases by satellite images. In Manfred Schroeder, Karsten Jacobsen, Gottfried Konecny, and Christian Heipke, editors, *Sensors and mapping from space 1999*, page 9, Hanover, Germany, 1999. ISPRS.
5. Volker Walter and Dieter Frisch. Automatic verification of GIS data using high resolution multispectral data. *International Archives of Photogrammetry and Remote Sensing*, XXXII(3):74–79, 1998.
6. Arbeitsgemeinschaft der Vermessungsvervaltungen der Länder der Bundesrepublik Deustchland. Amtliches topographisch-kartographisches informationssystem (ATKIS). `http://www.adv-online.de/english/products/atkis.htm`.
7. Kort & Matrikelstyrelsen. TOP10DK geometrisk registrering, specifikation udgave 3.2.0. Copenhagen, Denmark: Kort & Matrikelstyrelsen, May 2001.
8. Michael Kiefner and Michael Hahn. Image compression versus matching accuracy. *International Archives of Photogrammetry and Remote Sensing*, XXXIII(B2):316–323, 2000.
9. Rob Reeves and Michael Hahn. Modeling the influence of JPEG compression on DTM accuracy. *International Archives of Photogrammetry and Remote Sensing*, XXXII(3-4W2):161–168, 1998.
10. Michael Hahn and Michael Kiefner. Image compression and matching accuracy. *International Archives of Photogrammetry and Remote Sensing*, XXXII(3):444–451, 1998.
11. Thomas Knudsen, editor. *Proceedings of the seminar on remote sensing and image analysis techniques for revision of topographic databases*, National Survey and Cadastre—Denmark, Technical Report no. 13, Copenhagen, Denmark, 2000. National Survey and Cadastre—Denmark.

EXPLOITING SPATIAL AND TEMPORAL INFORMATION FOR EXTRACTING BURNED AREAS FROM TIME SERIES OF SPOT-VGT DATA

BRIVIO P.A. AND MAGGI M.

Remote Sensing Department., CNR, Milan, Italy

E-mail: brivio@irrs.mi.cnr.it

BINAGHI E. AND GALLO I.

Istituto per le Tecnologie Informatiche Multimediali, CNR, Milan, Italy

E-mail: ebinaghi@itim.mi.cnr.it

GREGOIRE J-M.

GVM Unit, Space Applications Institute, JRC-EU, Ispra, Italy

E-mail: jean-marie.gregoire@jrc.it

This paper reports on a new methodology for the detection of burned areas in coarse resolution satellite images. The work has been carried out in the framework of the GBA (*Global Burnt Areas*) 2000 project whose aim is the mapping of burned areas at a global scale from SPOT-VGT for the year 2000. The proposed methodology is based on the use of Multi-Layer Perceptron (MLP) neural network which allows to exploit not only the spectral information of the observed targets but also the spatial and temporal relationships of the phenomenon. The study area corresponds to the Northern part of the African continent, entirely covered by a mosaic of daily SPOT-VGT images. Validation of the burned area maps obtained at different dates has been conducted by the comparison with visual classification of low resolution data and with automatic classification of Landsat-ETM+ scenes acquired over the study area in correspondence of the SPOT-VGT sensor.

1 Introduction

Vegetation fires is a global phenomenon spread over all the vegetation types of the world, from equatorial rain forests to tropical savannas, from temperate to boreal forests [1]. Its impacts on ecological and environmental conditions are such of causing large scale and long lasting environmental changes. Biomass burning causes the release of large quantities of gases such as carbon dioxide and methane, which are in part responsible of greenhouse effect [2].

Although over the recent years the growing concern in scientific community about the importance of biomass burning in the problem related to climate change and environmental degradation [3], large uncertainties remain in quantifying the impacts of vegetation fires at all scales. There is therefore an increased request for a systematic documentation on biomass burning both from the scientists and policy makers.

Satellite observations allow a global coverage and maintain their systematic data gathering activities for a long period of time. Although no specific sensor was designed for this particular scope, a suite of sensors and satellite were used to document vegetation fire activity over a range of spatial scales and spectral domains. A wide range of studies have been done for the extraction of burned surfaces information from satellite remote sensing. Depending on the extension of the zone under the analysis different instruments can be used: high resolution sensors are used for specific regions and local detailed evaluation, coarse resolution sensors (AVHRR, ATSR, VGT) are used for global estimation.

This work proposes a methodology for mapping burned areas using SPOT-VGT images. The approach is a supervised classification scheme based on a Multi-Layer Perceptron (MLP) neural network which allows to exploit the spatial and temporal information for burned area mapping. A supervised classification scheme is used to detect burned areas in a time series of SPOT-VGT daily images acquired over the Northern part of the African continent.

2 Study area and satellite sensor data

In the frame of the Global Burnt Area (GBA-2000) project, a module of the more ambitious initiative Global Land Cover (GLC-2000), promoted by Joint Research Centre of the European Commission in cooperation with CNES, a global daily coverage of SPOT-VEGETATION (VGT) images were acquired covering one year.

SPOT-VGT is a sensor carried on board of SPOT-4 dedicated to studies of vegetation cover type and status at a global level. Spatial coverage of 2200 km swath width allows to image about 90% of the equatorial areas each day, the remaining 10% being imaged the next day; for latitudes higher than 35° (North and South), all regions are acquired at least once a day. The enhanced geometry characteristics of the VGT sensor assure a low spatial variation of pixel size from 1 km, at nadir, to 1.7 km for off nadir pixels. Main spectral characteristics of the VEGETATION sensor are summarized in Table 1.

Table 1. VEGETATION system spectral characteristics.

Band Name	b0 (Blue)	b2 (Red)	b3 (NIR)	swir
Wavelength [μm]	0.43 - 0.47	0.61 - 0.68	0.78 - 0.89	1.58 - 1.75

The study area covers the part of the African continent located North of the Equator (Figure 1). According to the IGBP vegetation map the major vegetation types present in the area include savanna and woody savanna, evergreen broadleaf forest and limited part of cropland. A mosaic of daily images acquired by SPOT-VGT is used to cover the entire area.

The VGT data set covers the period from October 1999 to March 2000, corresponding to the dry season when fire activity is more intense and spreads across all the major biomes in the area. The used daily VGT-S1 products are ground reflectance values, atmospherically corrected by CNES using the SMAC (Simplified Method for Atmospheric Correction) model [4].

Figure 1. Study area is the African continent North of the Equator. A mosaic of daily images acquired by SPOT-VGT is used to cover the entire area. Location of a set of Landsat Thematic Mapper frames (180 x 180 km) available for validation is also indicated.

3 Detection of burned area

Burned area mapping from Earth observation data has considerably lower heritage than detection of active fires, however several approaches have been proposed for mapping burned areas in different biomes at regional and continental scales exploiting the spectral capabilities offered by coarse resolution sensors, such as AVHRR of NOAA [5], ATSR of the European ERS [6], MVIRI of Meteosat and VISSR of the Japanese GMS [7] and VGT of SPOT [8]. These approaches range from the application of multiple tests on spectral values or derived indices, to temporal analysis of remotely sensed data, to traditional image segmentation techniques.

However to date a consensus burned area mapping algorithm has not been developed, although there is a general agreement that multi-temporal algorithms have the greatest potential for operational implementation.

3.1 Classification Strategy

The approach here proposed is a contextual classification strategy aimed at exploiting spatial and temporal information required for identifying burnt areas in low resolution imagery such as SPOT-VGT data.

The classification is based on Multi-Layer Perceptron (MLP) [9] which is used as a soft classifier considering the gradual activation values of the output neurons as final partial membership to classes. The use of the MLP model is motivated by the experimentally proven effectiveness in running as a soft classifier [10] and the well-documented capability in dealing with patterns described by complex features that in our case result from the modelisation of spatial and temporal context. Figure 2 shows the overall classification strategy receiving as input a time series of imagery from day d_j to day d_{j+k} and producing in output a burned area map M_j associated to the day d_j.

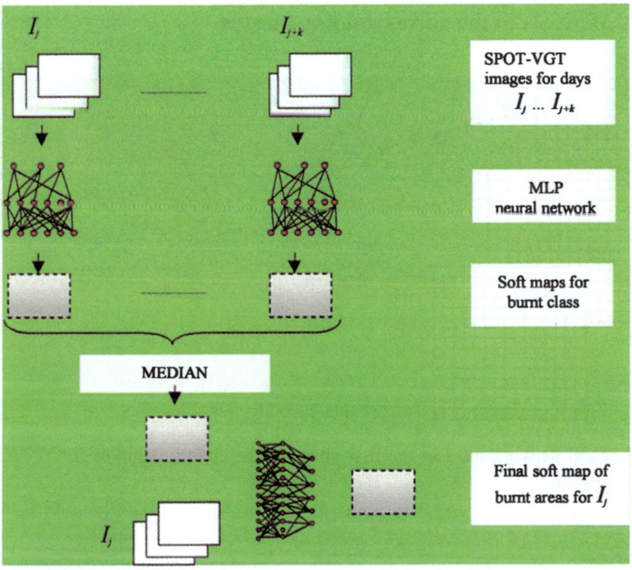

Figure 2. Diagram showing the classification strategy used for burned area mapping in SPOT-VGT time series of data.

The MLP model is inserted within a two stages classification strategy aimed to detect burned pixels in daily images. The approach of processing daily images was preferred to the compositing procedure used widely in the analysis of a set of multi-temporal images acquired by wide field of view polar orbiting sensors [11]. Compositing is a form of data fusion, that selects from a series of colocated pixels of different orbits the pixel that best satisfies some criteria allowing to obtain a single ideal data set with reduced atmospheric and cloud contamination and angular effects. However, these procedures that select the most suitable pixel form a set of measurements taken at different dates, often produce strong radiometric artifacts when compositing individual channels [12]. The use of daily images preserves the full range of information contained in the original radiometric measurements.

3.1.1 Stage 1

In the first stage each image of the time series $I_j \ldots I_{j+k}$ is classified separately according to a pixel-by-pixel classification scheme shown in Figure 3. The neural classifier receives in input pixel spectral values (b2, b3, swir) and computes the gradual membership to all the six classes conceived, Water, Burnt, Not Burnt, Medium Burnt, Cloud Shadows, Clouds. The result of the first classification stage is constituted by six soft maps for each image processed indicating the gradual memberships of pixels to the corresponding classes.

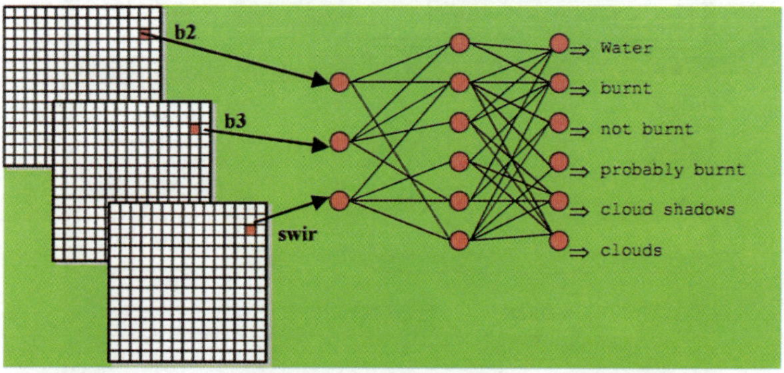

Figure 3. MLP neural network used in the first phase of the classification of SPOT-VGT daily images for burned area mapping.

3.1.2 Stage 2

The second classification stage is aimed to produce a soft classification of the daily image I_j. The first stage classification results are processed to prepare the input to the second stage classification. In particular, the series of soft maps related to burnt class are considered and fused together by applying the median operator, pixel-by-pixel. The rational behind this data fusion operation is that of exploiting temporal relationships in time series imagery. Values in the resulting median map synthesize information about the temporal variation of the pixels and play the role of reinforcing or attenuating the strength of membership to burnt class in the final map. Variability in spectral response due to no data conditions or clouds is automatically managed within this data fusion preprocessing stage contributing to make the classification output more reliable.

The second stage neural classifier is configured to receive in input I_j pixel spectral values and values of a 3x3 window drawn from the median map and centered on the coordinates of the pixel to be classified (Figure 4).

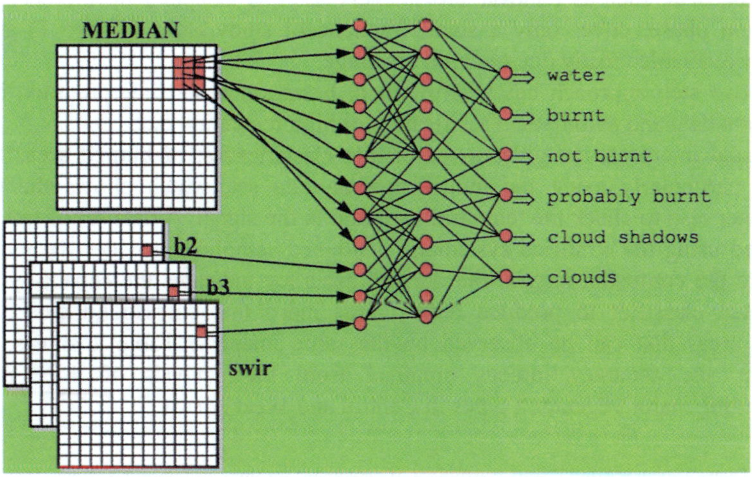

Figure 4. Diagram showing the MLP network architecture used for the final classification stage.

Figure 5 shows an example of the membership degree as result of the first phase and the final map of burned area detected. Extension of burned surface is reduced compared to the result of the classification first stage: temporal information conveyed by the median tends to diminish the presence of false alarms.

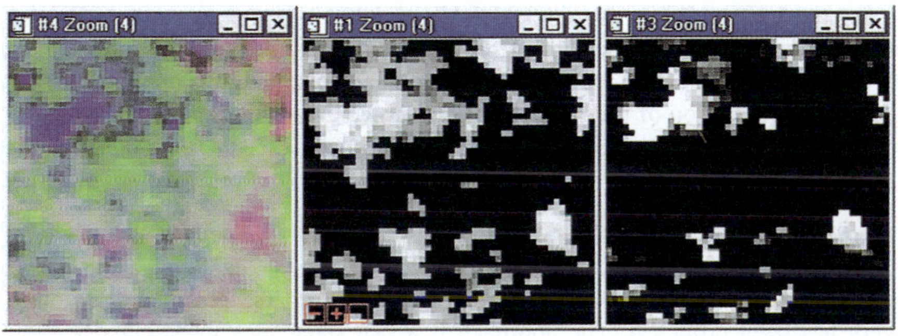

Figure 5. Spot-VGT colour composite (RGB: swir, b3, b2), map of the membership degree of burned class and final burned map

4 Analysis of the results

Evaluation of the results obtained from the classification of wide geographical areas covered by low resolution satellites is always a critic task for several reasons. High resolution data, such as Landsat TM imagery used as independent reference in the validation phase, cover only a small fractions of study areas (Figure 1) and are scarcely available mostly due to cloud coverage.

A two step approach has been defined, based on visual interpretation of low resolution data and automatic classification of high resolution data.

Visual interpretation gives a reasonable classification of burned areas on the image, particularly using the multi-temporal data set, which allows the photo-interpreter also to check the temporal variation in the signal. A data set of test pixels extracted using the stratified systematic unaligned sampling provides the means to compute the confusion matrix: the derived accuracy indicates the capability of the automated classifier to produce burned area maps that correspond closely with burned areas that can be observed on the same imagery. Table 2 shows, as an example, the accuracy indices obtained from the analysis of three regions representing major vegetation types in Central and West Africa.

Table 2. Overall accuracy and Kappa coefficient of agreement for three geographic regions.

Region	A	B	C
OA % - Kappa	82.3% - 0.57	83.8% - 0.67	87.0% - 0.73

Automatic classification of a set of Landsat TM images are used to produce an assessment of the surface area accuracy of the detected burned area, with limited location accuracy, on the basis of a regression line method between proportion of burned/unburned area in 15x15 km cells extracted from low and high resolution data. For the Landsat image acquired on 3 March 2000 over the area around Bangui (Central African Republic) the regression line shows a significant correlation ($R^2 = 0.60$) and an angular coefficient of 1.14. The main advantage of this approach is that it allows bias-correction of the low resolution burned area estimates, to improve the accuracy of the GBA's estimation of global areas burned.

5 Conclusions

A new methodology for detecting burned areas with coarse resolution satellite imagery has been proposed. The approach is a two stage supervised classification scheme based on a MLP neural network which allows to exploit the spatial and temporal information. The methodology has been applied to a time series of SPOT-VGT daily images acquired over the Northern part of the African continent, where burned surfaces appear very fragmented. Although the classification results appear to be satisfying, further efforts are foreseen to verify the robustness of the proposed methodology.

6 References

1. E. Dwyer, S. Pinnock, J. M. Gregoire and J. M. Pereira, "Global spatial and temporal distribution of vegetation fire as determined from satellite observations", *Int. J. Remote Sensing*, Vol. 21, no. 6&7, pp. 1289-1302, 2000.
2. J. S. Levine (ed.), *Global Biomass Burning. Atmospheric, Climatic and Biospheric Implications*, (Cambridge, MA: MIT Press), pp. 569, 1991.
3. P. J. Crutzen and J. G. Goldammer (eds.), *Fire in the Environment: the Ecological, Atmospheric, and Climatic Importance of Vegetation Fires.* (Chichester: Wiley and Sons), pp. 400, 1993.
4. H. Rahman and G. Dedieu, " SMAC: a simplified method for the atmospheric correction of satellite measurements in the solar spectrum", *Int. J. Remote Sensing*, Vol. 15, no. 1, pp. 123-143, 1994.
5. P. M. Barbosa, J. M. Gregoire and J. M. Pereira, "An algorithm for extracting burned areas from time series of AVHRR GAC data applied at a continental scale", *Remote Sensing Environ.*, Vol. 69, pp. 253-263, 1999.
6. H. Eva and E. Lambin, "Burnt area mapping in Central Africa using ATSR data", *Int. J. Remote Sensing*, Vol. 19, no. 18, pp. 3473-3497, 1998.
7. L. Boschetti, P. A. Brivio, J.M. Gregoire, "Detection of burned areas using geostationary satellite data in tropical environments", *Remote Sensing for Agriculture, Ecosystems and Hydrology*, M. Owe, G. D'Urso, E. Zilioli, Eds., Proc. SPIE Vol. 4171, pp. 156-165, 2000.
8. R. H. Fraser, Z. Li and J. Cilhar, "Hotspot and NDVI differencing synergy(HANDS): a new technique burned area mapping over boreal forest", *Remote Sensing Environ.*, Vol. 74, n. 3, pp. 362-376, 2000.
9. J. D. Paola and R. A. Schowengerdt, "A review and analysis of back-propagation neural networks for classification of remotely-sensed multi-spectral imagery", *Int. J. Remote Sensing*, Vol. 16, n. 16, pp. 3033-3058, 1995.
10. E. Binaghi, P.A. Brivio, P. Ghezzi, A. Rampini and E. Zilioli, "Investigating the behaviour of neural and fuzzy-statistical classifiers in sub-pixel land cover estimations", *Canadian J. Remote Sensing*, Vol. 25, n. 2, pp. 171-188, 1999.
11. J. Cihlar, D. Manak and M. D'Iorio, "Evaluation of compositing algorithms for AVHRR data over land", *IEEE Trans. Geoscience Remote Sensing*, Vol. 32, pp. 427-437, 1994.
12. E. Brems, G. Lissens and F. Veroustraete, "MC-FUME: a new method for compositing individual reflective channels", *IEEE Trans. Geoscience Remote Sensing*, Vol. 38, n. 1, pp. 553-569, 2000.

THE USE OF SOM-ENCODED TEXTURE SPECTRUM FOR CHANGE DETECTION

JONATHAN CHEUNG-WAI CHAN

Laboratory for Global Remote Sensing Studies, 2181 LeFrak Hall, University of Maryland, MD20742, USA
E-mail: jonchan@glue.umd.edu

KWOK-PING CHAN

Department of Computer Science and Information Systems, Hong Kong University, Pokfulam, Hong Kong
E-mail: kpchan@csis.hku.hk

We reported a methodology of combining Texture Spectrum with spectral features for land cover change detection in a complex urban environment. Texture Spectrum is a simple statistical texture describer and we improved the encoding procedure by Kohonen's Self Organizing Maps. Experiments with simulated data demonstrated that incorporating textures significantly improved change detection that are related to "built-up land". Experiments with real data showed that accuracy improved between 2.4% to 18.9% for certain change classes, but accuracy degradations were also found. Our results suggested that including spatial textures is potentially beneficial for detecting change in a comparatively homogeneous environment. In complex environments, texture analysis might be less effective because generation of textures representative of certain land cover types would be difficult due to noisy background.

1 Introduction

Texture analysis of remote sensing data is reportedly useful for land cover characterization. Intuitively, texture analysis should also be useful for detecting land cover changes if the change is adequately widespread to be texturally describable. To date there is no conventional way on operational use of textures for land cover change detection. This study investigated the use of SOM-encoded (an improved version of) Texture Spectrum [1] for detecting the nature of land cover change. A one-pass classification procedure described in [2] was implemented on both simulated and real data sets.

2 Encoded Texture Spectrum for change detection

Wang and He [1] proposed the use of statistical Texture Spectrum approach that extracts local texture attributes based on a running window of 3 by 3 pixels. The pixel in the middle is compared to each of its eight adjacent pixels. For the set V = $\{V_0, V_1, \ldots V_8\}$ where V_0 is the intensity value of the central pixel and V_i $\{i = 1, 2,$

… 8} represent the intensity of value of the eight surrounding pixels ordering in a clockwise manner. Texture Unit (TU) is built by forming a set of eight elements $\{E_1, E_2, \dots E_8\}$, where E_i ($i = 1, 2, \dots 8$) is determined by the formula:

$$E_i = \begin{cases} 0, \text{if } V_i < V_0 \\ 1, \text{if } V_i = V_0 \\ 2, \text{if } V_i < V_0 \end{cases}$$

for $i = 1, 2, \dots, 8$. As each element of the TU has three possible values, there is a total of $3^8 = 6{,}561$ possible Texture Unit Labels (TL). Eight different ways of ordering can be formed by starting from a different pixel of the surrounding pixels. Texture Spectrum, the measurement of TUs over a region of an image, is defined as the frequency distribution of all the TUs with the y-axis showing the occurrence frequency and the x-axis listings of the 6,561 labels of texture units. Land covers with different spatial heterogeneity will form different distributions of TUs and hence spectra with unique features.

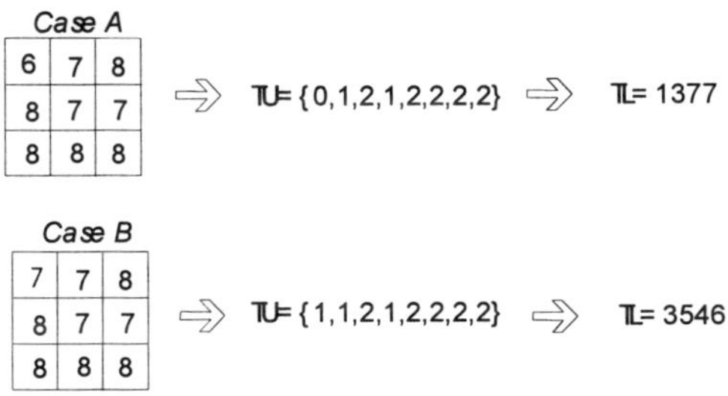

Figure 1. Original coding might have caused poor result because similar Texture Units (TU) are presented by a very different Texture Unit Labels (TL).

The advantage of texture spectrum is that it represents texture information in all eight directions and hence provides a more complete textural characterization than some of the conventional methods. Furthermore, construction of Texture Spectrum is computationally simple since only three values 0, 1 and 2 are used in Texture Units and the algorithm needed to build up spectra is also much simpler compare to other statistical methods. Texture Spectrum was applied to airborne SAR imagery for land cover characterization and encouraging results were reported [1]. The experiment in [1] used a 30 × 30 window but it was noted that appropriate window size should be determined by experiments. Our initial experiments found that the original coding of Texture Spectrum denotes similar TUs by very distant TLs meaning that similar texture patterns could produce very different spectra. For example, according to the label ordering, TL of {0,1,2,1, 2,2,2,2} is *1377* and that of {1,1,2,1,2,2,2,2} is *3564* (Fig 1). However, they are almost identical with only the first entry (top left) being different. We improved TL encoding with Self-Organizing Maps (SOM) [3] which has been widely experimented in studies for data compression and data mining. The entire set of Texture Unit Labels (6,561 units) is input to the SOM to form an arbitrary number of clusters that are represented by codebook vectors. The encoding procedure does not only generate more meaningful Texture Spectrum, since TUs in close features space were grouped together, but also trims the input dimensionality.

3 Data and Methodology

Figure 2. Multitemporal images of the study area – Tsing-yi Island, Hong Kong.

Two SPOT images of Hong Kong (Grid Reference System K/J: 386, 305) acquired on January 14, 1987 and February 5, 1995 were available for this study (Fig. 2). Geometric and atmospheric corrections were implemented to register the images. For our experiments, a sub-image of 227 × 204 pixels covering the entire Tsing-yi Island off the coast of Victoria Harbor was extracted. During that period of time, large-scale reclamation and land formation were carried out to provide land for housings and airport-related constructions. Since the visible red band gives good separation between water and land, Texture Spectrum was generated from band 2 (0.61-0.68 μm) of multi-date images. Four land cover types were identified for each period: built-up land, construction site, scrubland/grassland, and water surface. Figure 3 shows the typical texture spectrum generated from the 1987 image using a 30 by 30 window. Notice that each texture spectrum has distinct features. Instances for training and testing were randomly picked. Consequently, instances from the upper half of the image were used for training and the lower half for testing. The numbers of cases for each class were listed in Table 1.

Table 1. Classification scheme of change analysis. Listed are the number of cases for training (1st entries) and testing (2nd entries), and also the number of the class (#).

Time 2 / Time 1	Built-up land	Construction sites	Scrubland / grassland	Water surface
Built-up land	911, 103 #1	161, 37 #2	481, 36 #3	53, 20 #4
Construction sites	27, 32 #5	85, 26 #6	122, 3 #7	1, 0 #8
Scrubland/grassland	321, 17 #9	355, 8 #10	2160, 214 #11	3, 0 #12
Water surface	193, 20 #13	75, 4 #14	9, 0 #15	1441, 680 #16

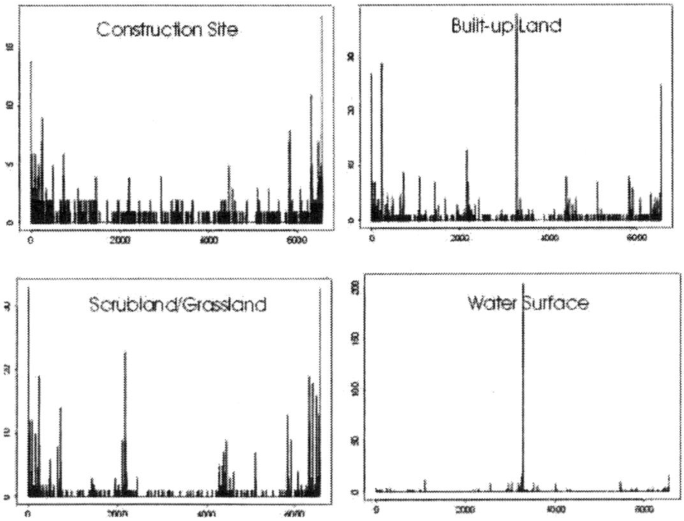

Figure 3. Texture Spectrum of typical land cover types using the original encoding method. Spectra are generated using a 30 by 30 window. The vertical represents occurrence frequency and horizontal Texture Unit Labels.

A one-pass change detection classification was adopted for our experiments [2]. Let (X) be a vector of a multi-temporal pixel in a $2N$-dimensional feature space $(X_{(t1)1}, X_{(t1)2}, \ldots, X_{(t1)N}, X_{(t2)1}, X_{(t2)2}, \ldots, X_{(t2)N})$ where $X_{(t1)N}$ is the input features N on time t_1 and $X_{(t2)N}$ is the those on time t_2. Suppose the number of land cover classes for time 1 and time 2 are m and n, respectively. Then, the maximum number of classes that can be generated is $m \times n$, or n^2 if number of land covers is the same for both dates. In this manner, an unknown pixel with a vector position (X) in $2N$-dimensional feature space is classified into a class providing nature-of-change information. This approach is preferable to conventional post-classification comparison in that it reduced the classification effort by half. The number of change classes is also easily controllable through training.

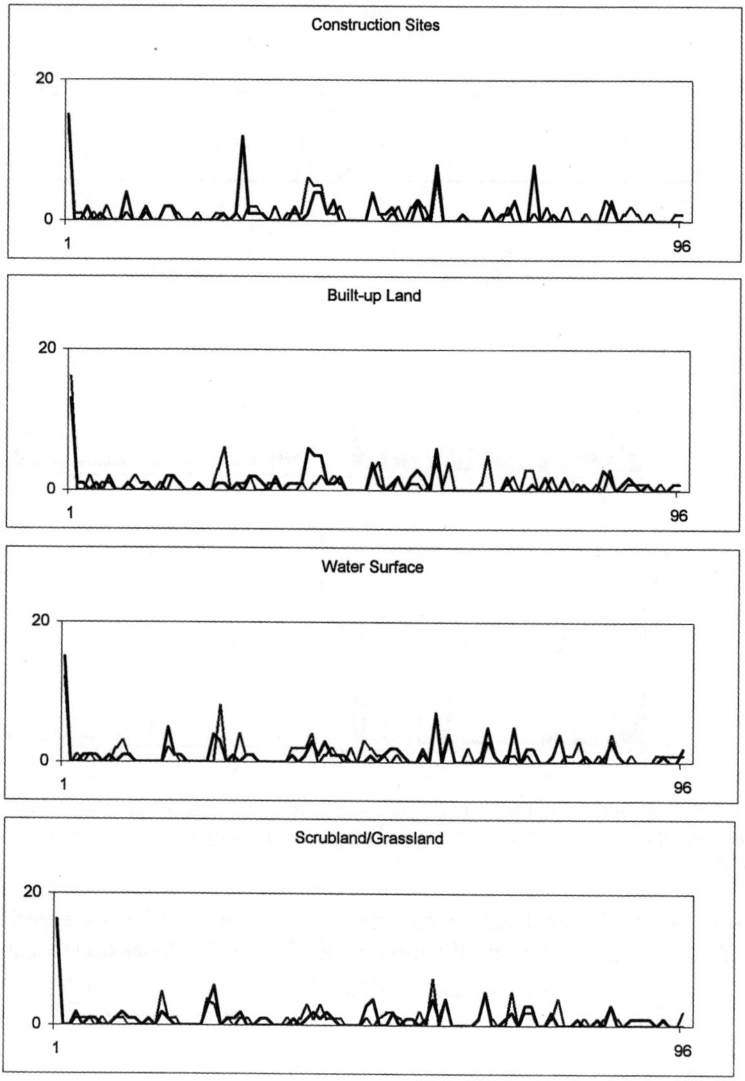

Figure 4. Texture Spectrum encoded by SOM method. Spectra are generated using a 10 by 10 window. The vertical represents occurrence frequency and horizontal Texture Unit Labels. The gray lines represent time 1 (1987) and the black lines time 2 (1995).

Since input dimension would be doubled for analysis of 2-date images, it will exert enormous burden on training costs if the original coding of texture spectrum was used, i.e. a feature space of 2 × 6561 dimensions. The use of SOM encoding method largely resolved this problem. The original coding was compressed into 96 labels. Figure 4 shows the typical form of the encoded Texture Spectrum of typical land cover types in time 1 and time 2. For two time periods, input dimension is 2 × (3 spectral bands + 96 encoded texture spectrum) = 198. We first generated simulated data to investigate the usefulness of Texture Spectrum for change detection. Simulated data maintained class homogeneity (only one class) within the texture-capturing window. It is not realistic compared to our study area where spatial pattern is rather sporadic and land use fragmentation is phenomenal. Several land cover types co-exist in a small window of 10 by 10 pixels is not uncommon. Then we tested the method using real data. The size of window was varied between 30, 15 and 10 by each side. A 10 by 10 window seems to work better and the results were presented. We have chosen Learning Vector Quantization [3] as our base learner.

Table 2. Results (%) of simulated data using different inputs. The last column shows the changes (%) after adding texture as compared to using spectral features alone.

Class	Spectral	Texture	Combined
1	57.6	58.4	4.0
2	70.4	84.8	13.6
3	74.4	96.8	4.0
4	84.8	100.0	9.6
5	80.0	52.8	-8.8
6	96.0	67.2	-2.4
7	95.2	89.6	Unchanged
8	94.4	64.0	4.0
9	83.2	56.8	-4.8
10	97.6	68.8	2.4
11	98.4	82.4	0.8
12	100.0	68.8	Unchanged
13	80.8	69.6	-4.8
14	100.0	88.0	Unchanged
15	100.0	92.0	Unchanged
16	100.0	95.2	Unchanged
Overall accuracy	88.3	77.2	89.4

4 Discussion and Conclusion

For the training of LVQ, a total of 320 codebook vectors were assigned. Training was repeated for 50,000 cycles and snapshots of every 5,000 cycles were recorded. The training states with best overall accuracy were presented. Results from simulated data show that texture alone attained 77.2% accuracy (Table 2). While spectral features achieve higher accuracies in 12 out of 16 classes as compared to Texture Spectrum, the latter achieved higher accuracies in 4 classes and have significant superior performance in classes 2,3 and 4, all related to the transformation from *built-up land* to other land types. In the case of class 3, *built-up land* to *grassland/scrubland*, accuracy for texture is 22% higher. Combined Texture Spectrum and spectral feature enhanced overall accuracy only marginally by 1.1%. However, identification of changes from *built-up land* to other land cover types clearly benefited from the addition of spatial textures. Class 2 (from *built-up land* to *construction sites*) improved by 13.6% and Class 4 (from *built-up land* to *water*) by 9.6%. It is, however, also observed that adding texture has degraded accuracies of other classes.

Table 3. Results (%) of change detection using different inputs. The last column shows the improvements (or degradations) led by adding texture as compared to using spectral features alone.

Class	Spectral	Texture	Combined
1	80.4	4.9	8.9
2	67.6		18.9
3	47.2		-11.1
4	80.0		-20.0
5	87.5		-3.1
6	96.2		3.9
7	100.0		Unchanged
8	—	—	—
9	47.1		5.9
10	87.5		Unchanged
11	97.2	98.1	2.4
12	—	—	—
13	90.0		10.0
14	100.0		Unchanged
15	—	—	—
16	98.7	93.2	0.6
Overall accuracy	92.8	70.8	94.4

Table 3 shows the results from real data. Note that classes 8, 12 and 15 did not appear in the test set. It is apparent that texture alone is not effective since class homogeneity is usually not maintained within the texture capturing windows. Noisy background that commonly occurred within the window has prevented the generation of a proper texture representation of the land cover labeled to the pixel-in-query. This explained why the class 11 (unchanged scrubland/grassland) and class 16 (unchanged water surface) have rather distinctive results. It is important to note that *Texture* result represents one training state that attained the highest overall accuracy. Hence, the learner was biased to maximize overall accuracy instead of class accuracy. Sub-optimal accuracy levels had shown non-zero accuracies for several other classes. Combining texture with spectral features improved overall accuracy slightly by 1.7%. However, more significant improvements were observed in several classes. Class 2 (from *built-up land* to *construction site*) has increased by 18.9%, class 9 (change from *scrubland/grassland* to *built-up land*) about 6%, and class 13 (from *water surface* to *built-up land*) 10%. Three classes experienced degraded accuracy and hence there is a tradeoff.

It is intuitive to use spatial texture for change detection. When applied to simulated data, SOM encoded Texture Spectrum has achieved an overall accuracy of 70%. Combining texture and spectral features benefited mostly the recognition of *built-up land* related changes. Our experiments with real data using Texture Spectrum have shown slight increase in overall accuracy. Combining textures with spectral features have benefited the rates of some classes but degraded others. The results suggested that texture analysis for change detection maybe more effective in a relatively homogenous environment. Future investigations should be directed to the problem of how to generate representative textures in a noisy environment.

Acknowledgement

This paper is a result of the first author's PhD study while he was affiliated with the Department of Computer Science and Information Systems at the Hong Kong University.

References

1. Wang L. and He D.C., A New Statistical Approach for Texture Analysis, *Photogrametric Engineering & Remote Sensing*, 56 (1990) pp. 61-66.
2. Chan, J. C-W, Chan K-P, and Yeh A. G-O, Detecting the nature of change in an urban environment – a comparison of machine learning algorithms, *Photogrametric Engineering & Remote Sensing*, 67 (2001) pp. 213-225.
3. Kohonen T., Self-Organizing Maps, (Springer, Berlin, 1995).

THE USE OF KOHONEN'S NEURAL NETS FOR THE DETECTION OF LAND-COVER TRANSITIONS

ANATOLY A. SAVELIEV, DMITRY V. DOBRININ

R&D Center ScanEx, 22/5 Lva Tolstogo st., 119021, Moscow, Russia

E-mail: saa@ksu.ru

A method proposed for multitemporal RS images analysis based on sequential SOM (self-organizing Kohonen neural nets) teaching. For multitemporal RS image comparison, change detection and results representation color seeds in feature space, generated by the SOM, are used.

1 Introduction and problem definition

In this work, it was shown that ordination features of Kohonen neural nets [1] make it possible to use them for the detection of land-cover transitions based on joint processing of the consequence of multi-temporal images made in short time intervals. The method proposed was used for the monitoring of spring dynamics in vegetation cover of the dry steppe in Kalmykia.

Remote sensing of land-cover transitions is dedicated to the solution of two main problems, i.e. to the revealing new thematic classes and to the detection of changes in the spatial positions of the classes already established. As a rule, these two problems are solving in remote sensing either separately or consecutively. Meanwhile, both of these two processes take place in the environment contemporary with each other and are of equal importance for the creation of correct dynamic model or prediction of the phenomenon investigating.

It should be explicitly mentioned that following conditions must be met for the sensing based on traditional processing algorithms:

1. Putting image to the standard conditions (atmospheric correction etc.)
2. Thematic calibration based on the data on surface observations (as an example, solving the problem if two spectral objects refer to the same or two just different phenomena);
3. Strategy choice assuming the priority of either pixel processing (monitoring of features dynamics) or working with spatial objects (structure dynamics) or the analysis of object moving as an example.

All these conditions required for sensing purposes become clear while Kohonen's self-organizing maps (SOM) are used.

In general, the dynamics in vegetation cover caused by the complex of anthropogenic factors and relating indirect effects can be considered in accordance with the following time scale:

1. long-term periodical (hundreds of years and more);
2. middle periodical (decennial);
3. short-term periodical (years);
4. seasonal;
5. phenological;

Only for first three cases the remote sensing can be conducted in relatively standard conditions (in one and the same season and time period, for similar sun angles and atmospheric conditions), for other cases this is absolutely impossible. Nevertheless, the importance of short-term land sensing is considered even higher than that if the identification of long-term changes. One of the possible reasons is that the long-term changes are always formed from a consequence of short-term effects.

In our case, i.e. for the dry steppe of Kalmykia, the monitoring of short-term periodic changes is of greater importance because of the high level of land degradation, the necessity in the planning of pasture loading and in the choice of the corridors for cattle migration.

2 Method Description

Let consider joint processing of two multispectral co-registered images and mark the set of band values of the first image as **x** and the set of the values of the same bands of the second image as **y**. Both sets are related to a single space point.

The concatenation of these set values with weights $<(1-\alpha)*\mathbf{x}, \alpha*\mathbf{y})>$ where $\alpha \in [0,1]$ is used for the neural net training. This procedure is first performed assuming $\alpha=0$. This corresponds to the training of the neural net on the first image only. Using Sammon mapping for the trained net we obtain the ordination of spectral classes for the first image followed by their interpretation and creation of thematic legend [2]. Then the procedure is repeated with α value gradually increasing up to 1 (the last one corresponds to the neural net training on the second image only). The analysis of newly obtained trained neural nets allows observing both the changes in the positions of existed classes in the ordination and the appearance of novel thematic classes.

Class ordination is rather stable toward changes in the sensing conditions so that the changes of the class position in this ordination as well as other changes (class amalgamation in one place of neural nets or the appearance of novel thematic classes in other places) facilitate joint thematic image interpretation and hence the monitoring of changes and their temporal progress.

3 Results

3.1 Description of data sets

The territory of Kalmykia Republic is located in the frames of four climatic zones, i.e. steppe, dry steppe, semi-desert and desert. Kalmykia climate is strictly continental, with hot and long summer and bleak but unstable winter. Small amount of sediments with high temperatures result in dryness of air and soil as well as in the frequency of droughts and hot winds. More than 40% of Kalmykia territory corresponds currently to the stages of strong and very strong desertification among them most part anthropogenic desertification. The investigation of dynamics in typological content of pastures showed that enormous pasturage results in the increasing contribution of halo- and xerophytes in the vegetation cover.

Two images, i.e. MSU-SK Resurs O1 ¹3, resolution 150ŏ150 m, obtained on May 10, 2000, and May 17, 2000, illustrating obviously the development of spring ephemeral vegetation in the dry steppe zone were chosen for processing (Figures 1-2). What was specific for this experiment, that the phenomenon of 'green wave' (consecutive changes in various phenological aspects depending on observation time) was clearly observed within the territory 130ŏ150 km processed in the images. In accordance with AVHRR NOAA data, this phenomenon is observed as a wavy increase in the vegetation index moving to the north with the rate about 15 km per day. On the space images of higher resolution this phenomenon cannot be considered as an integrated (i.e. with a single wave front). It falls into a lot of short-term periodical structures depending not only on the latitude and lightness but much more on the ecotop conditions (humification and salinization of territory, mechanical structure and humus content of soil, degree of pasture degradation, steppe fires).

3.2 Processing Results

Image processing was performed as described earlier, and the succession of the results are presented on the Figures 3-6. The following color legend was used in the figures to set the thematic color seeds [2]:

Ephemeral vegetation (developing only in spring and in autumn in the period of second flowering) and halophyte vegetation on seasonally moistering weakly salinated depressions and surrounds of solonchak low-lying lands (surrounds often the rice checks abandoned because of their salination). This kind of vegetation is developing often after the group of societies of semi-desert rhizome sedges.

Societies of semi-desert rhizome sedges.

Intensively developing close-up vegetation of mire and rice checks (similar on images only because rice checks are not in use currently and hence reed predominates).

Scattered vegetation surrounds the mire and rice checks not in use currently, vegetation of the moistered low-lying lands.

Solonchak and blowing sands.

Excluded (out of interest) regions

Color legend of other classes was obtained by interpolation of the color seeds using their distances on ordination surface [2].

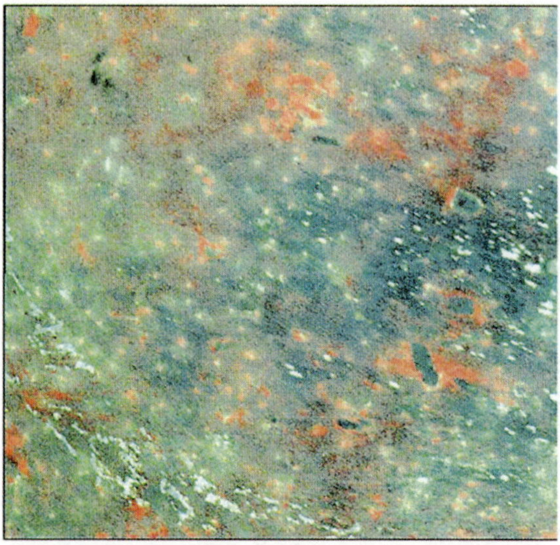

Figure 1. First image RGB- synthesis

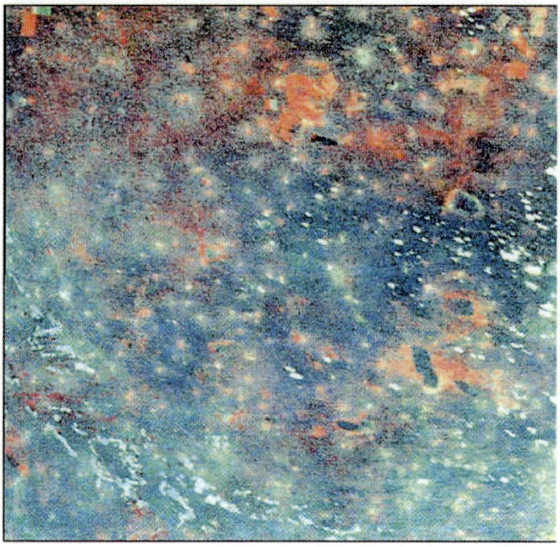

Figure 2. Second image RGB- synthesis

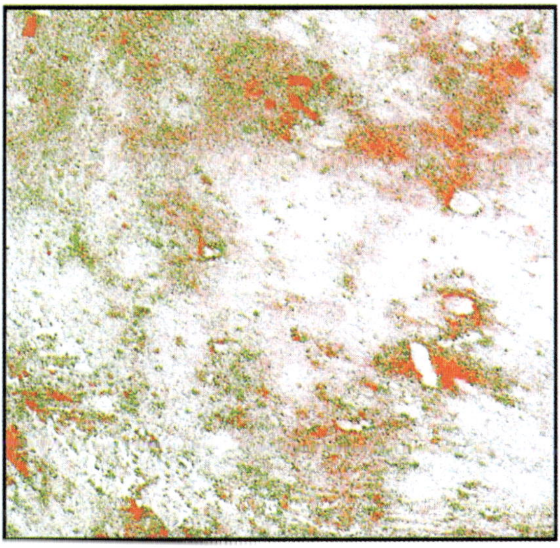

Figure 3. Classificatoin results for α=0.0

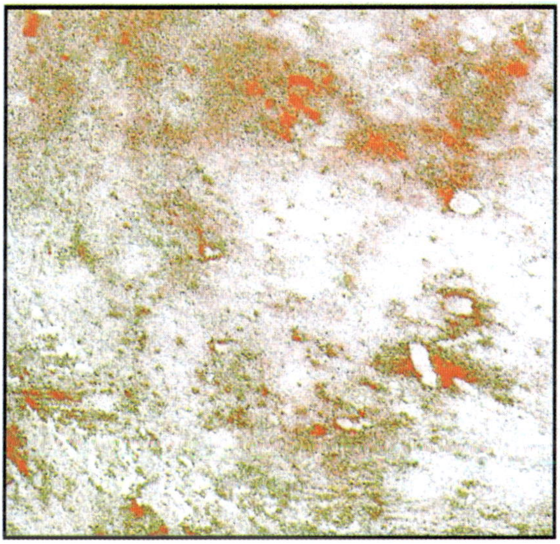

Figure 4. Classificatoin results for α=0.25

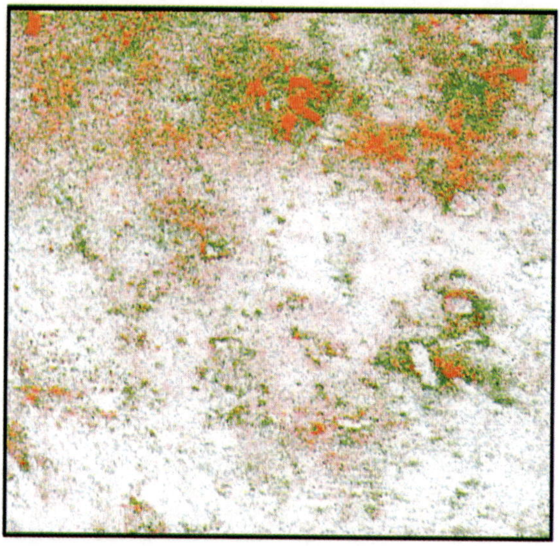

Figure 5. Classificatoin results for $\alpha=0.75$

Figure 6. Classificatoin results for $\alpha=1.0$

4 Discussion and conclusions

Although images used in this work were made in significantly different atmospheric and lightness conditions (Figures 1-2) and no atmospheric correction has been performed, method proposed made it possible to create single unified thematic legend (correspondence of the classes in neural-net ordination to their thematic interpretation) both for the classes of initial images and intermediate classifications, coincided on the same Sammon mapping plane.

The quality of the obtained thematic interpretation of the land-cover transition was then estimated by expert as a proper. This results shows that the method proposed could be used for the detection of short-term periodic land-cover transitions.

References

1. T.Kohonen. *Self-Organizing Maps*. Second Edition. Springer-Verlag, Heidelberg, 1997.
2. Saveliev, A.A., and Dobrinin, D.V., 1999. "Hierarchical multspectral image classification based on self organized maps". Proceedings of IGARSS'99, Hamburg, pp.2510-2512.

A HYPERSPECTRAL TOOLKIT FOR THE ANALYSIS OF MULTITEMPORAL HANDHELD SPECTRORADIOMETER DATA

ROGER L. KING, CHRIS RUFFIN, LORI MANN BRUCE,
JOHN VICKERY, AND NICOLAS YOUNAN

Remote Sensing Technologies Center, Mississippi State University, Box 9571, Mississippi State, MS 39762-9571, USA E-mail: rking@ece.msstate.edu

Remote sensing techniques have not been widely accepted in commercial agriculture due to the necessity of analyzing multi-temporal, high spatial resolution imagery within a short time to provide timely information to the end user. To advance the use of imagery collected with aerial and satellite hyperspectral sensors there must be experimental programs to verify and validate imagery over highly characterized field sites. These field sites require knowledge not only of the spatial nature of the targets, but also their spectral characteristics and how they vary over time. Typically, the spectral characteristics of the targets are collected periodically with one of several commercially available, handheld spectroradiometers. These units collect phenologic spectral information over a wide range of wavelengths and may measure over 1000 channels of data. To fully utilize these tools in ground reference research and in developing an understanding of the phenologic nature of the targets there is a need for specialized analysis tools. This paper will outline some techniques that have been investigated using hyperspectral data collected from both aerial and terrestrial based hyperspectral sensors, as well as describe the Matlab Hyperspectral Toolkit (MHTK) that was developed to perform analysis on handheld spectroradiometer data.

1 Introduction

The Remote Sensing Technologies Center (RSTC) at Mississippi State University (MSU) is investigating the application of high-resolution (spatial, spectral, temporal) imagery in the areas of precision agriculture, forest resource management, and environmental assessments of transportation systems. There are many facets of remote sensing that are particularly attractive to the characterization of these problem domains. Most are attractive because they hold the potential of superceding present ground based scouting practices that are expensive, time-consuming, and/or labor-intensive. In many cases, terrestrial-based characterization of these resources can be difficult to achieve on a large scale, as the targets of interest extend over a large spatial area. However, if the same characterization can be performed using remotely sensed data, then it becomes less burdensome to perform the characterization. For example, it would be impractical to manually perform terrestrial-based vegetation health characterization on a countywide scale, but the prospect of performing this characterization using remotely sensed data is more realistic. The problem until recently has been the lack of the appropriate technology needed to drive the high spatial and spectral resolution acquisitions necessary to provide data on a scale that are valuable to the management of the problem domains of interest.

There are a number of techniques being developed to perform the analysis of hyperspectral data [1,2,4]. Many of these techniques have been adapted from the area of spectroscopy, which is the study of light as a function of wavelength, and a wealth of signal processing and statistical approaches are also being considered. Hyperspectral remote sensing (imaging spectroscopy) departs from traditional spectroscopy in the sense that traditional spectroscopy occurs under very controlled conditions. Whereas, hyperspectral remote sensing normally occurs from airborne platforms in the Earth's atmosphere, from an orbiting satellite, or with a handheld spectroradiometer, and is subject to many factors that affect the data. Although, much emphasis has been placed on the more difficult problem of analyzing hyperspectral cubes, very little is recorded in the literature for producing information products form handheld imagery.

The importance of using handheld spectroradiometers for ground reference information has long been practiced by the remote sensing community, but analysis tools for deriving the information from the measured target has been lacking. This paper will focus on a set of tools being developed at MSU for deriving information products form handheld spectroradiometers. The mathematical and visualization engine for these tools is MATLAB.

2 The MATLAB Hyperspectral Toolkit

2.1 Rationale

Remote sensing is a data-oriented science and, due to the large number of channels associated with handheld hyperspectral remote sensing (typ. > 1000), often results in large data files. Consequently, a vital need in the data processing of handheld ground reference data is software that is tailored for the specific application. As new applications and technologies emerge, new software must be available that provides the means of manipulating and analyzing the data. Such a package should perform basic functions such as visualization, arithmetic operations on spectra, water band noise removal, as well as provide capabilities such as filtering, derivative spectroscopy, feature selection, and absorption band position detection. With these ideas in mind, the Matlab Hyperspectral Toolkit (MHTK) was designed to be an instructive example of a data analysis toolkit for generic hyperspectral data.

Specifically, the goal of the MHTK is to become a toolkit for the data analysis in the spectral dimension; it does not attempt to perform spatial analysis. However, the toolkit is designed in a modular fashion that allows for the extensibility necessary to perform complete analyses on a complete image. The toolkit contains the basic functionality to manipulate data using a set of built-in data analysis techniques that may be used on a variety of spectra. Presently, there are few such software tools to empower the remote sensing research community with hyperspectral data analysis capability.

Traditionally, remote sensing has been the result of subjective interpretation of a suite of images, but digital remote sensing empowers the research community with the capability of using quantitative analysis techniques in order to interpret the data. With these new capabilities, however, must come new computational tools in order to give researchers the flexibility they need to establish new algorithms, methods, and applications for the hyperspectral remote sensing. In that respect, the Matlab Hyperspectral Toolkit (MHTK) endeavors to provide the infrastructure that provides the capability, flexibility, and extensibility to ease the task of performing research with hyperspectral data.

2.2 *Toolkit design*

MHTK is written using the Matlab scientific computing environment to take advantage of the computational and graphical capabilities it provides. Since it operates on top of the Matlab environment, the toolkit operates on any of the computing platforms supported by Matlab such as Unix and Microsoft Windows. MHTK was designed to be modular. There are three main components to the MHTK system that provide this modularity: the functional routines, a set of graphical user interface routines, and a set of batch routines. The functional routines are the core of the toolkit and perform all analysis functions. They do so in a transparent way in order to give the functionality of being called either by the batch routines in order to process multiple data sets, the graphical user interface routines in order to process the data that is requested by the user of the GUI, or directly from the Matlab command line. This overall design makes MHTK modular and extensible. The GUI routines are a set of routines that perform the task of setting up the MHTK GUI main window, implement the dialog system, and interacting with the user in order to perform the desired task of managing data, filtering, and performing analyses. Finally, the batch routines of MHTK allow for the processing of large data sets that all require similar analyses to be performed. Figure 1 shows a functional block diagram of the MHTK software. The organization of the toolkit in this fashion provides excellent extensibility. Users may extend the toolkit by adding batch routines, GUI functions, or core MHTK routines.

2.3 *MHTK Functional Routines*

The core of the toolkit is a set of functional routines that perform analysis on the spectral data itself, such as those analysis techniques described in the chapter on hyperspectral data analysis. These routines operate on a data object containing among other things, a reflectance spectrum. The object is updated to include modifications to the spectrum itself such as filtering or water band removal, and is passed back to the calling function. Note that all the modifications are handled transparently, which brings forth the capability of calling the function from the

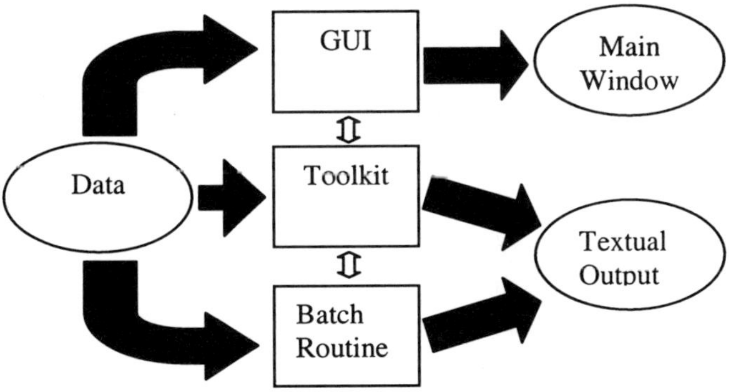

Figure 1. Overall toolkit structure

Matlab command line, used from another routine that is part of the graphical user interface, or from batch routines.

Within the functional routines of MHTK are the routines that handle data management. In order to provide modularity from the GUI, the data management system of the toolkit was abstracted to a level that simply requires the GUI function to call two functions that get the proper data from the workspace. One function gets the proper data from the workspace, and the other puts the data back into the workspace once it is modified and/or supplemented. This requires a set standard for the way data are handled. A single spectrum is treated as an object implemented as a data structure with a minimum of required meta-fields that represent the minimum level of data required to have declared an object. The only required field of a data object is the abscissa, which describes the exact points in wavelength where data was collected. Other data that are included in the object are additional data that add information to the object. In order for the data to become meaningful, an ordinate axis is defined. The data would normally describe the reflectance data, but the toolkit is designed for generality, and will work with other ordinates as well, such as radiance or absorbance.

2.4 *Graphical User Interface Routines*

MHTK also provides a graphical user interface that provides the functionality available in the core toolkit with the user-friendliness of a graphical user interface. These routines are completely separate from the core toolkit functions. In fact, a completely new graphical user interface could be written without spending any time modifying the core toolkit. Figure 2 shows the main GUI window. Within the window are three primary widgets: the spectrum list window, the spectrum plot window, and the window menus. The data list window shows a list of the components of the available data objects that are available to the user for

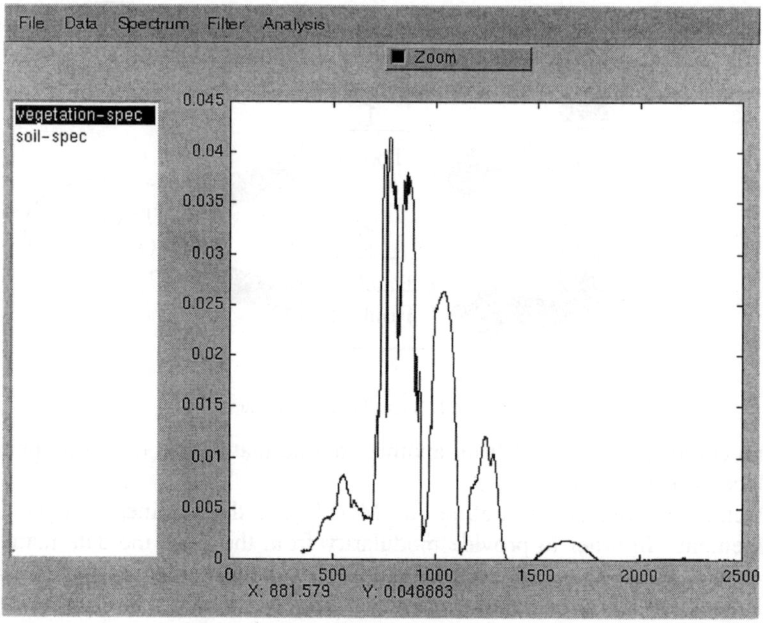

Figure 2. MHTK main window

visualization and analysis. The plot window is the main visualization capability of MHTK. As data are selected from the data list window, the data contained within that object is displayed in the spectrum plot object. This allows for rapid comparisons between different spectra in a variety of ways simply by selecting the desired data. In addition, each data object may contain information within the data object that may be plotted, such as the red edge position or the absorption band positions in the spectrum as determined by the analysis routines of the core toolkit.

The window menus contain most of the data manipulation and analysis capabilities of the GUI, and are the user's way of selecting what analysis to perform. In order to perform the same analyses on many different data objects, the user selects all the applicable data objects from the data list window, and selects the appropriate selection from one of the window menus. Thus, the menus are essentially the user's gateway into the functionality of the core toolkit. Most all of the menu selections call a single routine that brings up a dialog window that allows the user to select parameters applicable to the analyses being performed. Figure 3 shows a block diagram of this process. Every effort has been made to provide the user with reasonable default parameters, but it is essentially up to the user to enter the parameters most suitable for the goal of the analysis of the user's application.

The way the GUI functions are quite simple and intuitive. In order to perform analysis on one or multiple sets of data, the data are selected from the data list, which displays the data in the spectrum plot window. The user then selects what analysis or operation to perform, selects the appropriate parameters from the function's dialog, and the GUI takes care of collecting the selected data, calling the appropriate analysis function for each of the data objects, passing the modified data objects back to the workspace, changing the spectrum plot object to reflect any changes to the spectrum if applicable, and outputting the results of any analyses.

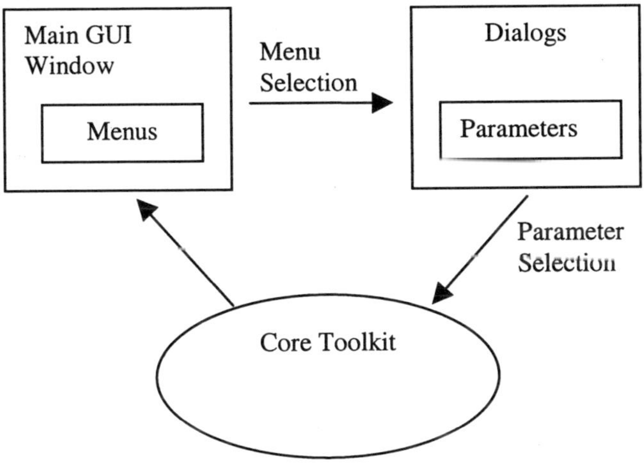

Figure 3. Graphical user interface

2.5 *Vegetation Indices Example*

Table 1 illustrates the utility of MHTK in performing comparative analysis of different spectra. Figure 4 shows two handheld spectroradiometer measurements of one cotton plant fertilized with a high nitrogen rate and the other with a low rate. From the figure it is obvious that the higher fertilized plant is a darker green and thus, has a lower reflectance in the green portion of the spectrum. With the use of MHTK and its built-in vegetation indices, it is very simple to compare these measurements among a variety of vegetation indices. For example, MHTK will integrate the spectra for various satellite channels and calculate a variety of indices based on a specific satellite's specific broadband channels. The red and near infrared bands built into MHTK for Landsat TM and SPOT are: Red = 630 – 690 nm NIR = 760 – 900 nm; SPOT: Red = 610 - 680 nm NIR = 790 – 890 nm, respectively. The vegetation indices in Table 1 are a combination of familiar ratio techniques and also several derivative techniques.

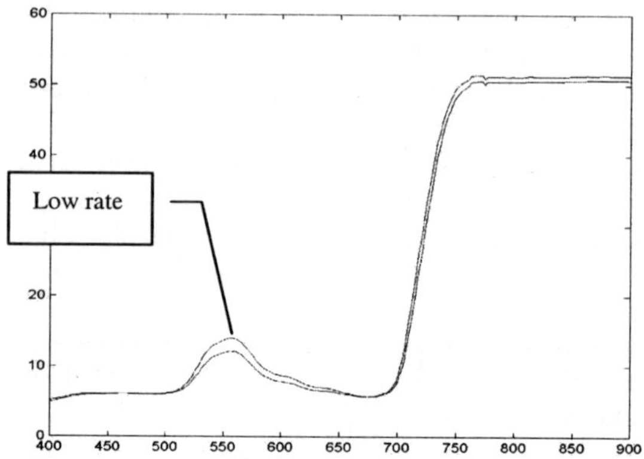

Figure 4. Spectra from low and high rates of nitrogen fertilization.

Table 1. MHTK derived vegetation indices from variable rate nitrogen fertilization.

Vegetation Index	Low rate N2	High rate N2
NDVI – Landsat TM	0.78267	0.78667
NDVI - SPOT HRV	0.77004	0.7778
DVI – Landsat TM	45.0281	44.5541
DVI - SPOT HRV	44.6189	44.2917
IPVI – Landsat TM	0.89133	0.89334
IPVI - SPOT HRV	0.88502	0.8889
RVI – Landsat TM	8.2025	8.3753
RVI - SPOT HRV	7.697	8.008
SAVI – Landsat TM	1.1639	1.1697
SAVI - SPOT HRV	1.1452	1.1565
Red Edge	698 nm	700.3 nm
1DL_DGVI	103.1122	98.85
1DZ_DGVI	96.6283	93.6026
2DZ_DGVI	4.8746	4.6576

2.6 Species Identification and Temporal Analysis

MHTK is also being expanded to include a variety of additional functionalities that will facilitate multiresolutional analysis, species identification, and feature extraction. A self-organized map has proven to be useful in species identification and in feature selection [3] and this capability is being incorporated into MHTK. Also, when detecting temporal changes in hyperspectral curves, one must consider resolution, or scale. The changes could be isolated within a narrow range of spectral bands or across a large range of spectral bands. In order to detect and quantify changes of both types, one must use an analysis technique that incorporates a multiresolution methodology. In order to avoid *ad hoc* approaches for determining the appropriate window sizes for analysis at varying spectral resolutions, MHTK uses a systematic approach for multiresolution analysis known as the discrete wavelet transform (DWT). The DWT is a mathematical procedure for transforming the spectroradiometer data from the spectral domain to the scalar domain; similar to the way the Fourier transform can be used to convert data from the time domain to the frequency domain. A major advantage of the wavelet transform, over the standard Fourier transform, is its properties of time-frequency localization. It can be used to decompose a hyperspectral curve according to scale, allowing one to separate the small-scale (detail) behavior from the large-scale (approximation) behavior of the hyperspectral signature. Presently, MHTK is being used to analyze multi-temporal ground reference data to determine the effect of band selection on some proprietary algorithms.

3 Conclusions

The Matlab hyperspectral toolkit is to be a complete system for the analysis and manipulation of handheld hyperspectral data used for ground reference. The toolkit includes analysis and manipulation operations crucial to absorption band characterization, red-edge detection, and both broadband and hyperspectral indices. The modular design allows for the extensibility that is needed in an ever-evolving field of hyperspectral remote sensing. The overhead of writing new data analysis routines is kept to a minimum by allowing the reuse of existing routines that perform data management, visualization, and filtering operations. In this way, the toolkit provides not only the analysis techniques that are presently available, but also the extensibility necessary to easily assimilate other analysis routines that may be developed. Since remote sensing research is about trying new methods and ideas, the toolkit provides an easy outlet for implementation of these ideas using a common infrastructure.

References

1. Bruce, L.M. and J. Li, Wavelets for computationally efficient hyperspectral derivative analysis. IEEE Trans. Geoscience and Remote Sensing **39** (2001) pp.1540-1546.
2. King, R.L., C. Ruffin, L. LaMastus, and D. Shaw, Classification of Weed Species Using Self-Organizing Maps. Proc. 2nd Intl Conf Geospatial Information in Agriculture and Forestry. Vol. II (Jan 10-12, 2000) pp. 151-158.
3. Lewis, M, V. Jooste, and A.A. de Gasparis, Discrimination of arid vegetation with airborne multispectral scanner hyperspectral imagery. IEEE Trans. Geoscience and Remote Sensing **39** (2001) pp.1471-1479.
4. Serpico, S.B. and L. Bruzzone, A new search algorithm for feature selection in hyperspectral remote sensing images. IEEE Trans. Geoscience and Remote Sensing **39** (2001) pp.1360-1367.

A TEMPORAL EXTENSION TO TRADITIONAL EMPIRICAL ORTHOGONAL FUNCTION ANALYSIS

ALLAN AASBJERG NIELSEN AND KLAUS BAGGESEN HILGER

Informatics and Mathematical Modelling, Technical University of Denmark, Building 321
DK-2800 Kgs. Lyngby, Denmark, Telephone +45 4525 3425, Telefax +45 4588 1397
{aa,kbh}@imm.dtu.dk, www.imm.dtu.dk/~{aa,kbh}

OLE BALTAZAR ANDERSEN AND PER KNUDSEN

National Survey and Cadastre, Rentemestervej 8
DK-2400 Copenhagen NV, Denmark, Telephone +45 3587 5050, Telefax +45 3587 5051
{oa,pk}@kms.dk, research.kms.dk/~{oa,pk}

This paper describes the application of temporal maximum autocorrelation factor analysis to global monthly mean values of 1996-1997 sea surface temperature (SST) and sea surface height (SSH) data. This type of analysis can be considered as an extension of traditional empirical orthogonal function (EOF) analysis, which provides a non-temporal analysis of one variable over time. The temporal extension proves its strength in separating the signals at different periods in an analysis of relevant oceanographic properties related to one of the largest El Niño events ever recorded.

1 Introduction

Empirical orthogonal functions (EOF) [5,10] analysis is often used in oceanography and other geophysical sciences to analyse temporal sequences of scalar (image) data. This paper deals with an extension of the EOF method, by applying maximum autocorrelation factor (MAF) analysis to two temporal sequences of scalar image data, namely global sea surface temperature (SST) and global sea surface height (SSH) one at a time. This type of analysis is typically carried out to maximise spatial autocorrelation. Here we maximise temporal autocorrelation. The data used to illustrate the analysis carried out come from the Ocean Pathfinder programmes set up by NASA/NOAA [7,9]. The data chosen represent relevant oceanographic properties related to one of the largest El Niño events ever recorded.

El Niño is a large-scale warm ocean event in the Pacific off the coast of Peru and Ecuador caused by eastward drifting toward the west coast of South America of the pool of warm waters normally residing in the western part of the Pacific. This event is not local but may influence weather conditions world-wide. This disruption of normal climate may have drastic socio-economic consequences [6]. The strength and frequency of El Niño events may be influenced by human activities.

2 Method

EOF analysis is a name often used in geophysical data processing for principal component analysis (PCA) [4,1]. Often the usual PCA assumption on variables with mean zero is replaced by an assumption of temporal means of zero. PCA finds linear combinations (called principal components, PCs) of the original data that maximise variance. PC1 contains maximum variance. Higher order PCs contain maximum variance subject to orthogonality or uncorrelatedness with lower order PCs. Maximum autocorrelation factor (MAF) analysis [11,2] finds linear combinations of the zero mean original variables that maximise autocorrelation between neighbouring observations. MAF1 is a linear combination of the original data that have maximum autocorrelation. Higher order MAFs have maximum autocorrelation subject to orthogonality or uncorrelatedness with lower order MAFs. Often the spatial autocorrelation is maximised, here we maximise the temporal autocorrelation. The temporal MAF (T-mode MAF or T-MAF) analysis is carried out first on the SST and then on the SSH data.

Linear combinations $a^T X$ with maximum (auto)covariance between a (zero mean) signal $X(t)$ and a temporally shifted version of the same signal $X(t+\Delta)$ are obtained by finding coefficients a that maximise $a^T\Gamma(\Delta)a$ where $\Gamma(\Delta)$ is the covariance between $X(t)$ and $X(t+\Delta)$. Introducing Σ as the dispersion matrix of the original variables $X(t)$ and Σ_Δ as the dispersion matrix of the difference between the original and the temporally shifted variables $X(t)-X(t+\Delta)$, we get for the autocovariance $a^T(\Sigma-1/2\,\Sigma_\Delta)a$ and hence for the autocorrelation $1-a^T\Sigma_\Delta a/2a^T\Sigma a$. Therefore the desired linear combinations are obtained by solving the generalised eigenvalue problem

$$\Sigma a = \lambda \Sigma_\Delta a$$

where a are (conjugate eigen-) vectors containing the desired weights to construct the variables. λ is the eigenvalue and $1-1/(2\lambda)$ is the autocorrelation.

The results from the analysis depend on Δ; here Δ is one month.

3 Data and results

The data used are global monthly mean values of 1996-1997 SST data from the NOAA/NASA Oceans Pathfinder AVHRR SST database [9] and global monthly mean values of 1996-1997 SSH data from the NASA/GSFC Ocean Altimeter Pathfinder database [7]. The SSH data are interpolated point observations from the TOPEX/Poseidon radar altimeter mission. The SST data come as 360 rows by 720 columns half degree data starting at 180° longitude, the SSH data come as 179 rows by 360 columns one degree data starting at 0° longitude. The SST data have been resampled to the SSH grid. The AVHRR instrument is influenced by cloud coverage whereas the radar altimeter provides uninterrupted data. Statistics for the SST analysis are calculated where SST have non-missing values for all 24 months and similarly for SSH.

Figure 1 shows the first six SST T-MAFs. The T-MAFs are stretched linearly from mean ± three standard deviations, the pseudo-colour scale goes from blue (minimum) over cyan, green, yellow to red (maximum). Figure 2 shows the first six SSH T-MAFs (same stretch and colouring).

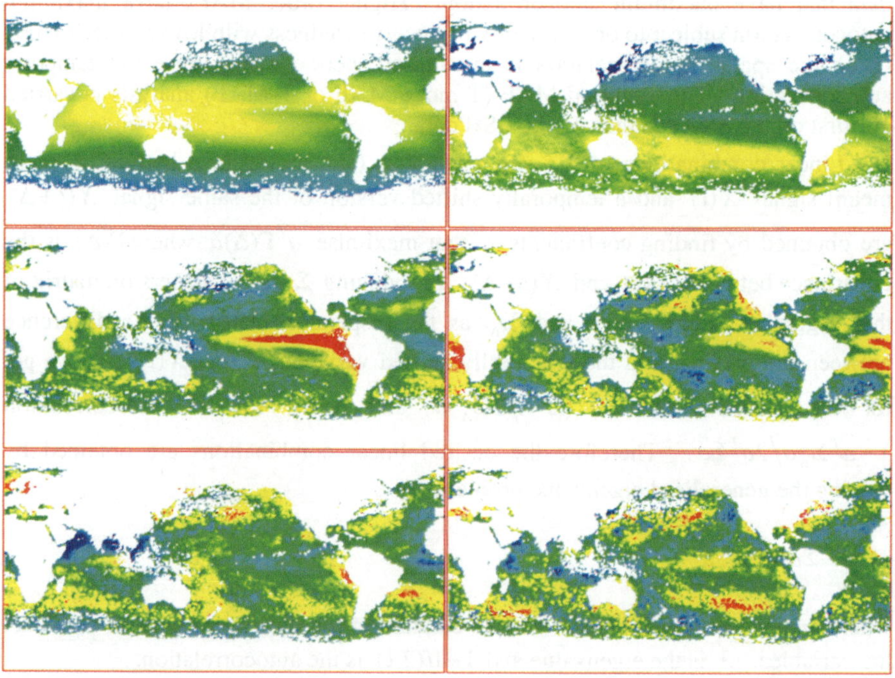

Figure 1. SST T-MAFs 1-6 row-wise.

Figure 3 shows the SST T-MAF autocorrelations (in black), and the correlations between the original data and the first six SST T-MAFs for both SST and SSH (correlations with T-MAFs 1-6 are shown in red, green, blue, cyan, magenta and yellow, respectively).

Figure 4 shows the SSH T-MAF autocorrelations (in black), and the correlations between the original data and the first six SST T-MAFs for both SST and SSH (correlations with T-MAFs 1-6 are shown in red, green, blue, cyan, magenta and yellow, respectively).

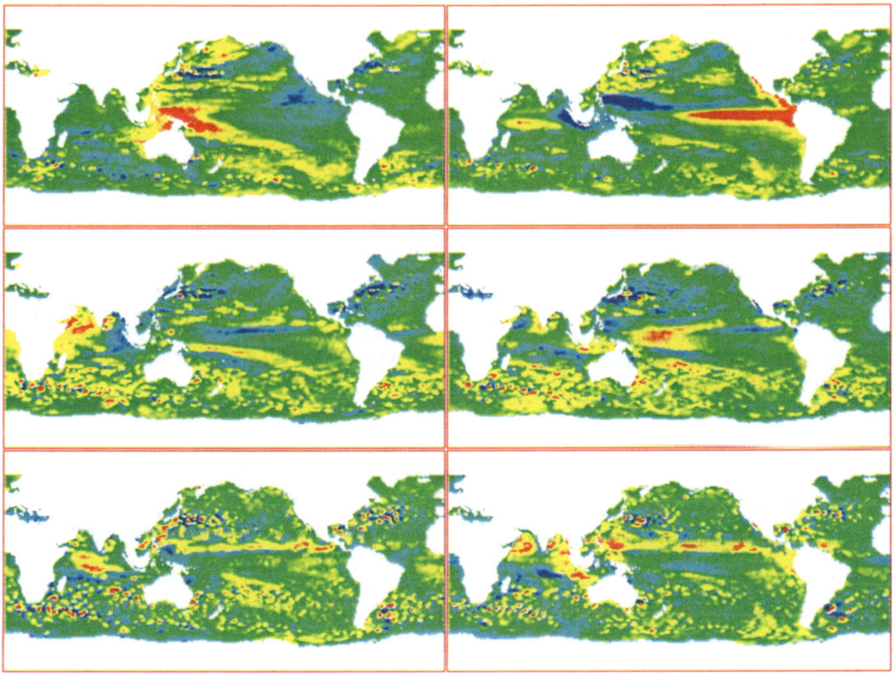

Figure 2. SSH T-MAFs 1-6 row-wise.

4 Discussion and conclusions

By maximizing the temporal autocorrelation we generally isolate features with the longest period in the first MAFs. This means that signals related to the mean and possible trend of the data will be captured first. Thereafter signals with shorter periods like annual signals and intra-annual variations will be captured.

Figure 3 shows that the first seven SST T-MAFs (especially T-MAFs 1-4) are highly autocorrelated temporally. SST T-MAF1 (shown in red) is highly (positively) correlated with all 24 months of SST data with a small semi-annual oscillation. Accordingly, the spatial structure of SST T-MAF1 shows the overall mean of the SST field having high temperatures in the tropics and cold temperatures away from the Equator. The SST T-MAF1 is nearly uncorrelated with all 24 months of SSH shown in indices 25-48. This corresponds well with the fact that the mean of the SSH field has been removed prior to the analysis.

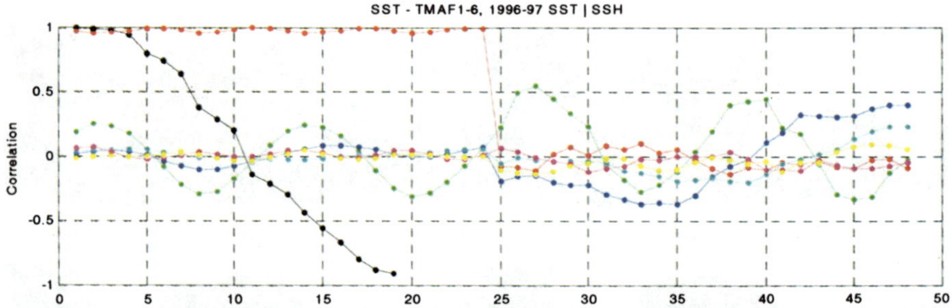

Figure 3. SST T-MAF analysis: autocorrelations (in black), and correlations between original SST (indices 1-24) and SSH variables (indices 25-48), and SST T-MAFs 1-6 (in red, green, blue, cyan, magenta and yellow, respectively).

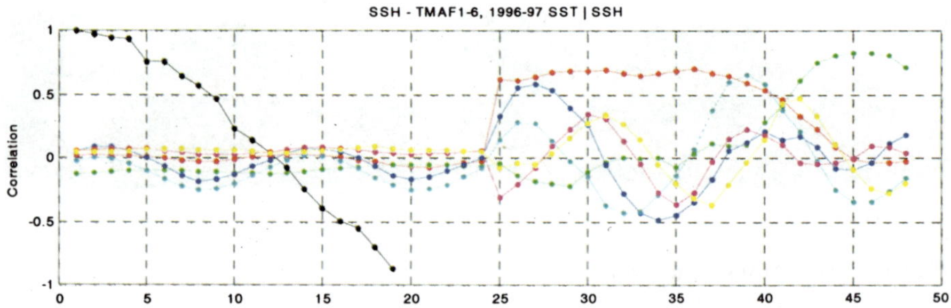

Figure 4. SSH T-MAF analysis: autocorrelations (in black), and correlations between original SST (indices 1-24) and SSH variables (indices 25-48), and SSH T-MAFs 1-6 (in red, green, blue, cyan, magenta and yellow, respectively).

Correlations between SST T-MAF2 and the 24 months of SST data (in green) exhibit strong annual oscillations [5] with positive correlations in the Northern Hemisphere's winter months and negative correlations in the summer months. The correlations with the 24 months of SSH are also high, but note that the extreme correlations occur one month later for the SSH than for the SST (e.g.

, minimum in month 8 and 20 for SST, but month 9 and 21 for SSH). The spatial structure of SST T-MAF2 shown in Figure 1 shows highs in the Southern Hemisphere and lows in the Northern Hemisphere and corresponds to an annual heating and cooling of the ocean. The apparent phase lag of one month between the highs of the SST and the SSH is explained by the fact that the SST represent the instant temperature of the sea surface, whereas the steric expansion causing the sea level to rise is a more integrated effect.

For both the SST T-MAF3 and the SST T-MAF4 (blue and cyan curves) the correlations also show some oscillations, but not on annual periods. This correlation pattern is furthermore dramatically disturbed in the second half of 1997.

This dramatic change in correlation is in accordance with the El Niño build-up during 1997. The connection to the El Nino is also captured by the corresponding spatial structure of SST T-MAF3. This exhibits a conspicuous high in the equatorial Pacific off the South American west coast. The spatial structure of the SST T-MAF4 is also El Nino related in the tropical Pacific, but some very interesting signals in the south Atlantic Ocean are also found.

The first seven to nine SSH T-MAFs (again especially T-MAFs 1-4) are highly autocorrelated temporally. SSH T-MAF1 is nearly uncorrelated with the SST, but highly (positively) correlated with the SSH data in 1996 and well into the first half of 1997. During the spring of 1997 this correlation drops to around zero where it stays for the rest of 1997. The corresponding spatial pattern of SSH T-MAF1 shows the warm and therefore high SSH "waters normally residing in the western part of the Pacific" mentioned in the introduction. By and large, SSH T-MAF2 exhibits an opposite correlation pattern: going from small oscillations close to zero to being very high in the second half of 1997. Accordingly, the spatial pattern of SSH T-MAF2 shows very high values in the equatorial Pacific off the South American west coast and also highs along the Mexican and south west U. S. coasts. This, again, complies with the El Niño phenomenon.

Generally all correlation structures between SSH T-MAFs and the 24 months of SSH data show large differences between 1996 and 1997. For T-MAF3 amplitudes are high and the annual oscillation is very conspicuous in 1996 with much smaller correlation values and oscillations in 1997. The spatial pattern of SSH T-MAF3 may show El Niño related edge effects with lows immediately to the north of the Equator and highs immediately to the south of the Equator in the Pacific.

For a joint analysis of SST and SSH data by means of canonical correlations analysis (CCA), see [8]. For a non-linear CCA of the same data, see [3].

In spite of the very short time span of the data and the associated risk of over-interpreting the results, simultaneous inspection of spatial patterns of the T-MAFs and the correlations between the original and transformed variables from the analysis gives good indications of a large anomaly in both the SST and SSH fields off the South American west coast taking place in the second half of 1997. This is in good agreement with established oceanographic knowledge on the build-up of one of the largest El Niño events on record.

Future analysis should include simultaneous maximisation of temporal and spatial autocorrelation. Also analysis of longer time series should be carried out in order to establish whether 1997 (and 1998) represent anomalous events in terms of global SST and SSH.

Acknowledgement

The Pathfinder SST data provision at JPL is due to J. Vazques, R. Sumagaysay and co-workers. The Pathfinder SSH data provision at GSFC is due to V. Zlotnicki and co-workers and the authors would like to thank B. Beckley for providing the data. This work is done as a part of the GEOSONAR project funded by the Danish National Research Councils under the Earth Observation Programme. GEOSONAR is headed by Dr. Per Knudsen, the National Survey and Cadastre, Denmark.

References

1. Anderson T. W., *An Introduction to Multivariate Statistical Analysis*, second edition, John Wiley, 1984.
2. Green A. A., Berman M., Switzer P. and Craig M. D., A transformation for ordering multispectral data in terms of image quality with implications for noise removal, *IEEE Transactions on Geoscience and Remote Sensing*, 26, 65-74, 1988.
3. Hilger K. B., Nielsen A. A., Andersen O. A. and Knudsen P., An ACE-based nonlinear extension to traditional empirical orthogonal function analysis, *MultiTemp 2001*.
4. Hotelling H., Analysis of a complex of statistical variables into principal components, *Journal of Educational Psychology*, 24, 417-441, 1933.
5. Knudsen P., Andersen O.B. and Knudsen T., ATSR sea surface temperature data in a global analysis with TOPEX/Poseidon altimetry, *Geophysical Research Letters,* 23(8), 821-824, 1996.
6. NASA Facts, *El Niño*, The Earth Science Enterprise Series, NF-211, 1999.
7. NASA/GSCF, http://neptune.gsfc.nasa.gov/ocean.html, Goddard Space Flight Center, National Aeronautics and Space Administration, Greenbelt, Maryland, USA.
8. Nielsen A. A., Hilger K. B., Andersen, O. A. and Knudsen P., A bivariate extension to traditional empirical orthogonal function analysis, *MultiTemp 2001*.
9. PO.DAAC, http://podaac.jpl.nasa.gov, Jet Propulsion Laboratory, National Aeronautics and Space Administration, Pasadena, California, USA.
10. Preisendorfer R. W., *Principal Component Analysis in Meteorology and Oceanography*, posthumously compiled and edited by C. D. Mobley. Developments in Atmospheric Science, 17, Elsevier, 1988.
11. Switzer P. and Green A. A., Min/max autocorrelation factors for multivariate spatial imagery, Technical Report, Department of Statistics, Stanford University, 1984.

AN ACE-BASED NONLINEAR EXTENSION TO TRADITIONAL EMPIRICAL ORTHOGONAL FUNCTION ANALYSIS

KLAUS BAGGESEN HILGER AND ALLAN AASBJERG NIELSEN

Informatics and Mathematical Modelling, Technical University of Denmark, Building 321 DK-2800 Kgs. Lyngby, Denmark, Telephone +45 4525 3422, Telefax +45 4588 1397 {kbh,aa}@imm.dtu.dk, www.imm.dtu.dk/~{kbh,aa}

OLE BALTAZAR ANDERSEN AND PER KNUDSEN

National Survey and Cadastre, Rentemestervej 8 DK-2400 Copenhagen NV, Denmark, Telephone +45 3587 5050, Telefax +45 3587 5051 {oa,pk}@kms.dk, research.kms.dk/~{oa,pk}

This paper shows the application of the empirical orthogonal functions/principal component transformation on global sea surface height and temperature data from 1996 and 1997. A nonlinear correlation analysis of the transformed data is proposed and performed by applying the alternating conditional expectations algorithm. New canonical variates are found that indicate that the highest correlation between ocean temperature and height is associated with the build-up of the El Niño during the last half of 1997.

1 Introduction

The aim of this paper is to present an extension to the traditional empirical orthogonal functions (EOF) analysis [10,17]. EOF is often applied in geophysical sciences to analyze temporal sequences. In the two-set case the traditional EOF analysis can be extended by the means of linear canonical correlation analysis (CCA) [8,4,1,14]. This paper presents a nonlinear correlation analysis of two-set data using the alternating conditional expectations (ACE) transform. ACE was originally proposed for nonlinear multiple regression [2]. The ACE transform is applied using the EOF analysis as a preprocessor. Two temporal sequences of scalar image data representing global sea surface temperature (SST) and sea surface height (SSH) are analyzed as these quantities are physically related. The data period covers 1996 and 1997 on a monthly basis. This particular period was chosen, as the build-up of one of the largest El Niño events [11] ever recorded occurred during the last half of 1997.

2 Proposed method

The EOF analysis is closely related to the principal components analysis (PCA) [7,1]. Often the usual PCA assumption on variables with mean zero is replaced by an assumption of temporal means of zero. A traditional CCA analysis involves a joint assessment of two sets to produce pairs of canonical variates (CVs) that are linear combinations of the original variables with maximal correlation. All the CV pairs are restricted to be orthogonal/uncorrelated. Generalizations of CCA to handle multiset scenarios can be made [9,13]. In this paper we address the problem of finding the pair of canonical variates that maximize the correlation in a more general manner. We do not restrict transformations to be simple linear combinations of the original variables, but we allow for nonlinear mappings of the data. The nonlinear CV transformations are found as conditional expectations of the original data, applying the ACE algorithm. Traditionally ACE is used to handle data that naturally separates into two sets often with only one variable in one of the sets. The algorithm can however be generalized to handle several variables in both sets and even multiset scenarios [6]. By allowing for nonlinear transformations, it is possible to determine CVs with higher correlations than in the linear CCA analysis. However, it becomes more challenging to interpret the CV sets, as the degree of freedom for the transformations is very high.

Consider a multivariate data set of P variables with gray levels $r_i(\boldsymbol{x})$, $i = 1,\ldots,P$ where $\boldsymbol{x}$ is the coordinate vector denoting a grid point of the sample. Let $\boldsymbol{r}(\boldsymbol{x}) = [r_1(\boldsymbol{x}) \ldots r_P(\boldsymbol{x})]^T$ represent a random signal variable and assume first and second order stationarity such that $\mathbf{E}\{\boldsymbol{r}(\boldsymbol{x})\} = 0$ and $\mathbf{E}\{\boldsymbol{r}(\boldsymbol{x})\, \boldsymbol{r}(\boldsymbol{x})^T\} = \Sigma$.

Determining the direction of maximum variation means finding the direction $\boldsymbol{a}$, with $\boldsymbol{a}^T \boldsymbol{a} = 1$, such that the linear combination $y(\boldsymbol{x}) = \boldsymbol{a}^T \boldsymbol{r}(\boldsymbol{x})$ possesses maximum variance. The principal components transform chooses P linear transformations $y_i(\boldsymbol{x}) = \boldsymbol{a}_i^T \boldsymbol{r}(\boldsymbol{x})$, $i = 1,\ldots,P$ such that the variance for $y_i(\boldsymbol{x})$ is maximum among all linear transforms orthogonal to $y_j(\boldsymbol{x})$, $j = 1,\ldots, i-1$. The variance is given by $\mathbf{V}\{\boldsymbol{a}_i^T \boldsymbol{r}(\boldsymbol{x})\} = \boldsymbol{a}_i^T \Sigma\, \boldsymbol{a}_i$. Thus, we see that the basis for the principal components is identified as the conjugate eigenvectors of the dispersion matrix. Let $v_1 = \mathbf{V}\{\boldsymbol{a}_1^T \boldsymbol{r}(\boldsymbol{x})\} \geq \ldots \geq v_P$ be the eigenvalues with the corresponding conjugate eigenvectors $\boldsymbol{a}_1, \ldots \boldsymbol{a}_P$, then $y_j(\boldsymbol{x})$ is the i'th principal component.

Consider two sets of random signal variables $\boldsymbol{r1}(\boldsymbol{x})$ and $\boldsymbol{r2}(\boldsymbol{x})$, each consisting of P1 respectively P2 variables. The alternating conditional expectations (ACE) algorithm chooses P = P1+P2 transformations $\{\varphi 1_1, \ldots, \varphi 1_{P1}\}$ and $\{\varphi 2_1, \ldots, \varphi 2_{P2}\}$ of the original variables such that

$$e^2 = \mathbf{E}\{[\, \Sigma\, \varphi 1_i(r1_i(\boldsymbol{x})) - \Sigma\, \varphi 2_j\, (r2_j(\boldsymbol{x}))\,]^2\,\} \,/\, \mathbf{E}\{[\, \Sigma\, \varphi 1_i\, (r1_i(\boldsymbol{x}))\,]^2\}$$

is minimized.

Applying ACE under the constraints that the new canonical variates must have variance one, ACE determines the transformations that maximize the correlation, $\mathbf{Corr}\{z1(\boldsymbol{x}), z2(\boldsymbol{x})\} = 1 - e^2 / 2$, between the new components $z1(\boldsymbol{x}) = \Sigma\varphi 1_i(r1_i(\boldsymbol{x}))$ and $z2(\boldsymbol{x}) = \Sigma\, \varphi 2_j\, (r2_j(\boldsymbol{x}))$. The new transformations need not be linear but can be even non-monotonic mappings. The ACE algorithm can be implemented to handle arbitrary mixtures of continuous ordered variables and categorical variables.

Optimal transformations exist and satisfy a complex system of integral equations [2]. Convergence to the optimal solutions is obtained through an iterative algorithm using only bivariate conditional expectations. Thus the ACE algorithm can be applied using a wide range of bivariate smoothers; we apply Friedman's local nonlinear supersmoother [4]. Note that suboptimal solutions can also be found applying multiple regression as suggested in [3].

3 Data and results

The data used are global monthly mean values of 1996-1997 SST data from the NOAA/NASA Oceans Pathfinder AVHRR SST database [16] and global monthly mean values of 1996-1997 SSH data from the NASA/GSFC Ocean Altimeter Pathfinder database [12,16]. The SSH data are interpolated point observations from the TOPEX/Poseidon radar altimeter mission. The SST data come as 360 rows by 720 columns half degree data starting at 180° longitude, the SSH data come as 179 rows by 360 columns one degree data starting at 0° longitude. The SST data have been resampled to the SSH grid. The AVHRR instrument is influenced by cloud coverage whereas the radar altimeter provides uninterrupted data. For a temporal analysis of the same data, see [15]. Statistics for the analysis are calculated where SST and the SSH have non-missing values for all 24 months.

In Figure 1 the first three principal components (PCs) are shown for both the SST and the SSH data. The first three SST-PCs explain approximately 99% of the total variation of the data. The first three SSH-PCs explain 59% of the variation in the data. By studying the principal components one can obtain valuable information about the dynamics of the ocean. In Figure 2 is included the correlations between the original SST and the SSH variables, and the PCs.

The PC transformed SST and SSH data are used as input to ACE. The orthogonal transformation is helpful when we wish to evaluate the results of the nonlinear correlation analysis. Inference problems are avoided that can occur when the original data are cross-correlated. The ACE algorithm applied generates nonlinear mappings of each input variable. The sum of all the transformed variables in each set determines the first ACE CV pair. In Figure 3 the first two ACE CV pairs are shown. The correlation for the first pair is 0.88 and for the second pair 0.63. The two highest correlations found using a linear approach are 0.76 and 0.72 [14]. In the figure is also shown the squared differences for each CV pair. The ACE analysis is performed on a complete set using all PCs for both SST and SSH. In Figure 4, for the first ACE pair, is shown the first three ACE transformations of the PCs together with a bar plot of the variances of all the ACE transforms. The bar plot may be useful when determining which transforms dominate each ACE CV.

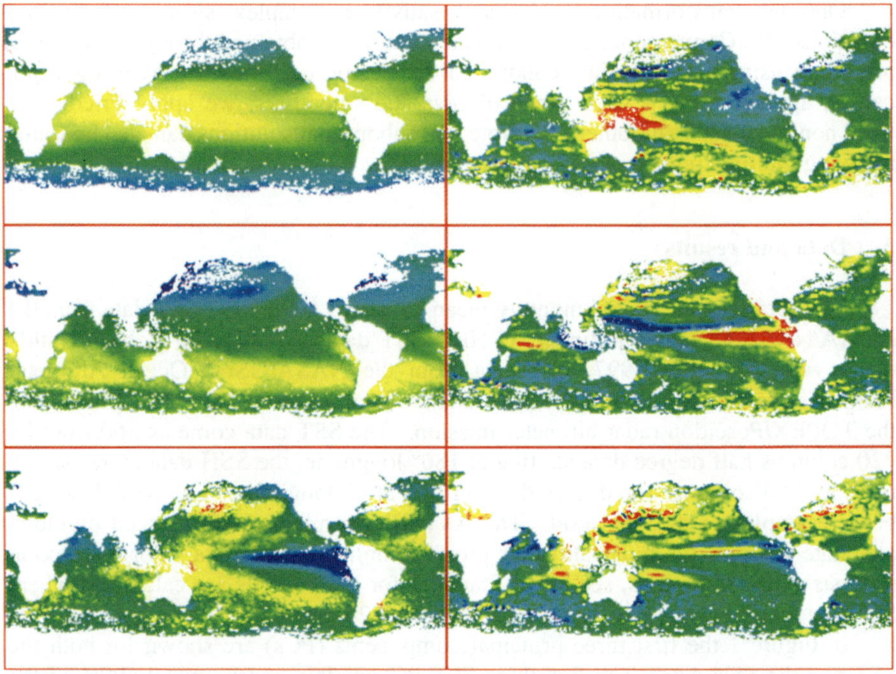

Figure 1. First column (top-down): The first three principal components of the SST data, the percentage of the variance explained by each component is 95%, 4%, and 0.2%. Second column: The first three principal components of the SSH data. The percentage of variance explained by each component is 29%, 23%, and 7%. The images are stretched linearly from mean ± three standard deviations and shown in pseudocolor, blue is minimum and red maximum.

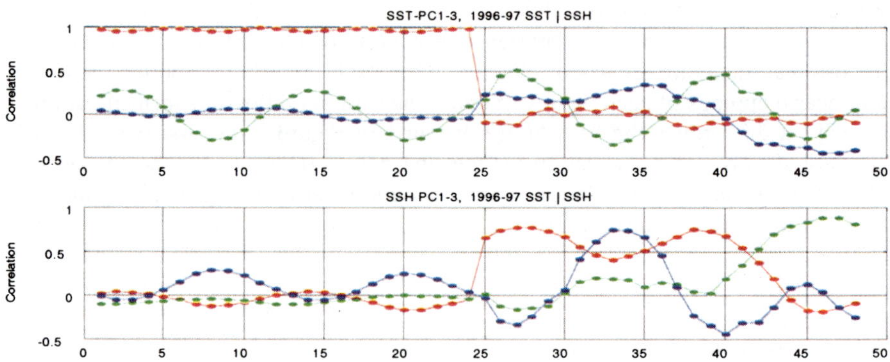

Figure 2. Top: The correlations between original SST (labeled 1-24) and SSH variables (labeled 25-48), and SST-PCs 1 through 3 (shown as red, green, and blue). Bottom: The correlations between original SST (labeled 1-24) and SSH variables (labeled 25-48), and SSH-PCs 1 through 3 (shown as red, green, and blue).

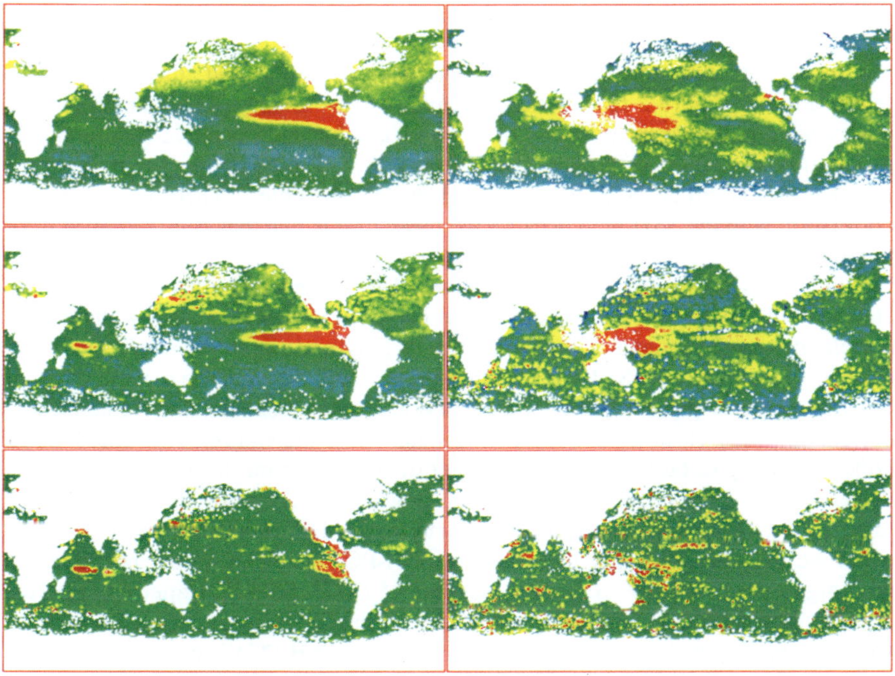

Figure 3. The columns contain the first respectively the second ACE CV pairs, and their squared differences. Each column (top-down): The ACE CV of the SST-PCs, the ACE CV of the SSH-PCs, and the squared differences. The images are stretched linearly from mean ± three standard deviations and shown in pseudocolor, blue is minimum and red maximum.

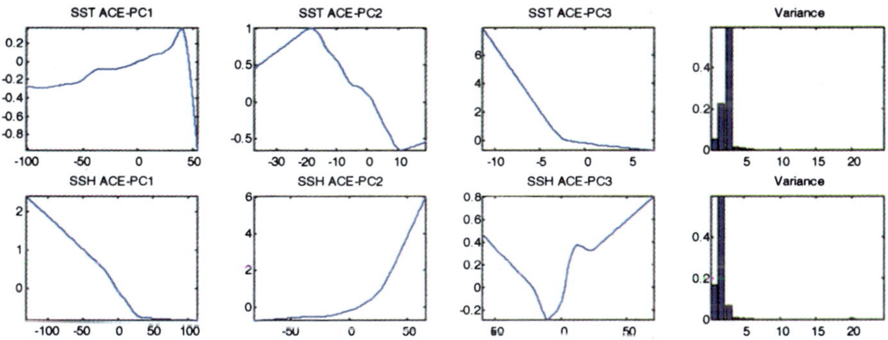

Figure 4. Top row: The first three ACE transformations of the SST-PCs - transformed value vs. input value. The last column contains a bar plot of the variances of all the ACE transformations for the SST PC1-24. Bottom row: The first three ACE transformations of the SSH-PCs - transformed value vs. input value, and a bar plot of the variances of the ACE transformations for the SSH PC1-24.

4 Discussion and conclusions

Inspecting the correlations between the SST-PCs and the sea surface temperature in Figure 2 reveals that the annual cycle is very dominant. SST-PC1 represents a temporal mean signal of the SST and has high positive semi-annually oscillating correlation with all of the SST months. The component represents the fact that it is warm at the Equator and colder near the poles. There is no apparent correlation with the SSH field because the mean of this field has been removed prior to the analysis.

In SST-PC2 we see highs in the Southern Hemisphere and lows in the Northern Hemisphere. In the correlations the annual cycle [9] is clearly present with positive correlations in the months from December to May and negative correlations from June to November. In SST-PC3 a strong signal off the equatorial west coast of South America is present. The component does contain some annual signal but has correlations very close to zero to the SST data. Near the end of 1997 where the El Niño is starting to build up we see a slight divergence from the zero into negative correlations. The negative correlations indicate that during the El Niño the temperature of the west coast of South America is high. Focusing on the El Niño event we see that especially the signal in SSH-PC2 is related to the phenomenon with high correlations to SSH in the last half of 1997. SSH-PC1 appears to be the mean height signal and is highly correlated with the SSH expect during the El Niño where it has negative correlations closer to zero. SSH-PC3 seems related to the same annual cycle as found for SST-PC2.

In Figure 4, for the first ACE CV pair, we see that the variance of the first two to three ACE-PCs dominate the SST and the SSH data. The remaining ACE transforms are all very close to zero. Inspecting SST-PC3 and SSH-PC2 in Figure 1, we would expect ACE to change the sign of one of the components when looking for maximum correlation. This is exactly what is happening with SST ACE-PC3 being transformed using a monotonely decreasing function and transforming SSH ACE-PC2 with a function that is monotonely increasing. Thus the ACE CVs focus on the El Niño event as the part of the signal containing the highest correlation between SST and SSH. The nonlinear transformations of the SST-PC1 and 2, and the SSH-PC1 and 3, appear to be included in the ACE CVs to explain additional spatial patterns present primarily within the Pacific Ocean. In the CVs in Figure 3 the pattern can be seen as a low signal in the western and southern part of the Pacific Ocean. Inspecting the ACE CVs for the first ACE pair the El Niño appears very strongly. Highs in the squared difference image show where the global model is less successful in correlating sea surface temperature to the sea surface height. This is particularly conspicuous near the west coast of Central and South America. There are also contributing regions in the Atlantic and Indian Oceans. Looking at the second ACE CV pair we notice an interesting signal in the western part of the Pacific Ocean. The signal is recognized as being related to the usual ocean temperature and height configuration. A full interpretation of the lower order ACE pairs is beyond the scope of this paper. The ACE analysis seems to be able to separate relevant ocean configurations from the temporal data. It looks for high correlations between the involved variables through nonlinear mappings, and finds components with higher correlations in comparison to those found by a linear analysis, in effect producing a more detailed decomposition of the multivariate data. Furthermore, the ACE analysis is purely data-driven and thus constitutes a useful exploratory tool for a data analyst when looking for insight into the structure of data.

The analysis presented in this paper is in good agreement with the established oceanographic knowledge on the build-up of one of the largest El Niño events on record.

Acknowledgement

The Pathfinder SST data provision at JPL is due to J. Vazques, R. Sumagaysay and co-workers. The Pathfinder SSH data provision at GSFC is due to V. Zlotnicki and co-workers. This work is done as a part of the GEOSONAR project funded by the Danish National Research Councils under the Earth Observation Program. GEOSONAR is headed by Dr. Per Knudsen, National Survey and Cadastre, Denmark.

References

1. Anderson T. W., *An Introduction to Multivariate Statistical Analysis*, 2nd edition, J. Wiley, 1984.
2. Breiman L. and Friedman J. H., Estimating optimal transformations for multiple regression and correlation, *Journal of the American Statistical Association*, 80(391), 580-598, 1985.
3. Buja A. and Kass R. E., Comment: Some observations on ACE methodology, *Journal of the American Statistical Association*, 80(391), 602-607, 1985
4. Cooley W. W. and Lohnes P. R., *Multivariate Data Analysis*, John Wiley and Sons, 1971.
5. Friedman J. H. and Stuetzle W., Smoothing of scatterplots, Technical Report ORION006, Department of Statistics, Stanford University, 1982.
6. Hilger K.B., *Exploratory analysis of multivariate single and multisets*, Ph.D. thesis, Informatics and Mathematical Modelling, Technical University of Denmark, in prep.
7. Hotelling H., Analysis of a complex of statistical variables into principal components, *Journal of Educational Psychology*, 24, 417-441, 1933.
8. Hotelling H., Relations between two sets of variates, *Biometrika*, XXVIII, 321-377, 1936.

9. Kettenring J. R., Canonical analysis of several sets of variables, *Biometrika*, 58, 433-451, 1971.
10. Knudsen P., Andersen O. B. and Knudsen T., ATSR sea surface temperature data in a global analysis with TOPEX/Poseidon altimetry, *Geophysical Research Letters,* 23(8), 821-824, 1996.
11. NASA Facts, *El Niño*, The Earth Science Enterprise Series, NF-211, 1999.
12. NASA/GSCF, http://neptune.gsfc.nasa.gov/ocean.html, Goddard Space Flight Center, National Aeronautics and Space Administration, Greenbelt, Maryland, USA.
13. Nielsen A. A., Multiset canonical correlations analysis and multispectral, truly multi-temporal remote sensing data, *IEEE Transactions of Image Processing*, accepted 2001.
14. Nielsen A. A., Hilger K. B., Andersen O. B. and Knudsen P., A bivariate extension to traditional empirical orthogonal function analysis, *MultiTemp 2001.*
15. Nielsen A. A., Hilger K. B., Andersen O. B. and Knudsen P., A temporal extension to traditional empirical orthogonal function analysis, *MultiTemp 2001.*
16. PO.DAAC, http://podaac.jpl.nasa.gov, Jet Propulsion Laboratory, National Aeronautics and Space Administration, Pasadena, California, USA.
17. Preisendorfer R. W., *Principal Component Analysis in Meteorology and Oceanography*, posthumously compiled and edited by C. D. Mobley. Developments in Atmospheric Science, 17, Elsevier, 1988.

A BIVARIATE EXTENSION TO TRADITIONAL EMPIRICAL ORTHOGONAL FUNCTION ANALYSIS

ALLAN AASBJERG NIELSEN AND KLAUS BAGGESEN HILGER

Informatics and Mathematical Modelling, Technical University of Denmark, Building 321
DK-2800 Kgs. Lyngby, Denmark, Telephone +45 4525 3425, Telefax +45 4588 1397
{aa,kbh}@imm.dtu.dk, www.imm.dtu.dk/~{aa,kbh}

OLE BALTAZAR ANDERSEN AND PER KNUDSEN

National Survey and Cadastre, Rentemestervej 8
DK-2400 Copenhagen NV, Denmark, Telephone +45 3587 5050, Telefax +45 3587 5051
{oa,pk}@kms.dk, research.kms.dk/~{oa,pk}

This paper describes the application of canonical correlations analysis to the joint analysis of global monthly mean values of 1996-1997 sea surface temperature (SST) and height (SSH) data. The SST data are considered as one set and the SSH data as another set of multivariate observations, both with 24 variables. This type of analysis can be considered as an extension of traditional empirical orthogonal function (EOF) analysis which provides a marginal analysis of one variable over time. The motivation for using a bivariate extention stems from the fact that the two fields are interrelated as for example an increase in the SST will lead to an increase in the SSH. The analysis clearly shows the build-up of one of the largest El Niño events on record. Also the analysis indicates a phase lag of approximately one month between the SST and SSH fields.

1 Introduction

This paper deals with an extension of the method of empirical orthogonal functions (EOF) [14]. EOF analysis is often used in oceanography and other geophysical sciences to analyse temporal sequences of scalar (image) data. In this paper we extend this type of analysis by applying canonical correlation analysis to two temporal sequences of scalar image data, namely global sea surface temperature (SST) and global sea surface height (SSH). This type of analysis can be extended further to more than two sets of data [6,10,11]. The data used to illustrate the analysis carried out come from the Ocean Pathfinder programmes set up by NASA/NOAA, [8,13].

The data chosen represent relevant oceanographic properties related to one of the largest El Niño events ever recorded. El Niño is a large-scale warm ocean event in the Pacific off the coast of Peru and Ecuador caused by eastward drifting toward the west coast of South America of the pool of warm waters normally residing in the western part of the Pacific. This event is not local but may influence weather conditions worldwide with potentially drastic socio-economic consequences.

2 Method

EOF analysis is a name often used in geophysical data processing for principal component analysis (PCA) [4,1]. Often the usual PCA assumption on variables with mean zero is replaced by an assumption of temporal means of zero. PCA finds linear combinations (called principal components, PCs) of the original data that maximise variance. PC1 contains maximum variance. Higher order PCs contain maximum variance subject to orthogonality or uncorrelatedness with lower order PCs. Canonical correlation analysis (CCA) [5,2,1] jointly analyses two sets of variables. It finds two sets of linear combinations (called canonical variates, CVs) of the zero mean original variables that maximise correlation between the two. The two CV1s are linear combinations of the original data (one from each set) that are maximally correlated. Higher order CVs are maximally correlated subject to orthogonality or uncorrelatedness with lower order CVs. The correlations obtained between corresponding CVs are termed canonical correlations.

The desired linear combinations $a^T X$ and $b^T Y$ of the two sets of variables X and Y are obtained by maximising the correlation

$$\rho = a^T \Sigma_{12} b \Big/ \sqrt{(a^T \Sigma_{11} a)(b^T \Sigma_{22} b)}$$

between the two. Σ_{11} and Σ_{22} are dispersion matrices of X and Y, respectively, and Σ_{12} is the covariance between them.

The desired linear combinations are obtained by solving the generalised eigenvalue problems

$$\Sigma_{12}\Sigma_{22}^{-1}\Sigma_{21} a = \rho^2 \Sigma_{11} a$$
$$\Sigma_{21}\Sigma_{11}^{-1}\Sigma_{12} b = \rho^2 \Sigma_{22} b$$

where a and b are (conjugate eigen-) vectors of the desired weights ($\Sigma_{21} = \Sigma_{12}^T$). The maximised ρ is called the canonical correlation; the corresponding $a^T X$ and $b^T Y$ are the canonical variates.

3 Data and results

The data used are global monthly mean values of 1996-1997 SST data from the NOAA/NASA Oceans Pathfinder AVHRR SST database, [13], and global monthly mean values of 1996-1997 SSH data from the NASA/GSFC Ocean Altimeter Pathfinder database, [8]. The SSH data are interpolated point observations from the TOPEX/Poseidon radar altimeter mission. The SST data come as 360 rows by 720 columns half degree data starting at 180° longitude, the SSH data come as 179 rows by 360 columns one degree data starting at 0° longitude. The SST data have been resampled to the SSH grid. The AVHRR instrument is influenced by cloud coverage whereas the radar altimeter provides uninterrupted data. Statistics for the CCA are calculated only where both variables have non-missing values for all 24 months.

Figure 1 shows the first six SST CVs resulting from an analysis where the 24 months of SST data are considered as one set and the 24 months of SSH data are considered as the other set. The CVs are stretched linearly from mean ± three standard deviations, the pseudo-colour scale goes from blue (minimum) over cyan, green, yellow to red (maximum).

Figure 2 shows the first six SSH CVs (same analysis, stretch and colouring).

Figure 3 shows the canonical correlations, and the correlations between the original data and the first three CVs for both SST and SSH.

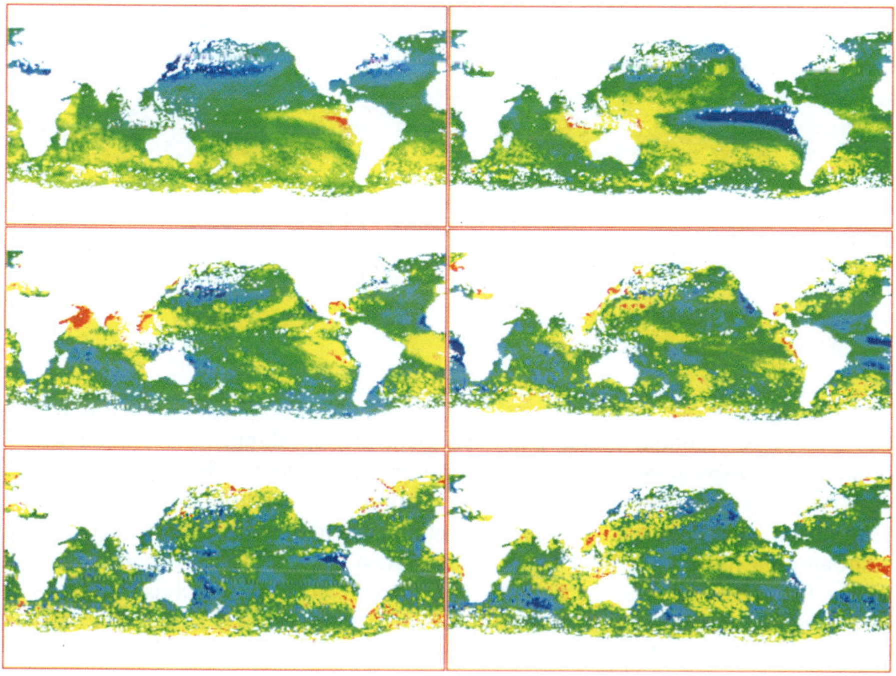

Figure 1. SST CVs 1-6 row-wise.

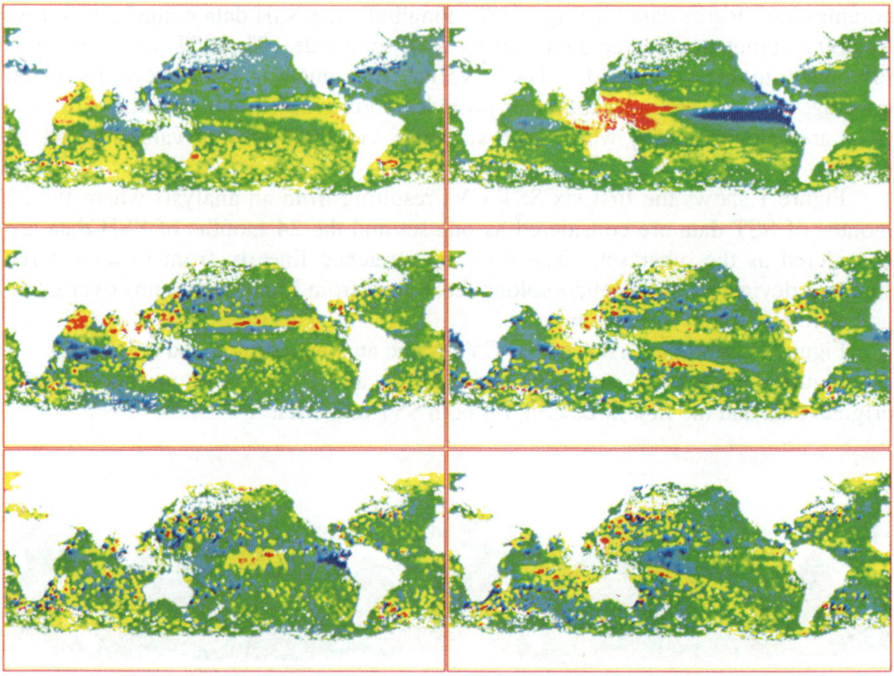

Figure 2. SSH CVs 1-6 row-wise.

4 Discussion and conclusions

The first pair of CVs has a correlation coefficient of 0.7587 and accordingly the two exhibit very similar spatial patterns. Generally speaking, we see highs in the Southern Hemisphere and lows in the Northern Hemisphere. This is in accordance with the observed maximally negative correlations between both SST and SSH and both CV1s in the Northern Hemisphere's summer months. A high in SSH CV1 and a strong high in SST CV1 off the equatorial west coast of South America is seen. This is in accordance with the El Niño phenomenon. The conspicuous oscillation is clearly at an annual period, but the correlations between CV1 and SST remain negative and do not alternate around zero which should be expected. The highest correlations occur in month 9 and 21 for both the SST and the SSH. Normally the highest correlation for the SST would occur one month earlier, but because of the joint analysis of the SST and SSH, this is dominated by the SSH. This suggests an analysis of SST and a temporally shifted version of SSH.

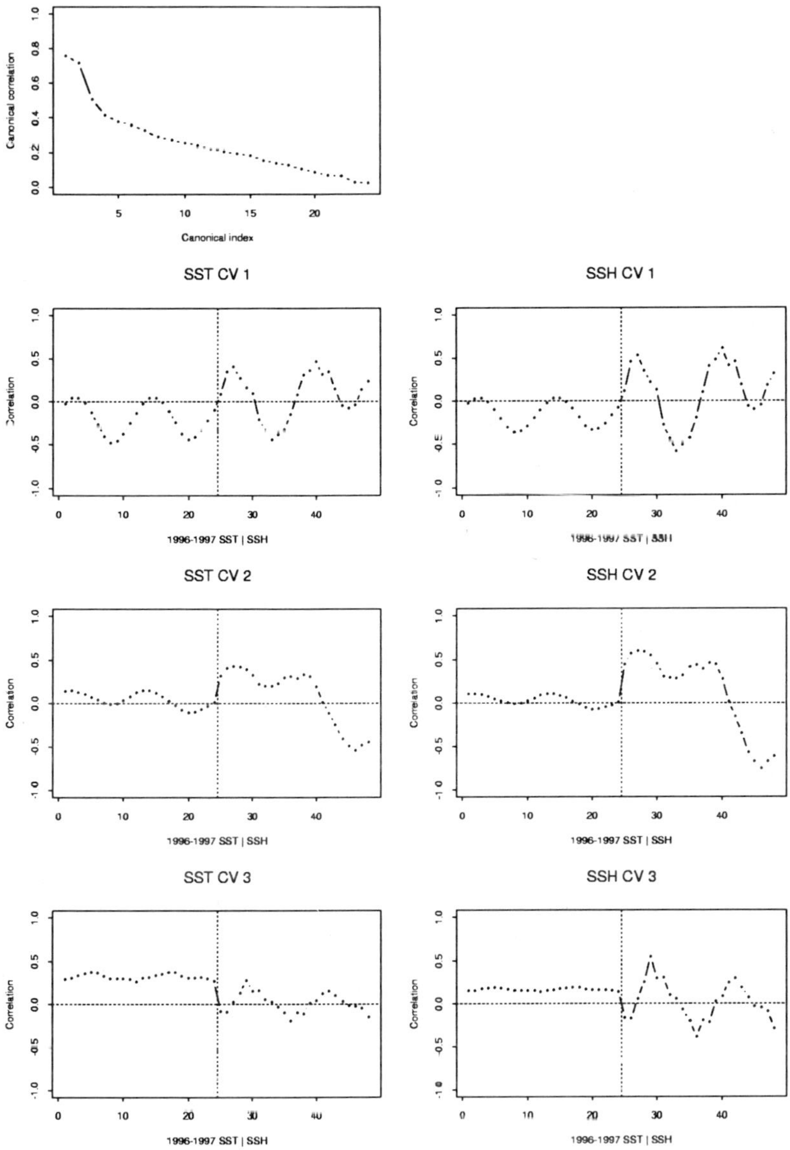

Figure 3. Canonical correlations (top), and correlations between original SST (indices 1-24, left of the dashed vertical line) and SSH variables (indices 25-48, right of the dashed vertical line), and SST CVs 1-3 (left column) and SSH CVs 1-3 (right column)

The second pair of CVs has a correlation coefficient of 0.7179 and again the two exhibit similar spatial patterns. For this CV pair we see strong lows in both SSH CV2 and SST CV2 off the equatorial west coast of South America which again is in accordance with the El Niño event. Also, a high in SST CV2 and a very strong high in SSH CV2 east of Indonesia and the Philippines is seen.

The correlations between the original data and the CVs indicate that both CVs 1 and CVs 2 reflect strong annual oscillations [7]. Especially for SSH and CVs 2 but also for SSH and CVs1 and to a lesser degree for SST and CVs2 these oscillations seem to be disturbed in the second half of 1997. This is agreement with the spatial patterns observed in especially CVs 2 and the El Niño build-up during the last eight months of 1997.

The third canonical correlation is 0.5061, and the fourth is 0.4148. This is still relatively high, but the fact that the first two canonical correlations are much higher than the remaining ones shows that the joint variation of the two geophysical fields over the two-year period is well explained by only two CVs. On the other hand the two highest canonical correlations are only 0.7587 and 0.7179, and therefore there are also variations in the two fields that are not well explained by a joint (linear) analysis. For a nonlinear analysis of the same data, see [3].

In spite of the very short time span covered by the data and the associated risk of over-interpreting the results, simultaneous inspection of spatial patterns of the CVs and the correlations between the original and transformed variables from the analysis gives good indications of an anomaly off the South American west coast taking place in the second half of 1997. This is in good agreement with established oceanographic knowledge on the build-up of one of the largest El Niño events on record.

Future analysis should include longer time series in order to establish whether 1997 (and 1998) really represent anomalous events in terms of global SST and SSH.

Investigations should also be made in which the SST field is shifted temporally to the SSH field in order to investigate if the correlations between the CV and the individual SST and SSH fields could be increased. The apparent phase lag of one month between the highs of the SST and the SSH is explained by the fact that the SST represent the instant temperature of the sea surface, whereas the steric expansion causing the sea level to rise is a more integrated effect.

Acknowledgement

The Pathfinder SST data provision at JPL is due to J. Vazques, R. Sumagaysay and co-workers. The Pathfinder SSH data provision at GSFC is due to V. Zlotnicki and co-workers, and the authors would like to thank B. Beckley for providing the data. This work is done as a part of the GEOSONAR project funded by the Danish National Research Councils under the Earth Observation Program. Dr. Per Knudsen, the National Survey and Cadastre, Denmark heads GEOSONAR.

References

1. Anderson T. W., *An Introduction to Multivariate Statistical Analysis*, second edition, John Wiley, 1984.
2. Cooley W. W. and Lohnes P. R., *Multivariate Data Analysis*, John Wiley and Sons, 1971.
3. Hilger K. B., Nielsen A. A., O. B. Andersen and Knudsen, P., An ACE-based nonlinear extension to traditional empirical orthogonal function analysis, *MultiTemp 2001*.
4. Hotelling H., Analysis of a complex of statistical variables into principal components, *Journal of Educational Psychology*, 24, 417-441, 1933.
5. Hotelling H., Relations between two sets of variates, *Biometrika*, XXVIII, 321-377, 1936.
6. Kettenring, J. R., Canonical analysis of several sets of variables, *Biometrika*, 58, 433-451.
7. Knudsen P., Andersen O. B. and Knudsen T., ATSR sea surface temperature data in a global analysis with TOPEX/Poseidon altimetry, *Geophysical Research Letters*, 23(8), 821-824, 1996
8. NASA/GSCF, http://neptune.gsfc.nasa.gov/ocean.html, Goddard Space Flight Center, National Aeronautics and Space Administration, Greenbelt, Maryland, USA.
9. NASA Facts, *El Niño*, The Earth Science Enterprise Series, NF-211, 1999.
10. Nielsen, A. A., *Analysis of regularly and irregularly sampled spatial, multivariate, and multi-temporal data*, Ph.D. thesis, IMM-, Technical University of Denmark, http://www.imm.dtu.dk/~aa/phd, 1994.
11. Nielsen A. A., Multiset canonical correlations analysis and multispectral, truly multi-temporal remote sensing data, *IEEE Transactions on Image Processing* accepted 2001.
12. Nielsen. A. A., Hilger K. B., Andersen O. B. and Knudsen P., A temporal extension to traditional empirical orthogonal function analysis, *MultiTemp 2001*.
13. PO.DAAC, http://podaac.jpl.nasa.gov, Jet Propulsion Laboratory, National Aeronautics and Space Administration, Pasadena, California, USA.
14. Preisendorfer R. W., *Principal Component Analysis in Meteorology and Oceanography*, posthumously compiled and edited by C. D. Mobley. Developments in Atmospheric Science, 17, Elsevier, 1988.

ENCAPSULATION OF DYNAMIC INFORMATION CAPTURED IN LONG SEQUENCES OF HYPER-TEMPORAL IMAGE DATA IN SINGLE TEMPORAL IMAGES - EXAMPLES, CHALLENGES AND DIRECTIONS FOR DEVELOPMENT

K. R. McCLOY

Danish Institute of Agricultural Research, Post Box 50, 8830 – Tjele Denmark

E-mail: Keith.mcccloy@agrsci.dk

Long sequences of image data, or hyper-temporal image data, capture information on how the environment that is recorded in those images responds to various dynamic environmental processes. Such information is essential in better understanding the inter-relationship between these processes in the environment, and in improving models of those processes. At present there are few suitable tools for the analysis of hyper-temporal imagery, for the purpose of extracting information on the dynamic processes recorded in the image data.

The paper will discuss a number of methods developed by the author to create temporal images, or images that capture dynamic change features, as well as methods developed to support the comparison of two independent sets of data, such as actual and modelled data. The paper will show how some of these developments are currently being used with long sequences of global NDVI data, and will then discuss other potential applications.

The paper will summarise with some more general findings in relation to the analysis of hyper-temporal image data.

1 Introduction

Long sequences of image data have existed in various formats for some time. Examples of such sequences include:

- The time series of AVHRR based land products supported by NASA at http://daac.gsfc.nasa.gov/CAMPAIGN_DOCS/LAND_BIO/GLBDST_Data.html [1, 7]. For this study data were used from July 1981 - June 1994 and from July 1995 to June 2000 to form a 216-image stack consisting of 18 annual cycles of monthly data. The images have been rectified, calibrated and corrected for some atmospheric effects. The images are calibrated using the Libyan Desert as a standard reflectance surface http://psbsgi1.nesdis.noaa.gov:8080/EBB/ml/niccal1.html. The data are then corrected for ozone absorption and Rayleigh scattering [3]. The data have not been corrected for water vapor or aerosol effects.
- The time series of AVHRR based Sea Surface Temperature held by NASA at http://podaac.jpl.nasa.gov/sst/

These technical developments in remote sensing are part of a much broader ferment of activity directed at better understanding global processes. There are two important components to all of these activities that are relevant to the work reported in this paper. The first is the use of models to investigate, better understand, and ultimately to be able to predict the response of processes to specific changes. The second is the use of image data in this process. This paper suggests that there needs to be closer integration of remote sensing and modelling if the power of both technologies is to be truly realised. This integration should take these forms:

- Provision of data to drive the models.
- Conduct of accuracy and reliability assessment.
- Comparative analyses to investigate the nature, extent, spatial, temporal and informational characteristics of the divergence between the model results and the estimates derived from image data.
- Depiction of spatio-temporal dynamic change so as to better understand the characteristics of that change.

The last two of these four roles are where the focus has been in the work being conducted at the Danish Institute of Agricultural Sciences (DIAS).

2 Current Activities at DIAS

2.1 Depicting Dynamic Processes

Three tools have been developed to depict the dynamic information captured in a temporal image sequence:

- Temporal Enhancements
- Temporal Variograms and Fourier Analysis
- Temporal Distance Images

2.1.1 Temporal Enhancements

A pixel profile is enhanced by weighting the values at each date in the profile by the weighting function shown in Figure 1, and summing the derived values [4]. This basic form of the enhancement was modified to apply the enhancement over a user-defined subset in the data, with the capacity to replicate that subset through the whole range of data in the multi-temporal image. The temporal enhancement shown in Figure 2 shows average NDVI values at different times of the year as different colours in accordance with Figure 1. The image has been analysed in relation to the climate map depicted as Plate 3 in [8]. This analysis shows that:

- The differences between the northern, southern and equatorial regimes are obvious. The rainforests of South America, The Congo Basin and New Guinea are dominantly white, indicating constant high values of NDVI throughout the year. Parts of the Amazon basin, Borneo, Sumatra and Madagascar exhibit light cyan colouring, indicating high annual NDVI values, with slightly greater NDVI values in the first half of the year. The humid rainforests, and monsoon forests of Bangladesh, north eastern India, Burma, south east China, Laos and northern Vietnam are dominantly yellow, with some isolated areas of red, indicating greatest NDVI values in September-October, and isolated areas with greatest NDVI values in November–December. This coincides closely with the monsoon rainfall and cloud patterns of this area, where peak rainfall is usually in June-July, and ceases by December. The pink to magenta areas of southern Africa and South America indicate peak NDVI in the November-February period, coinciding with peak rainfall and temperatures in these areas. On the African east coast, in the vicinity of northern Mozambique and southern Tanzania, this peak NDVI shifts to later in the season, typically to late December- January, as it also does in eastern Brazil, in the vicinity of Recife. The green of the USA, Yugoslavia, Korea, north eastern China, eastern Russia and northern Japan match the growing seasons in this part of the world, where peak temperatures and rainfall usually occurs in July.
- Areas of desert are uniformly dark grey to black, whilst areas of sparse or intermittent vegetation are grey in colour. Intermittent vegetation can include agriculture if the fields of crop are interspersed with areas of bare soil. Mountainous areas also appear as grey.
- Some of the boundaries in the image closely correspond to the boundaries shown in the map, whilst others do not. There are some boundaries in the image that do not appear in the map.
- The boundaries around the monsoon rainforests of India, Bangladesh, Burma, China, Laos and northern Vietnam do not appear to support similar boundaries in the climate map.

2.1.2 Temporal Variograms and Fourier Analysis

Temporal variograms were constructed for each pixel using the temporal data of 216 monthly images in accordance with the standard form of the variogram equation [2]. The variograms were then subjected to the Fourier transformation, with the Fourier amplitude image at three frequencies being shown in Figure 3. The relationship between Frequency and Interval or wavelength is given for this data set by:

Frequency * Wavelength = 256

The Fourier amplitude image shows the weights (y- axis) for the frequencies that are displayed along the x-axis. The Fourier image is symmetrical about the central term. The Fourier profiles show that:

- All surfaces have a constant term, the magnitude of which is related to the average value in the variogram.
- Many surfaces are cyclic in nature, and the dominant sine wave frequencies in these cyclic patterns will be shown as a spike in the Fourier. The profiles show two spikes for annual and six-monthly cycles.
- Some of the data contain longer-term cycles, including three, five seven to eight and eleven-year cyclic patterns.

Choosing channel 124 as red, 107 as green and channel 85 as blue means that most of the information on cyclic patterns in the image data will be displayed in the image. Such an image is thus close to the goal of a temporal image, and will be treated as such in this paper. Such an image is displayed in Figure 3. It shows areas of:

- **red** have high channel 124 values and low channel 107 and 85 values. This is consistent with surfaces that either have high central variances or have a long term cyclic pattern centred on channel 124, and no annual or six-monthly cycles. Such surfaces can be found in the northern Amazon rainforest, parts of the rainforest of west Africa, Borneo and New Guinea, as well as for most of the Arabian desert, parts of the Iranian and Chinese deserts and the south western deserts of the USA.
- **gray** to dark gray have low Fourier values at all channels, and no cyclic patterns. Such areas are common in parts of the Saharan and Arabian deserts, as well as parts of the rainforest areas of Sumatra, Borneo and New Guinea.
- **green** are dominated by the annual cyclic pattern of vegetative growth. Such areas are common in China, central India, the Sahel, southern and south eastern Africa, the northern USA and northern Japan.
- **blue** are dominated by six-monthly cycles of vegetative growth. Such areas exist in northern India, Bangladesh, the Mekong delta of Vietnam, the Horn of Africa, the northern part of the Congo basin and the southern coast of the USA. The Horn of Africa and the

Figures

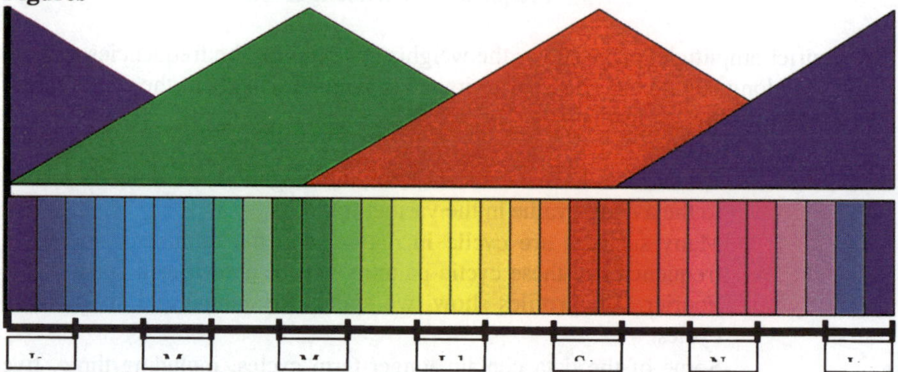

Figure 1. The temporal colour enhancement algorithm and the resulting temporal colour assignment.

Figure 2. Temporal global enhancement depicting seasonal vegetation dynamics.

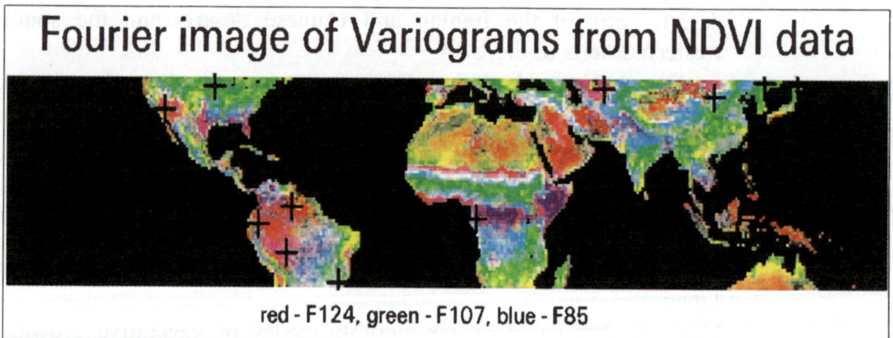

Figure 3. Fourier image derived from the three frequencies that dominate the dataset representing (i) no cyclic pattern (red), (ii) Annual patterns (green) and (iii) six monthly patterns (blue).

northern Congo basin exhibit significant magenta amongst the blue, indicating high central values as well as some six-monthly cycling.

- **magenta**, such as the southern fringe of the Saharan desert, southern Nigeria, the southern half of the Amazon basin and south western Texas, exhibit some six-monthly cycles, and high central variance values.
- **cyan** have both annual and six-monthly cycles, and these areas can be found in the southern part of the Congo basin, south-central Brazil, central India and parts of inner China.
- **yellow** indicate annual cycling with high central values, as occurs in extensive parts of the Saharan desert, northern Australia and eastern Brazil.

The images thus depict cycling patterns in the vegetation around the globe.

2.1.3 Temporal Distance Images

A distance image depicts the distance of the pixel response from a model response at the date of this image, where this distance can be Euclidean or Mahalonobis types of distance metrics. To calculate such a distance the model response has to be determined. This can be done from a numerical model, or by using a training area in the image as being representative of that model or surface condition. A temporal distance image is an integration of the distance images over time, where the most common form of integration would be the average value.

Pixels that have a short distance in a Temporal Distance Image will have profiles through time that are close to the model. Pixels that have large distances will have profiles that are a long way from the model. Such a temporal distance image is a black and white image, with black representing short distances, and white representing large distances. Three such temporal distance images can be combined in the three different primary colours, and provide comparative information on the relative distances to the three cover types (Figure 4).

Whilst the average distance is used as a first order image, it can mask significant differences in the distance that occur through time. Thus a pixel with a moderate distance across the whole set of images may end up with a similar temporal distance value as another pixel that has small values most of the time, but the occasional very large value. Development of a distance metric that can deal with this type of issue would be an advantage.

2.2 Comparing images to models

This comparative work has been done using the global NDVI data set in comparison with the Lund-Potsdam-Jena (LPJ) vegetation model [6]. The LPJ model has been developed to investigate the impact of climate change on vegetation. It uses different plant groups to investigate the propagation of plant groups with global climate.

2.2.1 Theoretical basis for the study

The data are taken to approximate a sine wave in this development. If these data sets are compared along the **t** axes, then this removes the **t** parameter from the resulting relationship between the data sets. The shape of the resulting figure depends on the amplitude, phase and phase shift of both source data sets as shown in Figure 5. Such figures can be:

- A line when the difference between the time series is due to changes in amplitude or mean value.
- An ellipse when the difference is due to a phase shift between the time series. This type of figure can be reduced to a line by comparing the values at the phase shift between the signals (Figure 5©).
- A figure of 8 when one time series has wavelength double the other time series.
- More complex figures when the wavelengths of the time series vary in different ways. Whilst such variations occur in these data sets, they have not been considered in this development.

2.2.2 The data used in the study

Two data sets are used in the study. The actual data are based on the NDVI data referred to in the introduction. The data were converted to estimates of FPAR using the approach of [5]. Dr Wolfgang Lucht of the Potsdam Institute for Climate Impact Research supplied the modelled data, computed using the LPJ model [6].

2.2.3 Conduct of the Analyses

Two types of analyses were conducted. The first type compared a time series from the AVHRR based data set against another time series taken from the sáme data. The second comparison was between the AVHRR based data set and the modelled data set as derived by the LPJ model. The analyses were conducted in two stages. In the first stage, the data for each annual cycle were analyzed. The data were analyzed at each step or phase difference in the annual cycle to find the regression with the highest R-squared value. The gain, offset, R-squared value and the phase difference of this solution was retained for this cycle. Similar values were retained for each annual cycle.

The second stage was to simplify this data set for display and analyses. The data set derived in stage 1 was used to find, in stage 2, for each pixel:

Figure 4. Composite of three Temporal Distance images. Areas with NDVI values like desert are depicted in cyan, areas like rainforest in yellow, and areas like Mediterranean climates in magenta. Intermediate areas will have intermediate colours between these limits. The enhancement used is critical. In this image the enhancement is designed to clearly delineate "core areas", or areas with short distances to the cover type of interest.

Figure 5. Modelling of the comparison of sine waves. (a) the source sine waves, (b) Comparison of the reference wave with the other three waves and (c) the effects of stepping one wave relative to the other when comparing the reference wave against the phase shifted wave.

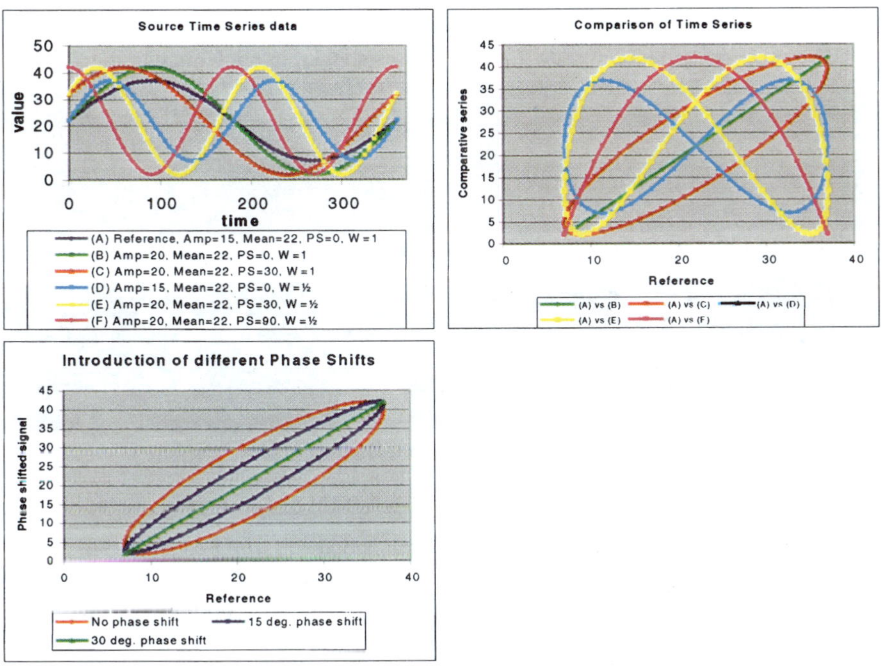

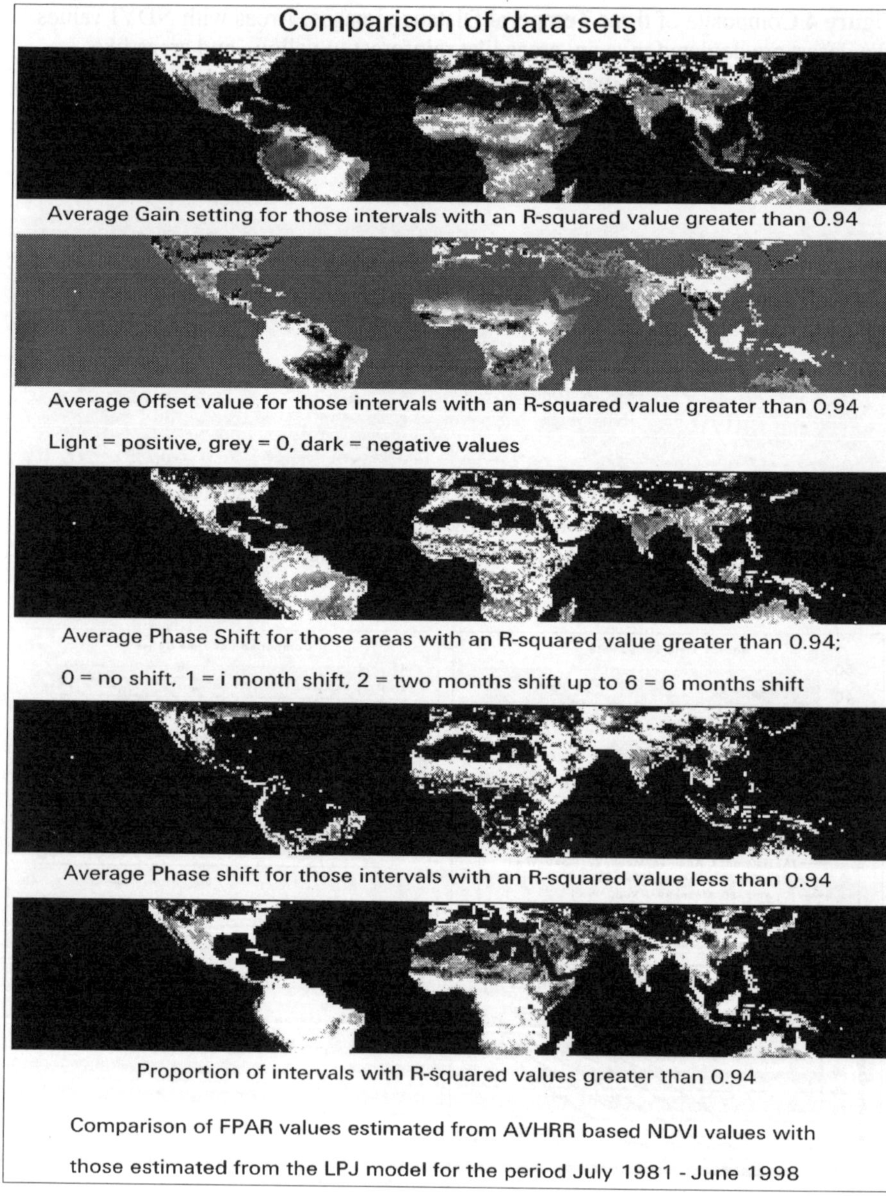

Figure 6. Comparison of FPAR values derived from NDVI with those estimated by the LPJ model for the period 1981 – 1998.

- The proportion of the cycles (years) that met the R-squared threshold value as being a good fit to the linear model, the proportion of these that had this value with a step of zero (No phase shift between the data sets) and the proportion with a step larger than 0.
- The proportion with a step value of zero was assigned to red, with the intensity a function of the proportion value.
- The proportion with a step value greater than 0 was assigned to green, with the intensity being this proportion value.
- The proportion of cycles that did not meet the R-squared value criteria was assigned to blue, with the intensity of blue representing the step value.

2.2.4 Results of the Analyses

The comparison of the two AVHRR data sets showed that most areas have time series that are of similar characteristics and they most pixels had R-squared values above 0.94, with the mode being 0.98. Areas of approximately constant FPAR values, such as rainforest and desert, can exhibit phase shifts. The largest anomalies occurred along coastlines.

The comparison between the FPAR data derived from the NDVI and the LPJ model showed that all of the theoretical shapes occurred in the data. The comparison also showed that the fluctuations in the trajectories were less pronounced possibly because the LPJ model should take weather patterns into account, whereas these will contribute to the variability found in the first comparison. Most pixels had R-squared values above 0.94, with the mode being 0.98, but the distribution was flatter than that from the first comparison. Whilst the annual trajectories were usually consistent in shape, they were not always consistent in position. It suggests that there may be a discrepancy at some locations in the estimation of FPAR by the model in comparison with the estimate derived from the NDVI data.

3 Concluding Discussion

3.1 Data stability

Current methods of atmospheric correction cannot deal adequately with local effects, due primarily to water vapour and aerosols. A satellite based technique is required, so as to ensure stable data at a reasonable cost. There is also the need to ensure data stability from sensor to sensor.

3.2 Data formats

At present there exist no established formats for temporal image sequences. We have established a Temporal Band Sequential (TBSQ). This format is identical to standard BSQ if there exists either a temporal sequence with one band of data, a single multispectral image, or where the combination of bands and images can be considered as one dimension in the data set. Equivalent temporal formats could be established to match the other data formats.

3.3 Image processing software

At present none of the existing image processing systems can deal with temporal data sets.

3.4 The depiction of temporal information and the surrenders that have to be made

Temporal image data will usually contain six dimensions. Collapsing this down to one image (3 dimensions of data for black and white image data, 3+ dimensions for colour imagery) will require surrenders. There are a number of potential forms that these surrenders could take, and which may not involve significant difficulties:

- **Assume that the landcover is known.**
- **Assume that the condition of the landcovers being observed is known.**
- **Assume that the analyst can retain a mental model of the process.**

The work conducted so far at the Danish Institute of Agricultural Sciences indicates that we can either:

- **Create very general information**, or information that provides a very general overview of the situation. This is valuable in itself, but it is not sufficient, and it seems to be surrendering a lot of the information content in the data.
- **Create very specific information**. This again can be very valuable, but it is often difficult to know what specific information is required, is the best to use and indeed is available from the analysis.

3.5 The nature of the changes that can be detected in image data

There are three forms of temporal dynamic change;

- **Where the change in time is related to change in location.**
- **Where the change in time is related to change in condition.**
- **Where the change in time is related to both changes in location and condition.**

4 Acknowledgements

This work was done as part of the "Image Segmentation and Temporal Change" project funded by the Danish Research Agency. Data used by the author in this study include data produced through funding from NASA's Earth Observing System Pathfinder Program in co-operation with National Oceanic and Atmospheric Administration. The data were provided by the Earth Observing System Data and Information System (EOSDIS), Distributed Active Archive Centre at Goddard Space Flight Centre, which archives, manages, and distributes this data set. The provision of modelled FPAR data by Dr Wolfgang Lucht is appreciated.

References

1 Agbu P.A., and James M.E., 1994, The NOAA/NASA Pathfinder AVHRR Land Data Set User's Manual, Goddard Distributed Active Archive Center, NASA, Goddard Space Flight Center, Greenbelt, MD.

2 Cressie N.A.C., 1993, *Statistics for Spatial Data* (John Wiley and sons, New York).

3 James M.E. and Kalluri S.N.V., 1994, The Pathfinder AVHRR land data set: An improved coarse resolution for terrestrial monitoring, *Int. J. Remote Sens.*, **15** (17): 3347-3363.

4 McCloy, K.R., 2000, Creation of temporal images from sequences of multi-temporal image data, Proceedings, XIXth Congress of the International Society for Photogrammetry and Remote Sensing, Amsterdam, The Netherlands.

5 Myneni R., 1997. Estimation of Global Leaf Area Index and Absorbed PAR Using Radiative Transfer Methods, IEEE Transactions on Geoscience and Remote Sensing, Vol. **35**, No. 6., pp.1380-1393.

6 Sitch S., Prentice I.C., Smith, B., Cramer W., Kaplan J., Lucht W., Sykes M., Thonicke K., and Venevsky S., LPJ - A coupled model of vegetation dynamics and the terrestrial carbon cycle. In *The role of vegetation dynamics in the control of atmospheric CO2 content*, Dissertation, Sitch, S., Lund University, Lund, Sweden.

7 Smith P.M., Kalluri S.V.N., Prince S. D., and DeFries R., 1997. The NOAA/NASA Pathfinder AVHRR 8-km land data set, Photogrammetric Engineering & Remote Sensing **63** (1), 12-32.

8 The Times World Atlas, Comprehensive Edition, 1993.

PERFORMANCE ASSESSMENT OF MULTI-TEMPORAL SAR IMAGE FILTERING

E. TROUVÉ, M. CIUC, F. BUJOR AND PH. BOLON
Laboratoire d'Automatique et Micro-Informatique Industrielle (LAMII/CESALP)
Ecole Supérieure d'Ingénieurs d'Annecy - Université de Savoie
B.P. 806 - 74016 Annecy Cedex - FRANCE
E-mail: {trouve | ciuc | bujor | bolon}@esia.univ-savoie.fr

N. CLASSEAU AND J.P. RUDANT
Laboratoire de Géomatériaux - Institut Francilien de Géosciences
Université de Marne-La-Vallée - Cité Descartes
5, bd Descartes, Champs-Sur-Marne, 77454 Marne-La-Vallée Cedex 2 - FRANCE
E-mail: {classeau | rudant}@univ-mlv.fr

DA. COLTUC
Dept. of Applied Electronics - University "Politehnica"
Iuliu Maniu 1-3 - Bucharest - ROMANIA
E-mail: coltuc@atm.neuro.pub.ro

Multi-Temporal SAR Image filtering is useful for many applications which require speckle reduction. Several multi-temporal filters are now available with very different characteristics. In this paper, performances of three filters are assessed with respect to both objective and subjective criteria. Objective criteria include measures of bias, noise reduction and spatial resolution preservation, whereas subjective criteria evaluate the accuracy of manual detection of geographical features. Results are presented on multi-temporal SAR images from ERS satellites.

1 Introduction

An increasing number of applications in geosciences are now using multi-temporal SAR images as a regular source of information: forest monitoring [1], land cover classification [2]... Most of these applications use pre-filtering in order to reduce speckle, since speckle noise makes automatic analysis or operator photo-interpretation difficult. Classical bi-dimensional (2D) speckle filters operate image by image, the result being a trade–off between noise reduction and spatial resolution degradation [3].

Multi-temporal data creates a new opportunity to develop filters which benefit from the additional information brought by the third dimension (the time axis). This paper presents an evaluation of three multi-temporal speckle filters, the principles of which are briefly recalled in Section 2. A strict comparison of filter performances is not possible since they all require parameter adjustment which depends on data and application. The objective of this paper is to provide SAR image end–users with measurements of filter main characteristics in order to guide their choice according to the expected results. This is done by proposing two kinds of criteria:

- objective criteria presented in Section 3. Filters are evaluated in terms of radiometry preservation (bias), noise reduction (equivalent number of looks) and spatial resolution preservation (false alarms and detection probabilities for a classical edge detector [4]).
- subjective criteria presented in Section 4. Since many applications still rely on image photo-interpretation, it is important to evaluate filtering results with respect to their visual aspect. In order to obtain a quantitative evaluation, trained human operators are asked to manually extract simple geographic features. Results are compared with a reference map and a visual interpretation of original images.

2 Multi-temporal speckle filters

In this section, a brief review of three multi–temporal filters is presented. These particular filters have been selected in order to cover different approaches to multi–temporal SAR image filtering. Results obtained by the selected filters are further used in the performance assessment stage.

2.1 Texture–Compensation Multichannel Filter

The Texture–Compensation Multichannel Filter (TCMF) proposed by Bruniquel and Lopès [5] is a purely–temporal method based on optimality criteria. The basic idea is to calculate the filter output at a given pixel as a linear combination of its temporal neighbors. The weights of the linear combination are derived so that the output variance is minimal. One single filtered image is synthesized, the others being computed by local correction with a term taking the local texture into account.

2.2 3D Adaptive–Neighborhood Filter

The 3D Adaptive Neighborhood Filter (3D–ANF) proposed by Ciuc et al. [6] is a spatio–temporal filter, that uses both spatial and temporal neighbors of the current pixel to derive the filter output. Multi-temporal sequences are treated as 3D volumes, with two spatial and one temporal coordinates. For every "seed" pixel, a 3D adaptive neighborhood is determined by means of a modified region growing algorithm. The goal is to retain in the adaptive neighborhood only pixels coming from the same distribution as the seed distribution. Their values are then used to estimate the seed filtered value.

2.3 Time–Space Filter

The Time-Space Filter (TSF) proposed by Coltuc et al. [7, 8] is an homomorphic filter based on the use of unitary transforms. It combines two powerful tools: a Discrete Cosine Transform (DCT) applied on the time axis to represent the data in the frequency domain and a spatial adaptive filter applied on the resulting DCT channels. Since "temporally" uncorrelated noise spreads in all the DCT channels and most of radiometric information concentrates in the "zero" frequency channel, the former is left unfiltered in order to preserve the spatial resolution whereas the noise is strongly reduced by the filtering applied in the other channels.

2.4 Data

These three filters are evaluated on a set of 6 multi-temporal SAR images from satellites ERS-1 and ERS-2 acquired between 1992 and 1995 during descending passes over French Guyana. Data are classical 3-look precision images (PRI) co–registered by translation without resampling to avoid artifacts. A 2300×2400 window was selected in order to include wide homogeneous regions of tropical forest which are stable in time and human made features (rice fields, roads...) with spatial contrast and temporal evolutions. The different filtering results are illustrated in Figure 1 on two 512×512 sub-images.

3 Evaluation by statistical criteria

Multi-temporal filtering techniques perform a trade–off between noise reduction and preservation of spatial resolution and temporal information. The best tuning depends on the application: minimizing the degradation of spatial resolution is a priority for applications requiring feature detection whereas preserving the mean radiometry in each image is very important for applications using the ground reflectivity. The following statistical criteria are chosen in order to illustrate filters behavior with respect to these different goals.

3.1 Bias and noise reduction

Mean radiometry preservation and noise reduction are evaluated by measuring bias and equivalent number of looks L_{eq} on homogeneous regions. In these regions, the speckle probability density function (pdf) of intensity images is known as a Gamma distribution with mean $\mu_s = 1$ and variance $\sigma_s^2 = 1/L_{eq}$. The relative bias B is measured by computing, at each pixel, the ratio between the original image and the local mean of the corresponding filtered image. If the filter output is biased, this ratio image contains the speckle divided by the bias. In this case, the ratio image mean is equal to $1/B$.

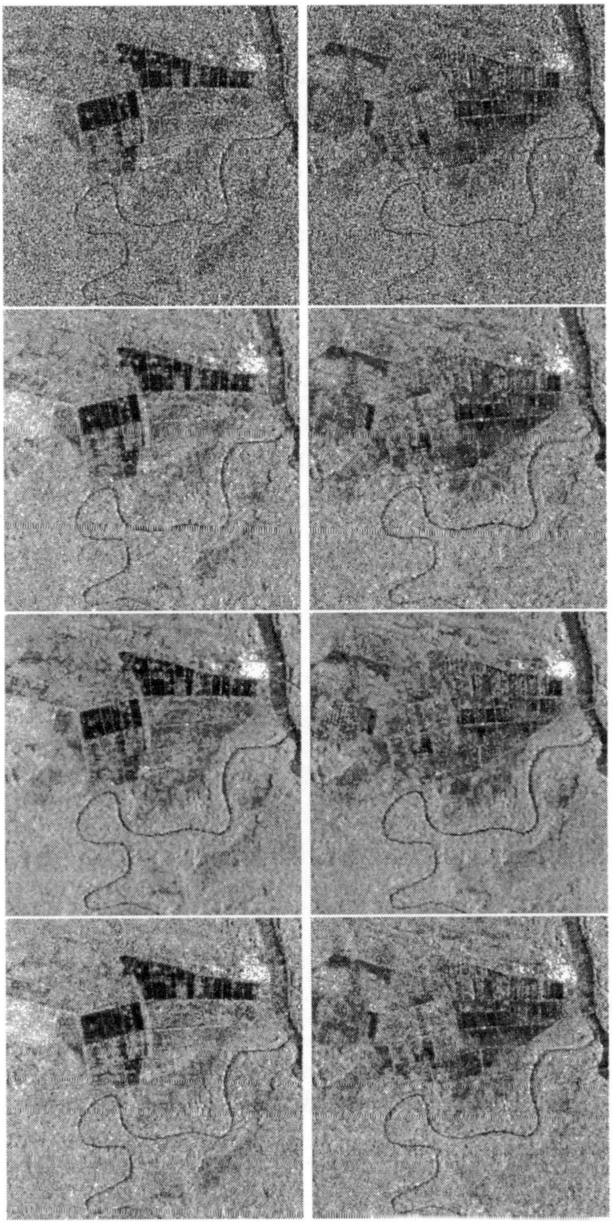

Figure 1. 2 out of a set of 6 multi-temporal SAR images, original and filtered

The equivalent number of looks is measured by computing the squared coefficient of variations of intensity images $C_I^2 = \sigma_I^2/\mu_I^2$ which is equal to $1/L_{eq}$ in homogeneous regions. Since precision images (PRI) are equivalent to 3-look images, the local value of C_I^2 computed on 9×9 windows in the original image is used to build up a mask of homogeneous regions: only pixels with C_I^2 less than 1/3 (about 50% of the 2300×2400 pixels) are considered to measure the ratio mean value and the coefficient of variation.

Results obtained by applying this procedure are presented in Table 1. No significant bias can be observed in the filtered images. For TCMF and TSF which are mainly "temporal" filters, the final number of looks is equal to 11, which correspond to a gain of about 4, when 6 images are used. The equivalent number of looks obtained by the 3D-ANF is much larger thanks to spatial neighborhoods. The difference between these two kinds of filter is increased by the fact that only pixels from homogeneous areas (where the 3D–ANF can compute large filtering regions) are taken into account.

Table 1. Bias and noise reduction assessment

	TCMF	**3D-ANF**	**TSF**
B	0.984	0.999	1.039
L_{eq}	10.95	56.91	11.01

3.2 Spatial resolution preservation

Spatial resolution preservation is evaluated through the performances of a classical edge detector: the ratio of local means [4]. Ratio histograms are measured within two regions defined by the masks presented in Figure 2: on the sides of a river visible in Figure 1 to obtain the detection pdf, and on forest homogeneous regions to obtain the false alarm pdf. Filtering effects are assessed by means of operational curves (detection probability versus false alarm probability when the threshold applied on the ratio output ranges in [0, 1]).

Results obtained on the original set of 6 images and after applying the 3 different filters are presented in Figure 2. After filtering, the detection pdf slightly moves to ratio values closer to 1 because of spatial resolution degradation, whereas the false alarm pdf concentrates near 1 thanks to the noise reduction. Operational curves show that the 3 filters increase the detection performances for any false alarm rate. The stronger filtering of the 3D–ANF appears with the positive effect on the false alarm pdf and the negative effect on the detection pdf. But the benefit from the speckle reduction seems to be more important than the loss of spatial resolution in terms of operational performances.

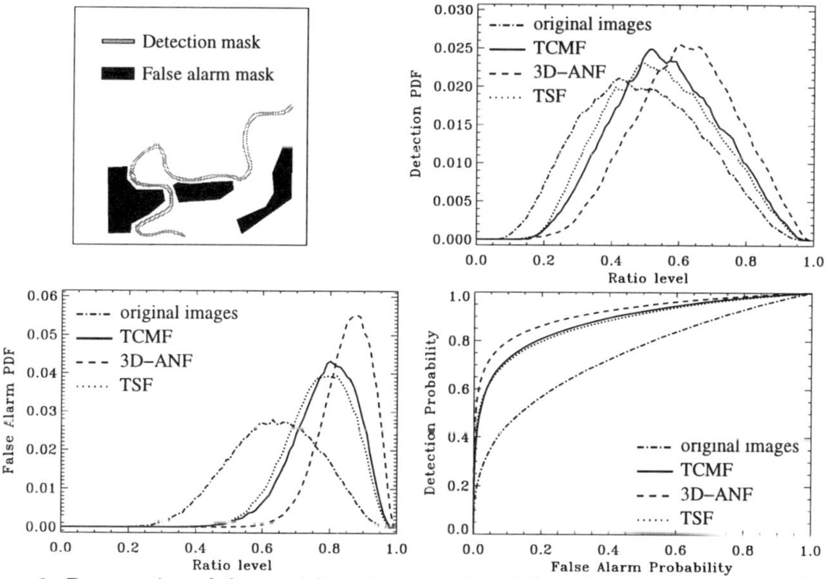

Figure 2. Preservation of the spatial resolution evaluated by using the ratio edge detector and masks to measure detection and false alarm pdf's and to derive operational curves;

4 Evaluation by visual interpretation

The previous statistical criteria are derived from signal processing theory and provide objective measures. But they do not always reflect the subjective point of view of photo-interpretors who often "like" or "dislike" some filtering effects. In order to transform these opinions into quantitative measures, a performance assessment strategy based on image visual interpretation and Geographic Information Systems (GIS) is presented in this section.

4.1 Study area and references

The studied area, located in French Guyana near the town of Mana ($53°46W$, $5°40N$), is characterized by the presence of river, rice cultures, roads and swamp savanna. During data acquisition period, the area is supposed to have no significant change except seasonal evolutions. Filtering results are evaluated by comparing visual interpretation with two references: the *map* which is a vector layer set, extracted from a 1: 25000 scale topographic map (Institut Geographique National), and the ***original data visual interpretation (ODVI)*** guided by the map. The choice of geographic features and their representation by geographic primitives have been established by a selection of features jointly visible in the map and in the original SAR images (cf. Figure 3). For each feature, references and criteria are described in Table 2.

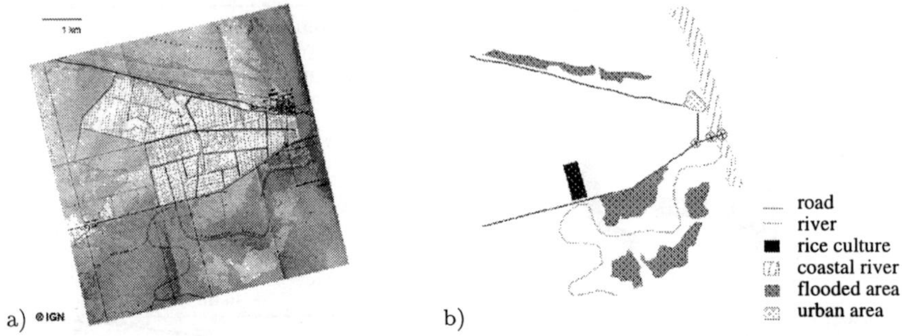

Figure 3. a) Reference map; b) Geographic features to detect by photo-interpretation

Table 2. References and criteria. *ODVI*: visual interpretation of the SAR image guided by the map; *map*: 1: 25000 scale map vector layer.

primitive	criteria	reference(s)
point	visibility and distance to the reference	*ODVI*
line	% of detection and distance to the reference	*ODVI*
surface	% of intersection and % of false alarm	*map* and *ODVI*

4.2 Evaluation method and results

GIS topological tools are used to calculate distances between punctual or linear features and to compare object surfaces. Results obtained for each filter are presented in Table 3. Operator's inaccuracy is supposed to be the same for each interpretation. Difficulties appear to locate some punctual objects (bridge extremity, traffic circle ...) and to distinguish one-pixel-wide linear objects from other targets which are spatially and radiometricaly similar. For surfaces, results obtained from the *map* reference are less discriminant than with the *ODVI* reference since the boundaries of some geographic features change with precipitation, particularly those in flooded savanna.

Photo-interpreters observe that TCMF doesn't preserve temporal information for some buildings which are not identified in one of the images and appear after the filtering process. The 3D–ANF offers a good preservation of each image information. Nevertheless, some punctual objects seem to have disappeared and could not be detected. Images are over-filtered and some areas are extremely smooth. On the other hand, this filter offers a very good radiometric contrast between riparian vegetation and dense forest and could be very useful to detect some geographic entities. The preservation of temporal evolution is very good for the TSF filter. From the operator point of view, textural information, contrast and boundaries are well preserved. The good results obtained on lines (74% detected) and surfaces (74% detected and only 13% of false alarm) make it useful for visual interpretation.

Table 3. Performance assessment by visual interpretation using *map* and *ODVI* as reference; ND: not detected; false-alarm(A/B): complement(A∩B) - B

		TCMF		3D-ANF		TSF	
Feature	**Measure**	map	ODVI	map	ODVI	map	ODVI
point	distance (m)	72	42	96+ND	30+ND	82	30
line	% detection	64	-	65	-	74	-
surface	% intersection	71	81	62	71	64	74
	% false-alarm	28	23	23	16	21	13

5 Conclusions

Two approaches to performance assessment for multi-temporal SAR filters are presented in this paper. The results obtained for three filters are significantly different when evaluation is performed with statistical criteria or by photo-interpreters. The main result is that performance criteria must be chosen according to the applications. Statistical measures are interesting to assess filter efficiency for automatic image processing whereas visual interpretation measures are interesting for manual exploitation of the filtered data.

Acknowledgments

The authors wish to thank GdR ISIS, (Information Signal Image and viSion) and the PNTS (Programme National de Télédétection Spatiale) for supporting this project and the European Space Agency for providing images under project ID539.

References

1. S. Quegan, T. Le Toan, J.J. Yu, F. Ribbes, and N. Floury, Multitemporal ERS SAR analysis applied to forest mapping. *IEEE Trans. Geosci. and Remote Sensing*, 38(2):741–753, 2000.
2. G. Chust, D. Ducrot, J. Bruniquel, and J.L. Petrus, Capabilities of ERS sensor for Mediterranean vegetation detection using multitemporal data. In L. Guerriero F. Posa, editor, *Proc. EOS/SPIE on SAR Image Analysis, Modeling and Techniques III*, vol. 4173, pp. 291–302, Barcelona, Spain, 2000.
3. J.S. Lee, I. Jurkevich, P. Dewaele, P. Wambacq, and A. Oosterlinck. Speckle filtering of synthetic aperture radar images: A review. *Remote Sensing Reviews*, 8:313–340, 1994.
4. R. Touzi, A. Lopes, and P. Bousquet. A statistical and geometrical edge detector for SAR images. *IEEE Trans. Geosci. and Remote Sensing*, 15(3):764–773, 1988.
5. J. Bruniquel and A. Lopes. Multi-variate optimal speckle reduction in SAR imagery. *Int. J. Remote Sensing*, 18(3):603–627, 1997.
6. M. Ciuc, Ph. Bolon, E. Trouvé, and H. Trébossen. Multitemporal SAR image filtering using 3D adaptive neighborhoods. In S. Serpico, editor, *Proc. EOS/SPIE on Image and Signal Processing for Remote Sensing VI*, vol. 4170:1–12, Barcelona, Spain, 2000.
7. D. Coltuc, E. Trouvé, F. Bujor, N. Classeau, and J.P. Rudant. Time-space filtering of multitemporal SAR images. In *IGARSS 2000*, Honolulu, USA, pp. 2909–2911.
8. D. Coltuc, E. Trouvé and Ph. Bolon. Bias correction and speckle reduction in time–space filtering of multitemporal SAR images. In *IGARSS 2001*, Sydney, Australia.

Monitoring and Management of Natural Resources

USING TEMPORAL CHANGE OF THE LAND COVER SPECTRAL SIGNAL TO IMPROVE BURNT AREA MAPPING

DANIELA STROPPIANA AND JEAN-MARIE GRÉGOIRE

EC-Joint Research Centre, Space Applications Institute-Global Vegetation Monitoring Unit

TP 440, Via Fermi 1, I-21020 Ispra (VA), Italy

Phone: +39 0332 785769, Fax: +39 0332 789073, E-mail: daniela.stroppiana@jrc.it

An algorithm has been developed for burnt area mapping using SPOT-VEGETATION images over Northern Australia. The methodology applied is of a supervised classification type and is based on Classification And Regression Trees (CART). Burnt area maps from VGT data have been produced for the period May 25 - July 03, 1999. The performance of the algorithm was assessed using Landsat TM images. In the area common to the two datasets a total amount of 2774 km^2 was estimated to be burnt from VGT, of which 64% was confirmed burnt according to TM burnt area maps. The two datasets showed also a good spatial agreement of the burnt areas.

1 INTRODUCTION

Vegetation fires play an important role in the emission of aerosols and gaseous compounds contributing to global warming and in various land use practices that can result in land cover changes. Biomass burning can be considered a global phenomenon affecting boreal, tropical and temperate ecosystems. The estimation of the area burnt is a prerequisite for assessing the effects of vegetation fires on the tropospheric chemistry and on the ecosystems. Several algorithms have been developed for mapping burnt areas at regional and continental levels with low spatial resolution images acquired by sensors such as the Advanced Very High Resolution Radiometer (AVHRR) (e.g. [1]) and the Along Track Scanning Radiometer (ATSR) (e.g. [2]) on board orbital satellites.

The images acquired by the VEGETATION (VGT) sensor on board the SPOT 4 platform were used for mapping burnt areas in Northern Australia for the period May 15 to July 15 1999 by [3]. The algorithm was based on an image-wide threshold approach and the results showed that VGT is a promising sensor for burnt area mapping purposes at regional level. In this work we present an improvement to the algorithm described in [3] obtained by introducing a temporal change detection component in the classification. The main objective was to overcome misclassification between burnt areas and other targets with a similar spectral signature such as mixed water/land pixels, sparse and dry vegetation.

1.1 The study site

The study site is located in the Northern part of the Australian continent (125°-146° E; 10°-17.7° S) where tropical savannas (discontinuous trees over grass; [4]) is the predominant ecosystem. The density of the vegetation cover decreases southwards as described by the vegetation map in figure 1 [5].

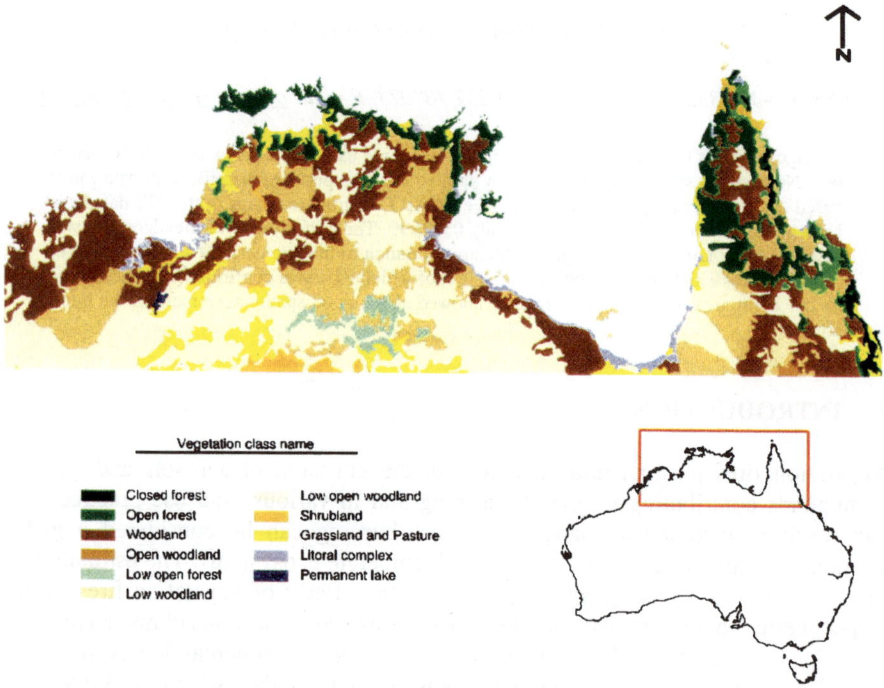

Figure 1. The vegetation cover in the study area.

This region is characterized by two marked seasons. The wet season covers the period between November and March and witnesses 90% of the annual rainfall leading to an abundant production of fine fuel (grass). The vegetation starts curing with the onset of the dry season (April) and fires, mainly lit by land managers and pastoralists, occur generally between April and October. The intensity of the fires increases through the dry season as the vegetation dries out and winds get stronger.

2 The remotely sensed datasets

2.1 SPOT VEGETATION

The VGT time series is composed of daily S1 images for the period May 15 to July 15, 1999. Daily VGT images are acquired in four spectral bands (table 1), with a 1 km spatial resolution (nadir), for the descending node, at 10.30 am local solar time.

Table 1. The spectral domain covered by the VGT bands.

VGT band	Spectral domain	Wavelength
b0	BLUE	0.43 - 0.47 μm
b2	RED	0.61 - 0.68 μm
b3	NIR	0.78 - 0.89 μm
MIR	SWIR	1.58 - 1.75 μm

The S1 product, provided by CNES (Centre National d'Etudes Spatiales), is corrected for misregistration of the different channels, the detectors are calibrated for each spectral band and the product is resampled to a "Plate-Carrée" projection (Latitude-Longitude). An additional atmospheric correction is performed using the Simplified Method for Atmospheric Correction (SMAC) [6]. Hence, the pixel value is a measure of the ground reflectance.

Ten-day composites were derived from daily VGT images by applying a minimum value technique to the NIR band. The ten-day periods covered by the composite images derived with the available dataset are summarized in table 2.

Table 2. The compositing periods.

Start date (1999)	End date (1999)
May 15	May 24
May 25	June 03
June 04	June 13
June 14	June 23
June 24	July 03
July 04	July 13

2.2 Landsat TM

Two Landsat Thematic Mapper (TM) scenes (105/069 and 105/070), acquired on June 9 1999, were used for the accuracy assessment. The TM images were geolocated with Ground Control Points taken from the best available digital topographic maps (scale 1:250,000) with 1.8 TM pixels accuracy. A burnt area map was derived by visual inspection and assumed to be the ground truth.

3 Methodology

3.1 The tree classifier

The algorithm has been developed based on the Classification And Regression Trees (CART) theory [7] that has already been applied successfully for land cover and burnt area mapping using satellite images (e.g. [8], [9]). CART is a supervised classification in which a training set extracted from the satellite images is used for deriving a decision tree. The rules extracted from the decision tree are composed of a set of thresholds applied to variables such as the spectral bands. In this work the main objective is the classification of VGT composite images into burnt/unburnt surfaces and so the variables were selected amongst those that best discriminate between these two targets. We used the VGT spectral bands, the ALBEDO index and their temporal difference. The temporal difference is computed between the VGT composite image to be classified (identified as at time 2) and either the previous one (time 1) or the following one (time 3). The product is a classification of a VGT composite image mapping areas that burnt since the previous one. The variables used in the decision tree are the following ones:

- ALBEDO (time 2) = (RED + NIR) / 2
- NIR (time 2)
- SWIR (time 2)
- D1NIR = NIR (time 2) – NIR (time1)
- D2NIR = NIR (time 3) – NIR (time 2)
- D2SWIR = SWIR (time 2) – SWIR (time 1)

The classification rules are summarised in table 3. A pixel is assigned to the "burnt class" if it satisfies the conditions of any the classification rules outlined in table 3.

Table 3. The classification rules in a temporal change detection approach.

Rule	NIR	SWIR	ALBEDO	D1NIR	D2NIR	D2SWIR
a	≤ 0.167	> 0.072		≤ -0.041		
b			≤ 0.142	> -0.041		≤ -0.341
c	≤ 0.147			> -0.041 and ≤ -0.034		> -0.341
d	≤ 0.255			> -0.034	≤ -0.324	> -0.341
e	≤ 0.158	> 0.134		> -0.034 and ≤ -0.024	> -0.324	> -0.341

3.2 *The masks*

A set of thresholds (table 4) has been applied for the individuation of those targets for which the classification rules are not tested. Besides those "bad" pixels for which the rules are not applicable, the objective is the reduction of either the likelihood of misclassification (water bodies) or the computation time (clouds).

Table 4. The conditions applied for creating the masks.

Target	Flag	Conditions (ground reflectance)
No data	1	BLUE = 0 or RED = 0 or NIR = 0 or SWIR = 0
Negative value	2	BLUE (or RED or NIR or SWIR) = 0.0005
High SWIR	3	SWIR > 0.500
Water	4	SWIR ≤ 0.025
Cloud-free	5	BLUE ≤ 0.075 and SWIR ≤ 0.090

A pixel identified as "no data" is characterised by a value equal to zero in any VGT of spectral band. The "negative value" flag is associated to those pixels with an extremely low reflectance (ρ=0.0005, Digital Number=1) due to an over correction of the atmospheric effects. The image processing performed at VITO (Mol, Belgium) implements an atmospheric correction of the raw data for producing deliverable product (such as the dataset used in this work) and if the correction leads to a negative reflectance the pixel value is set to a digital number equal to 1 (DN=ρ*2000). Also pixels with high SWIR reflectance due to deficient detectors are masked out because the classification rules are not applicable. Finally the water and cloud masks are applied.

4 Results

4.1 *The burnt area maps*

The classification rules presented in table 3 were applied to the composite images in order to derive the burnt area maps. Since the algorithm is based on temporal change detection between composites, burnt area mapping was not feasible for the first and last periods presented in table 2. The burnt area maps derived for the remaining compositing periods were summed in a synthesis map (Fig. 2) representing the areas classified as burnt between May 25 and July 03, 1999.

Figure 2. The synthesis burnt area map for the period May 25-July 03, 1999.

4.2 Accuracy assessment

The performance of the algorithm was assessed on the burnt area map derived from one VGT composite image (May 31 – June 09, 1999) using burnt area polygons derived from Landsat TM images (June 09, 1999). The VGT composite was derived for the above period in order to maintain the contemporaneity between the date of acquisition of the TM scenes and the last day of the VGT composite. The resulting burnt area map depicts those areas that burnt since the previous composite (May 21 – May 30) hence this validation exercise allowed only the assessment of the commission error (areas classified as burnt from VGT but not actually burnt according to TM). In the area covered by both the datasets the total amount of area burnt as derived from the VGT map is 2774 km^2. The comparison with the TM showed that 64% was correctly estimated.

Note that the amount of area burnt derived from TM and classified as unburnt in the VGT map is not significant because the algorithm maps the areas burnt between two dates.

The burnt area polygons were also overlaid to the VGT maps for a visual evaluation of the spatial agreement between the two datasets. Figure 3 presents two zoom windows over the study site showing the results.

Figure 3. Example of a burnt area map derived from VGT and the burnt polygons extracted from TM images for two zoom windows over the study area (93 km x 93 km).

5 Discussion and conclusions

This research presents an algorithm developed for mapping burnt areas using SPOT-VGT images in a study area located in Northern Australia. We implemented a temporal change detection approach in a Classification Tree methodology for the classification of ten-day VGT composite images. The algorithm has been applied to ten-day composite images for mapping areas burnt during the period May 25-July 03, 1999. The performance of the algorithm was assessed on one burnt area map (compositing period May 31 – June 09, 1999) using two Landsat TM scenes. The agreement between burnt areas derived from VGT and TM maps is 64%. A more exhaustive accuracy assessment will be performed using TM scenes acquired over the same area for different dates. Indeed this approach will allow us the assessment of both the omission and commission errors.

The visual evaluation of the burnt area maps showed a very good spatial agreement between the two datasets. The results confirmed that images acquired by the VGT sensor can be used for burnt area mapping over large areas. Although a more extensive validation exercise is required, this algorithm, developed by introducing a temporal change detection condition, overcomes the misclassification errors produced by an image-wide threshold approach.

6 Acknowledgement

The pre-processing and visual analysis of Landsat TM images was done by Marta Maggi. D.S. was supported by a doctoral grant from the European Commission's TMR programme.

References

1. P.M. Barbosa, J.-M. Grégoire, and J.M.C. Pereira, "An algorithm for Extracting Burned Areas from Time Series of AVHRR GAC Data Applied at a Continental Scale," *Remote Sensing of Environment*, 69, 253-263, 1999.
2. H.D. Eva, and E.F. Lambin, "Burnt area mapping in central Africa using ATSR data," *International Journal of Remote Sensing*, 19, 3473-3497, 1998.
3. D. Stroppiana, M. Maggi, J-M. Pereira, D. Graetz, J-M. Grégoire, J. Silva, A. Sà, P. Henry, V. Gond, and E. Bartholomé. Detection and mapping of burnt areas and active fires in tropical woodland ecosystems with the VEGETATION sensor: the SMOKO-FRACTAL case study over Northern Australia, proceedings "VEGETATION 2000, 2 years of operation to prepare the future", Belgirate, Italy, April 3rd – 6th, 2000.
4. Gillison, A.N. (1994), Woodlands. In: *Australian Vegetation* (2nd edn) (ed. R.H. Groves), Cambridge University Press, Cambridge, pp.227-255.
5. Atlas of Australian Resources, Australian Surveying and Land Information Group, Department of Administration Services, Canberra 1990.
6. H. Rahman, and G. Dedieu, G.SMAC: a simplified method for the atmospheric correction of satellite measurements in the solar spectrum. *Int. J. Remote Sensing*, Vol. 15, No.1, 123-143, 1994.
7. L. Breiman, J.H. Friedman, R.A. Olshen, and C.J. Stone, *Classification and Regression Trees*, Belmont, CA: Wadsworth Int., 1984.
8. M.C. Hansen, R.S. Defries, J.R.G. Townshend, and R. Sohlberg, "Global land cover classification at 1 km spatial resolution using a classification tree approach," *International Journal of Remote Sensing*, 21 (6,7), 1331-1364, 2000.
9. J.M.C. Pereira, M.J.P. Vasconcelos, and A.M. Sousa, "A ruled-based system for burned area mapping in temperate and tropical regions using NOAA/AVHRR imagery". In: Biomass burning and its inter-relationships with the climate system, edited by J.L. Innes, M.Beniston, and M.M. Verstraete (The Netherlands: Kluwer Academic Publishers), 2000.

LAND COVER AND SOIL LOSS MULTITEMPORAL ANALYSIS: AN APPLICATION OF GEOINDICATORS IN THE PANTANAL WETLANDS (BRAZIL)

DISPERATI L.(1), SALVINI R.(1), CIALI A.(1),
FANTOZZI P.L.(1), CARMIGNANI L.(1),
FIORI A.P. (2), BOCCI M. (3) & PARANHOS FILHO A.C. (2)

(1)Dipartimento di Scienze della Terra, Università di Siena, Via Laterina 8, Siena, Italia; tel. +39 0577233836; fax: + 39 0577233880; e-mail: pantanal@unisi.it

(2)Universidade Federal do Paraná, Departamento de Geologia, Jardim das Americas, Curitiba, PR, Brazil

(3)Via Sansedoni, 7, Siena, Italy

1 Introduction and problem definition

In the last decades economic growth of developing countries involved population growth and need of new productive spaces: new urban areas, industries and agricultural lands were set up, often causing environmental troubles. Increase of "human pressure" without any evaluation concerning sustainable development makes the environment weak, reducing its productivity in the medium and long term. In South America, Brazil were strongly affected by such problems in the last three decades. Especially in the states located inside the Amazonas and Upper Paraguay rivers basins, natural areas were transformed into agricultural lands through deforestation. Forest and savanna were progressively cleared using both mechanical devices and fire, the latter being widely utilised because efficient and cheap. Forest clearing is one of the cause of desertification of tropical areas, global warming and the main cause of accelerated soil erosion. Especially in the highlands of the Mato Grosso do Sul State (Fig.1), after forests clearing accelerated erosion developed and it is now threatening some urban areas and prevents the exploitation of wide lands. As a consequence, in the neighbouring Pantanal wetlands, one of the largest continental alluvial fan of the world, the solid discharge increased, causing accelerated deposition in the channel beds, widening of floodable lands and reduction of productive territories. These facts suggest that geoindicators like land cover and soil erosion changes (Berger & Iam, 1996), represent key parameters for sustainable development evaluations and land management.

In this work a multitemporal (years 1966, 1985, 1966) land cover database was created for a wide region of the Pantanal and the neighboring *planaltos* (highlands) (Fig.1) by utilising topographic data integrated with remotely sensed information. In a pilot area of the *planaltos* affected by appreciable land cover changes, GIS analyses and the USLE (Wischmeier & Smith, 1978) allowed to estimate the effects of changes on soil loss.

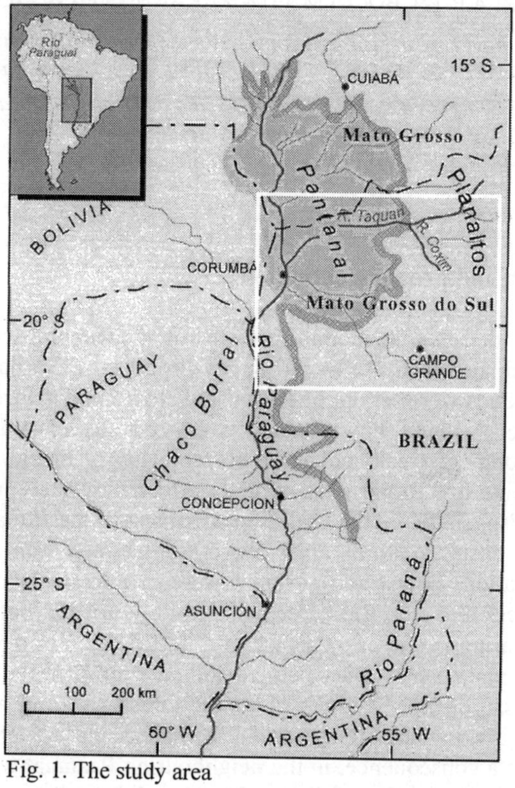

Fig. 1. The study area

2 Proposed Method

Aim of this work was to build multitemporal land cover databases by using topographic maps, aerial photos, geographic databases from literature (PCBAP, 1997) and satellite images. The authors carried out fieldwork in the study area since 1997 in order to support processing and analysis of existing data and remotely sensed images. Eleven topographic maps sheets at a scale of 1/250,000 and twenty-two sheets at a scale of 1/100,000, based on aerial photos of the year 1966, were rasterised, geo-referenced and vectorised for all their contents in order to create a

GIS. Then, elevation and hydrography themes were utilised as input in order to produce the Digital Elevation Model (DEM). The Topogrid function of the ESRI© ARC/INFO software was used. This function is based on an interpolation algorithm designed for the creation of hydrologically correct DEMs (connected drainage structure and correct representation of ridges and streams). Also the Triangulated Irregular Network (TIN) algorithm was utilised in those areas characterised by very sparse input data due to flat morphology (the Pantanal wetlands); in these cases linear interpolation provided the best approximation of the elevation data. The TIN was afterward converted to the grid format. The grid is a raster format implying the choice of the cell size. Accuracy of the DEM is strictly dependent on both scale and density of the input data, as well as the cell size. For these reason a study was performed in order to choose the cell size representing the trade-off between data redundancy and output accuracy.

The aerial photographs, after rasterisation, were oriented and interpreted utilising digital photogrammetric workstations. Their implementation guaranteed a better land cover monitoring of some *planaltos* area at the fly time.

Landsat TM satellite images related to the dry season (July-October) of the years 1985 and 1996 were acquired, pre-processed (radiometric and geometric correction) and processed. Supervised classification (maximum likelihood algorithm) integrated with segmentation techniques were utilised. Soils data were used as reference for segments identification. The raster data, after classification, were converted to vector format to in order to create preliminary land cover data bases of every date. Photo-interpretation check of the preliminary database allowed to assign the correct land cover class to those polygons representing areas which could not be identified through their spectral features. The CORINE Land Cover project legend (Heymann et al., 1994) was adopted. Some changes were made to the legend in order to take account of some peculiar characters of the study area. The numerical and hierarchical structure of the legend resulted very useful for data management, querying and analysis. Then, the detection of changes was performed by "post-classification comparison"; topologic intersection of the land cover GIS allowed to create databases describing type, location and extent of changes among the different epochs. Through integrated analysis of the DEM, the ground slope grid (calculated from the DEM), the database of geology (Radambrasil, 1982; PCBAP, 1997) and visual interpretation of Landsat TM images, the study area was split in physiographic provinces (Fig. 2), i.e. areas each other characterised by different geologic, geomorphologic, hydrologic and soils features (West, 1995).

Statistical evaluation of land cover changes and their visual representation were performed by means of charts (Fig. 3) and maps for the whole study area and at the scale of physiographic provinces.

Analysis of potential soil loss was carried out through the USLE (Wischmeier & Smith, 1978) in a watershed falling in the *planaltos* area (Upper Coxim River watershed, Fig. 2). The watershed experienced notable land cover changes after the '60s and it is actually occupied by extensive agriculture. Both the soil map and the

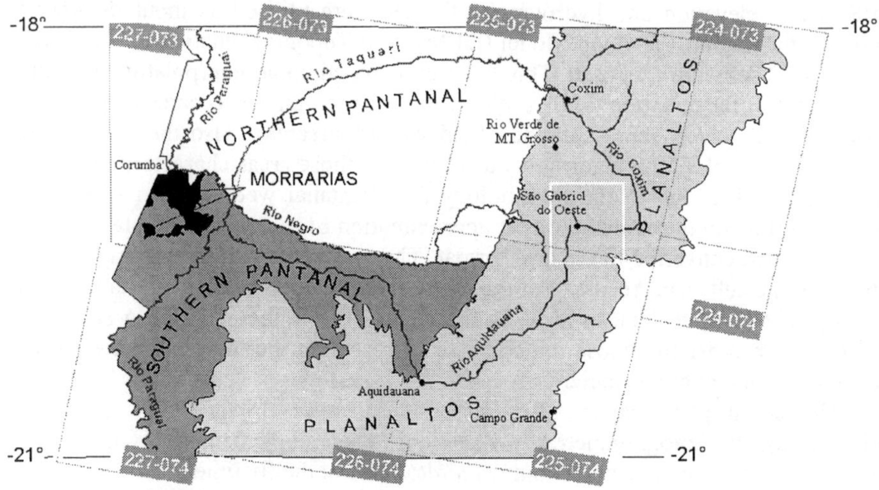

Fig. 2. Physiographic provinces of the study area. Grey squares and labels indicate the utilised Landsat TM scenes and the path-row identification numbers respectively. The white box indicates the area where the USLE was applied (Upper Coxim River Watershed).

rainfall erosivity factor (R) were assumed from PCBAP (1997); texture and organic matter determinations on soil samples collected in the watershed during two fieldwork missions (1998, 1999) allowed to calculate the soil erodibility factor (K) applying the nomograph from Wischmeier et al. (1971); the slope lenght and steepness factor (LS) was calculated from the DEM, the cropping management and erosion control practice factor (CP) was taken from Stein et al. (1987) and it was applied to the land cover database built for the years 1966, 1985 and 1996.

3 Results

Fig. 3 shows the extent of the land cover classes of the whole study area following the CORINE land cover legend. Data related to years 1966, 1985 and 1996 are represented. From 1966 to 1985, due to forest clearing, the extent of natural areas (class 3) reduced from 90,000 km^2 to 70,000 km^2. If areas changed from any land cover class to class 3 (re-naturalisation) in the same time span are neglected, the total extent of deforestation can be estimated, and it resulted to be 32,000 km^2. As a term of reference, this area is slightly bigger than the area of Belgium. Hence this result indicates that forest clearing was, at least till 1985, a major factor inducing environmental changes. From 1985 to 1996 the extent of natural areas passed from 70,000 km^2 to 61,000 km^2, while the total extent of deforestation resulted to be 15,000 km^2. Fig. 3 also highlights that in both the

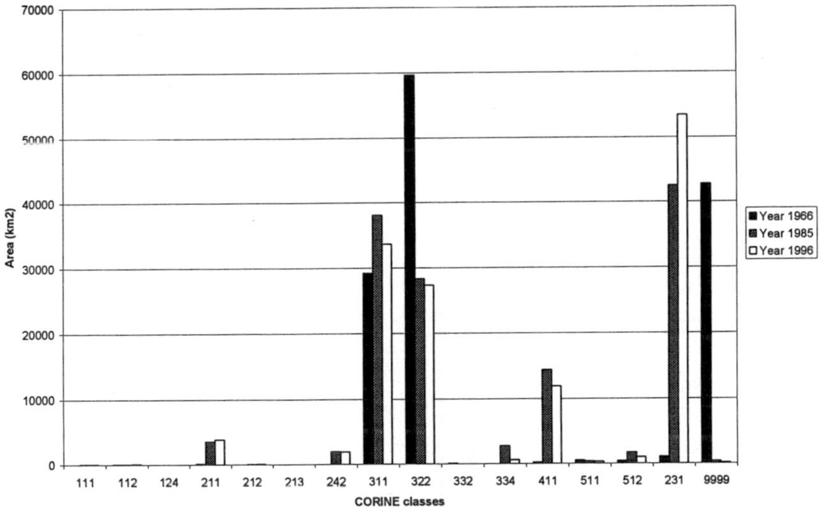

Fig. 3. Extent of the CORINE land cover classes for the years 1966, 1985 and 1996, for the whole study area. Classes 3xx refer to "natural and semi-natural areas", classes 2xx and 2xxx to "agricultural areas". Fourth level classes 2311, 2312 and 2313 were introduced in this work: 2311- grass savanna (*Campo limpo*, *Campo sujo*, *Caronal*, etc.): grass layer occasionally interrupted by rickety shrubs and small stemless palm trees;□ 2312 - grass-shrub savanna (*Savana parque*): soil predominantly covered by grass with rickety trees and shrubs; occasionally tree tufts occur; in the Pantanal this ecological type occurs mainly on periodically flooded plains; 2313 - burnt areas over plant formations of the classes 2311 and 2312. Class 9999 refers to unclassified areas.

considered time periods, no significant growth of urban areas occurred, so most of natural forest changed to pastures and agricultural lands. The same analyses performed at the scale of physiographic provinces allowed to identify regions where higher increase of human pressure occurred. As an example, multitemporal land cover data of both the Planaltos and Southern Pantanal physiographic provinces (Fig. 2) in Fig. 4 are compared.

The charts show that the Planaltos province experienced the widest changes: about 40% (23,000 km^2) of natural and semi-natural areas were cleared from 1966 to 1985. On the contrary, in the Southern Pantanal province minor land cover changes occurred (about 13% of class 3 was cleared): indeed this province is flooded many months per year, so resulting poorly suitable for human colonisation. After 1985, in the Planaltos province about 30% of natural areas changed to agricultural lands and pastures, suggesting that human pressure did not decrease.

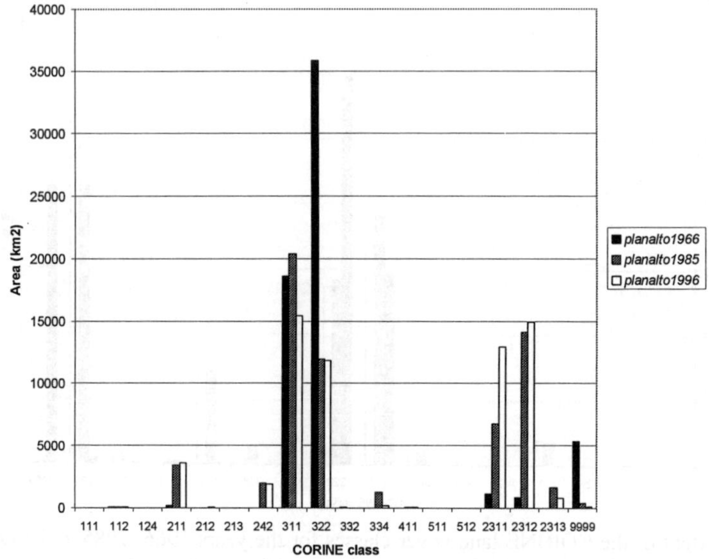

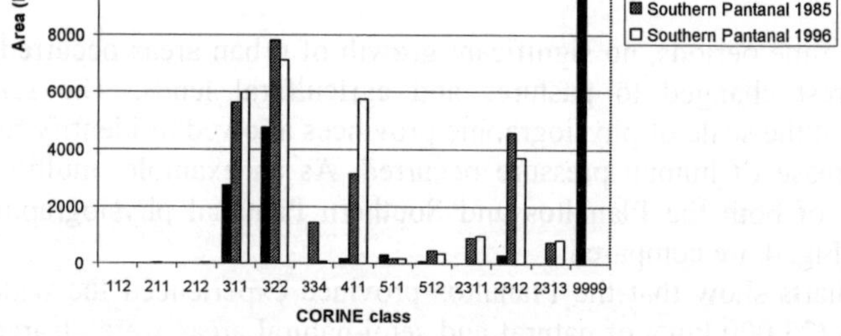

Fig. 4. Extent of the CORINE land cover classes for the years 1966, 1985 and 1996, for the Planalto (at the top) and the Southern Pantanal (at the bottom).

The results from the USLE show that in the Upper Coxim River watershed the detected land cover changes induced a meaningful growth of the estimated soil loss from 1966 to 1985 (Fig. 5a). The increase of potential soil loss (A) was larger: a) in the deforested steep areas of the eastern side of the basin; b) at the bottom of the slopes, close to watercourses, in the *planaltos* occupied by arable lands. The multitemporal map of soil loss change related to the period from 1985 to 1996 (Fig. 5b) highlights large areas where soil loss decreased. The average change of A (ΔA) for the whole watershed, in this time span, is about zero ($\Delta A_{1996\text{-}1985} \approx 0.1$ t·ha^{-1}·y^{-1}). In contrast, ΔA related to 1966-1985 is very big ($\Delta A_{1985\text{-}1966} \approx 15$ t·ha^{-1}·y^{-1}). These results indicate that human activities caused a strong increase of soil loss up to about 1985, while more recently human impact on soil erosion became steady in this area.

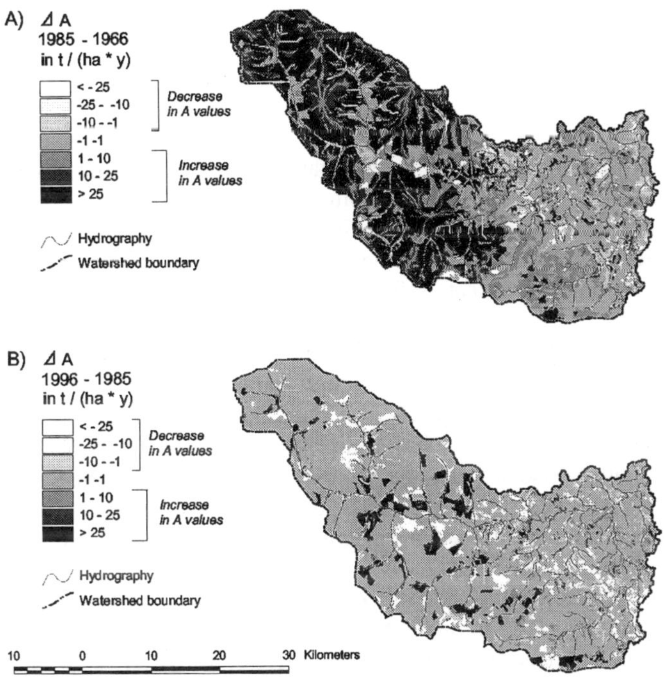

Fig. 5. Multitemporal analysis of potential soil loss (A) for the Upper Coxim River watershed: a) time span 1966-1985 ($\Delta A_{1985\text{-}1966}$); b) time span 1985-1996 ($\Delta A_{1996\text{-}1985}$).

4 Discussion and conclusions

The study area underwent wide land cover transformations from 1966 to 1985: wide extent of natural and semi-natural lands (forest and *cerrado*), underwent clearing to be replaced by agricultural and zoo-technical areas. From 1966 to 1985 about 32,000 km^2 of natural areas were cleared in the whole study area, and 23,000

km^2 were deforested in the Planaltos physiographic province alone. This province experienced the widest land cover changes of the study area. After 1985, deforestation continued even though Landsat images and fieldwork allowed to recognize, in the *planaltos*, natural vegetation within several previously deforested areas.

The USLE showed that in the Upper Coxim River watershed the detected land cover changes, related to the period 1966-1985, induced a meaningful growth of the estimated soil loss. These data are in agreement with: a) large gully erosion phenomena (*voçorocas*) observed during the fieldwork; b) increase of solid discharge detected in the Taquari and Coxim rivers from 1977 to 1997 (Padovani et al., 1998).

5 References

1. Berger A.R. & Iam W.J. (eds): Geoindicators - Assessing rapid environmental changes in Earth systems, Balkema Rotterdam, 480 pp, 1996.
2. Y. Heymann, C. Steenmans, G. Croisille, and M. Bossard, “CORINE land cover project - Technical guide”, pp. 136, European Commission Directorate General Environment, uclear Safety and Civil Protection ECSC-EEC-EAEC, Brussels, 1994.
3. C.R. Padovani, N.O. Carvalho, S. Galdino, and L.M. Vieira, “Produção de sedimentos da Alta Bacia do Rio Taquari para o Pantanal”, III Encontro de Engenharia de Sedimentos, pp. 18-24, Comissão de Engenharia de Sedimentos Regional de Minas Gerais – CNPq, Belo Horizonte, 1998.
4. PCBAP – Plano de conservacão da bacia do Alto Paraguai. Ministerio do Meio Ambiente dos Recursos Hidricos e da Amazonia Legal. Secretaria da Coordenacão dos Assuntos do meio ambiente. Programma Nacional do Meio Ambiente. Brasilia, 1997.
5. D.P. Stein, P.L. Donzelli, F.A. Gimenez, E.L. Ponçano and F. Lombardi Neto, “Potencial de Erosão Laminar, Natural e Antrópica na Bacia do Peixe-Paranapanema”. Anais do Simpósio Nac. De Controle de Erosão, 4, pp. 105-135, ABGE/DAEE, Marília São Paulo, 1987.
6. West T.R.– Physiographic Provinces and Engineering considerations. In: Geology applied to Engineering, Chapter 13, 261-280, 1995.
7. W.H. Wischmeier, C.B. Johnson, and B.V. Cross, “A soil erodibility nomograph for farmland and construction sites”, Journal of Soil and Water Conservation, 26/5, 189-193, 1971.
8. W.H. Wischmeier, and D.D. Smith, “Predicting rainfall erosion losses - a guide to conservation planning”, Agr. Handbook 537, pp. 58, U.S. Dept. Agr., Washington, 1978.

NATURAL RESOURCE IN SOUTHERN AFRICAN DRYLANDS: DETERMINING SPATIAL AVAILABILITY AND VARIABILITY USING ATSR2 TIME SERIES.

FRANCE GERARD
Centre for Ecology and Hydrology Monks Wood, Abbots Ripton, PE28 2LS, UK
Email: ffg@ceh.ac.uk

STEPHEN PLUMMER
ESA-ESRIN, Via Galileo Galilei, Casella Postale 64, 00044 Frascati, Italy
Email: Stephen.Plummer@ESA.int

PETER NORTH
Department of geography, University of Wales Swansea, Singleton Park Swansea SA2 8PP, Wales, UK
Email: p.r.j.north@swansea.ac.uk

ANDRE VAN ROOYEN
Desert Margins Programme of South Africa, ARC, PO Box 2983, Kimberley 8300, South Africa
Email: andrevr@imed.co.za

Effective management of natural resources (NR) is important in the drylands as local populations rely heavily on these resources. To evaluate the range of programmes that are in operation in a successful manner, the natural spatial and temporal variability of vegetation response to climate has to be separated from the effects of the different land management strategies on degradation and NR availability. This work explores the use of multi-temporal remotely sensed data to achieve the separation between climatic and management effects.
A NDVI time series of nine test sites, representing different states of degradation, were produced from Along-Track Scanning Radiometer images for a four year period. The cumulative NDVI, which is strongly related to NPP, was calculated over 12 months periods and the whole four year period (48 months). The cumulative NDVI of the 12 months periods are strongly affected by local climate variations and consequently show no apparent relation with the level of rangeland degradation. The cumulative NDVI of the 48 months period at the other hand shows some evidence of an increase in value with a decrease in rangeland degradation.

1 Introduction

In the drylands, local populations are most vulnerable to changes in the availability of natural resources (NR). The UN Convention to Combat Desertification views NR management as important in both effective dryland environmental management and poverty alleviation [1]. A variety of community based (bottom-up) NR management programmes and national action (top-down) programmes have been developed to combat desertification (land degradation). However, not all are necessarily successful [2].

Land degradation refers to the reduction or loss of available biomass, biological productivity and complexity caused by the over-utilization of the vegetation [1]. It is difficult to assess objectively and different actors may operate at different spatial scales. A variable and local climate is characteristic of the south-African drylands and leads to a naturally spatially and temporally variable plant phenology. To evaluate the range of programmes that are in operation in a successful manner, the natural spatial and temporal variability of vegetation response to climate has to be separated from the effects of the different land management strategies on degradation and NR availability.

The methodological problem is to find a compromise between the need to capture the phenological behaviour of bushveld, in response to the spatially and temporally variable rainfall, and the need to identify indicators of degradation which may be identifiable only at certain temporal scales. The difficulty with the latter is the establishment of links between indicators of degradation and observations of vegetation state at a given point in time and scale.

Field measurements of vegetation state are generally of high quality but they are limited in scope and scale. The use of satellite imagery can provide the spatial and temporal scales required. At the regional scale it is generally accepted that monitoring of vegetation behaviour can only be accomplished through repetitive measurements of biophysical attributes, their seasonality and spatial variability [3].

Extensive research has shown that the Normalised Difference Vegetation Index (NDVI) used in optical remote sensing has a strong relationship with a number biophysical attributes, such as the amount of absorbed photosynthetically active radiation, leaf area index and fractional vegetation cover [4]. Other research has shown that the integration of NDVI time series (i.e. ΣNDVI) is directly related to the net primary production (NPP) of the vegetation [5][6]. A number of studies have attempted to use NDVI time series to characterise the spatial variability and phenology of vegetation in semi-arid environments [7][8].

The objective of this work is to explore the use of remotely sensed data to examine whether the climatic variation and the response of the vegetation can be separated from the effects of the different management strategies. Along-Track Scanning Radiometer (ATSR2) NDVI time series are used to look at the wider spatial and temporal vegetation cover variability in two south African study areas. The study areas represent a transition zone in climate characterised by a south–north gradient in rainfall and different land management strategies comprising domestic stock ranching, communal land grazing, game ranching and wildlife management.

2 Data

2.1 Ground data

Transect data were acquired for each of 82 sites distributed from the Mier region, Republic of South Africa (lat: -26 20 000, long: 20 15 000) to north-east of Ghanzi, Botswana (lat: -21 24 500, long: 22 06 000) representing:

- an arid area in Kgalakgadi District, southwest Botswana and the adjacent Northern Cape Province, RSA (Fig. 1a);
- a semiarid area in Ghanzi and Ngamiland Districts, west Botswana, and adjacent Gobabis, Hereroland and Bushmanland in Namibia (Fig. 1b).

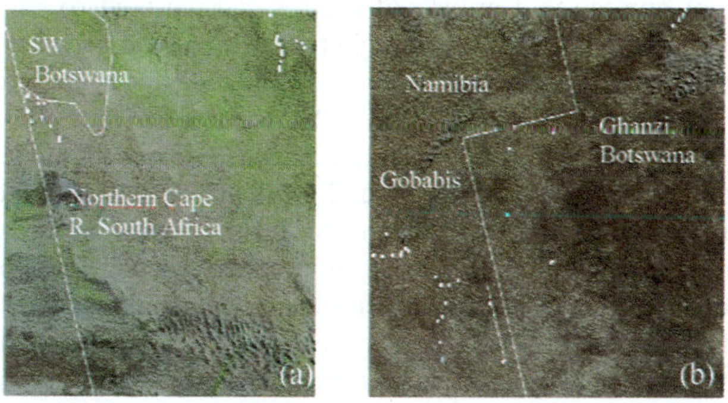

Figure 1. The location of 82 sites distributed from the Mier region, RSA (a) to North-east of Ghanzi, Botswana (b)

These data were processed to estimate the fractional cover of perennial grass, annual grass, litter and bush. Bush type and mean size are also available. The location of the centre of each transect was determined using GPS for subsequent registration with the imagery. Each site was assigned a qualitative measure of degradation (D: degraded, M: moderate, G: Good) based on the lifelong experience of local farmers and scientists [9]. Degraded sites normally show a mixture of a dense black rigozomatous bush (*Rhigozum*) and an annual grass (*Schmidtia kalaharensis*) in the interdunes and no vegetation on the dunes. Good sites have a good mixture of perennial and annual grasses, little black *Rhigozum* bush and in some cases other palatable bush species. Moderate sites show a reduction in perennial grass cover, resulting in a higher proportion of bare ground which is subsequently invaded by *Rhigozum* and *Schmidtia*.

2.2 Remotely sensed data

Generally cloud-free Along-Track Scanning Radiometer (ATSR-2) images were acquired over the period September 1996 to September 2000. Using GPS taken at each site, the corresponding ATSR-2 red (670nm) and near infra-red (870nm) Top of Atmosphere (ToA) reflectance values were extracted for each site and preliminary NDVI values calculated. NDVI outliers were removed by implementing the BISE method developed for AVHRR NDVI time series [10]. These data are available approximately weekly for the 4-year period. Figure 2 shows examples of the NDVI variation for several sites over time.

3 Proposed method

The outcome of rain is rapid growth of grasses causing a sharp rise in the NDVI which continues while the grass is actively green. The period between onset and end of this high NDVI indicates the longevity while the area under the 'growth' curve is associated with NPP [5][6]. The integral of NDVI time series (ΣNDVI) is used in this study to detect differences in production between degraded, moderate and good rangeland sites. The annual phenology and NPP of the sites is dependent on the local rain patterns which vary spatially and with time appearing anywhere between December and April (Fig. 2). However, if over a timespan of several years the spatial occurrence of rain is randomly distributed, the cumulative NPP (~ΣNVDI) is expected to decrease for sites which are more degraded, independent of the annually variable local climate. To test this, nine neighbouring sites were selected (three Degraded: D1, D2 , D3; three Moderate: M1, M2, M3 and three Good sites: G1, G2, G3) and their ΣNDVI calculated for periods of twelve months and 48 months, the total period for which ATSR-2 imagery was available.

4 Results and Discussion

Figure 3 shows the ΣNDVI for the nine selected sites. When the ΣNDVI is calculated over 12 month periods no apparent trend is visible between ΣNDVI and level of degradation (Fig. 3a). For example, for the moderate sites M2 and M3 the ΣNDVI of the first 12 month period is lower than the ΣNDVI of the degraded sites and the ΣNDVI of the fourth 12 month period is higher than the ΣNDVI of the good sites. The good sites show ΣNDVI that are higher and lower than the ΣNDVI of the degraded sites. The third 12 month period show low ΣNDVI for all sites indicating a long period of drought. The ΣNDVI calculated over the 48 month period, shows some evidence of an increase in value with improving rangeland state (Fig.3b), although the period is probably not sufficient to eliminate the effects of local climate, in particular extreme localisation of drought or rain. A high degree of

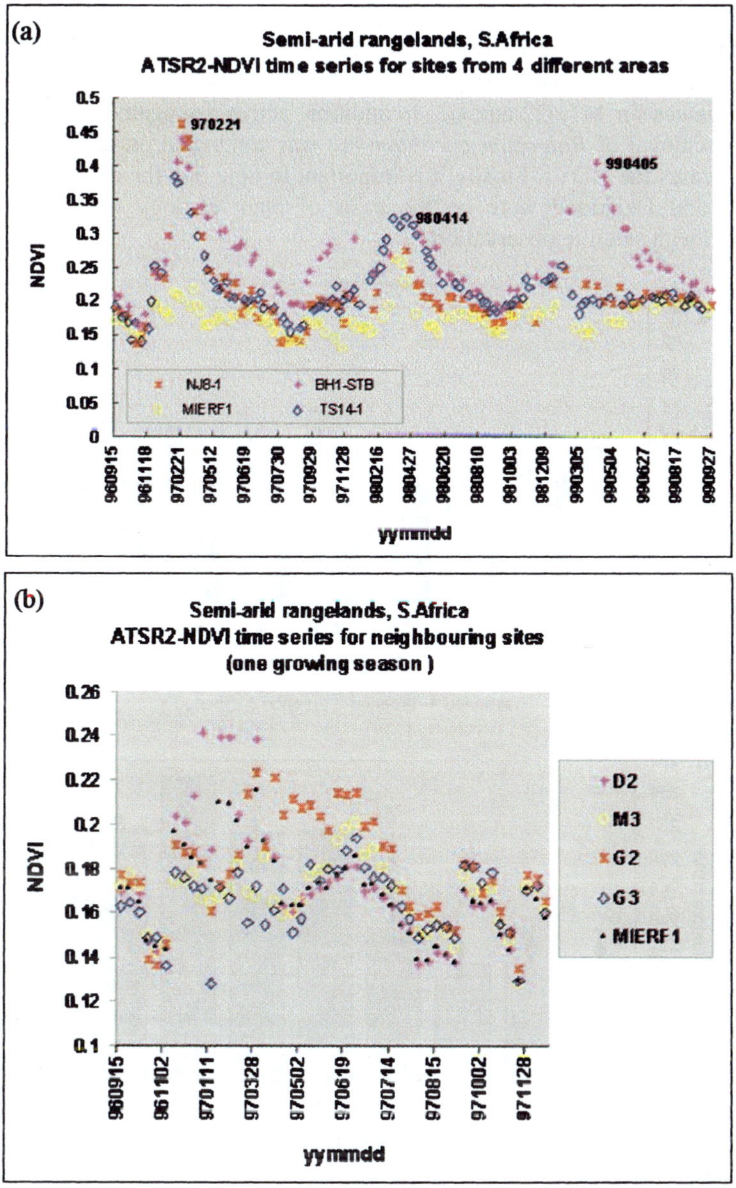

Figure 2. Sites revealing a trend in NDVI difference with climatic gradient which is primarily a function of rainfall (a). Deviation away from this line, particularly for sites in the same latitude-longitude zone, may indicate differences in management practice (b).

variability in the proportion of cover components (bare soil, bush, annual and perennial grass) within each rangeland categories and variable soil colour (red to brown) and spectra could explain the higher ΣNDVI values for D2 and lower ΣNDVI values for M1, G2 and G3. In addition, active management over the period (manual removal of *Rhigozum trichotomum*) was conducted on the degraded sites. This increases the NDVI. Finally it is important to note that the definition of Good, Medium and Degraded were on the basis of plant ecology which may not be correlated with satellite observations.

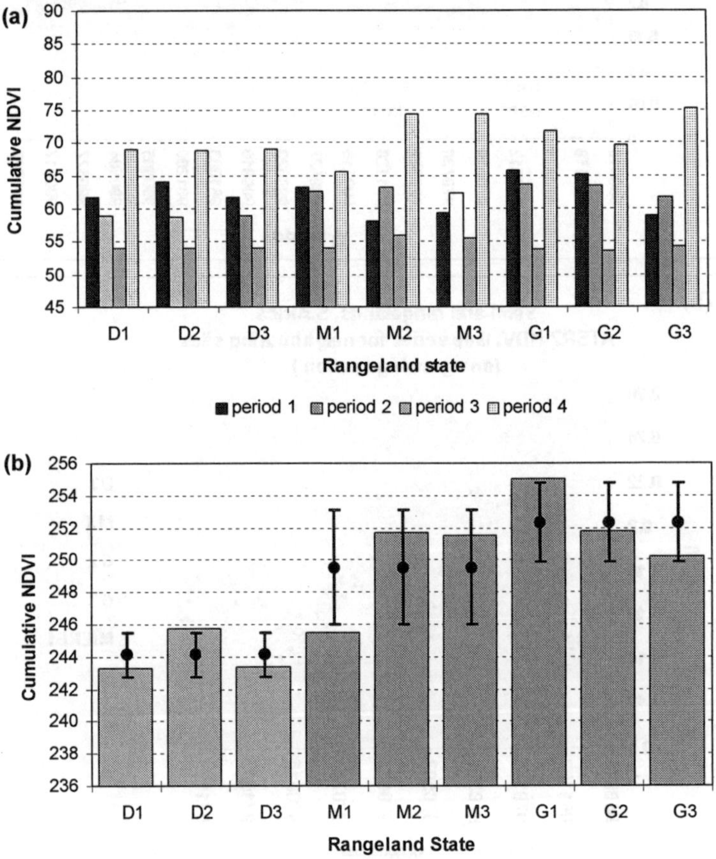

Figure 3. The cumulative NDVI calculated for (a) four 12 months periods: 15 Sept'96 - 15 Sept'97, 15 Sept'97 - 15 Sept' 98, 15 Sept'98 – 15 Sept'99 and 15Sept'99 – 15 Sept'00 and (b) the full 48 month period from 15 Sept'96 to 15 Sept'00 for nine sites ranked as good (G), moderate (M) and degraded (D) rangeland.

5 Conclusions

ATSR-2 NDVI time series are used to examine whether the climatic variation and the response of the vegetation can be separated from the effects of the different management strategies. The short term (12 months) local climate variability and its effect on the ΣNDVI appear to exceed the effects caused by variations in rangeland state. The behaviour of the cumulative ΣNDVI calculated over a longer term (48 months) suggests a decoupling of the effect of local climate variability from the effect of rangeland state.

6 Acknowledgement

This work was conducted as part of the Poverty and Natural Resource Utilisation in South Africa (PANRUSA) project funded by the UK Department for International Development (DFID). ATSR-2 data were provided through the UK Natural Environment Research Council and are copyright to ESA/BNSC/NERC/CLRC. The authors would also like to thank Dr. Chasca Twyman for helping with vegetation sampling and the Oxfam for access and help in the Omaheke Region of Namibia, particularly Dr. Jeremy Muller and Mr Albert Papama.

7 References

1. UN, "United Nations Convention to Combat Desertification in those countries experiencing serious drought and/or desertification, particularly in Africa. Text with annexes. Geneva, UNEP/CCD, 1995
2. R. K. Hitchcock, " Indigenous peoples, resource management and traditional tenure systems in African dryland environments". In Social aspects of sustainable dryland management. S. D. Chichester, John Wiley, 1995
3. E. Lambin, "Monitoring forest degradation in tropical regions by remote sensing: some methodological issues", vol. 8, pp. 179-190, 1999
4. J. Cihlar, L. St.-Laurent and J. A. Dyer, "Relation between the normalized difference vegetation index and ecological variables", Remote Sens. Environ., vol. 35, pp. 279-298, 1991
5. G. F. Asrar, E. T. Kanemasu, R. D. Jackson and P. J. Pinter, "Estimation of total above-ground phytomass using remotely sensed data", Remote Sens. Environ., vol. 17, pp. 211-220, 1985
6. E. O. Box, B. N. Holben and V. Kalb, "Accuracy of the AVHRR vegetation index as a predictor of biomass, net primary productivity and net CO2 flux", Vegetatio, vol. 80, pp. 71-89, 1989

7. S. Ringrose, T. MusisiNkambwe, D. Coleman, D. Nellis and C. Bussing, "Use of Landsat Thematic Mapper data to assess seasonal rangeland changes in the southeast Kalahari, Botswana", Environmental Managment, vol. 23, pp. 125-138, 1999
8. Y. Richard and I. Poccard, "A statistical study of NDVI sensitivity to seasonal and interannual rainfall variations in Southern Africa", Int. J. Remote Sens., vol. 19, pp. 2907-2920, 1998
9. A. van Rooyen, personal communications.
10. N. Viovy, O. Arino and A. S. Belward, "The Best Index Slope Extraction (BISE): A method for reducing noise in NDVI time-series", Int. J. Remote Sens., vol. 13, pp. 1585-1590, 1992

MODIS 250m AND 500m TIME SERIES DATA FOR CHANGE DETECTION AND CONTINUOUS REPRESENTATION OF VEGETATION CHARACTERISTICS

JOHN R.G. TOWNSHEND [1,2,3]
E-mail: jtownshe@geog.umd.edu
RUTH S. DEFRIES [1,2]
XIWU ZHAN [1]
ROBERT A. SOHLBERG [1]
MATTHEW C. HANSEN [1]
CHARLENE DIMICELI [1]
MARK L. CARROLL [3]
JILL C. EASTMAN [1]

[1] *Department of Geography, University of Maryland, 2181 LeFrak Hall, College Park, MD 20742*
Tel: 301-405-4050; Fax: 301-314-9299; http://glcf.umiacs.umd.edu

[2] *Earth System Science Interdisciplinary Center, University of Maryland, College Park, MD*

[3] *Institute for Advanced Computing Studies, University of Maryland, College Park, MD*

This paper addresses land cover and land cover change applications using spectral time series data from NASA's Earth Observing System (EOS) Moderate Resolution Imaging Spectroradiometer (MODIS) instrument onboard the Terra satellite. Special attention is given to the comparative advantages of MODIS over polar orbiting remote sensing heritage instruments such as the Advanced Very High Resolution Radiometer (AVHRR). MODIS provides spectral bands that are very similar to the Landsat series, in combination with a daily global repeat cycle. Spatial resolution, calibration, and geolocation are all significantly superior to previous data sources. MODIS is providing data of extraordinary value for global change and land management applications.

1 Time Series as a Tool for Land Cover Analyses

Time series of remote sensing data as they relate to land cover research are utilized for two primary purposes. The first is to derive representations of land cover components. This is achieved by analyzing the spectral trajectory of a pixel through the growing season. Such information can be used to derive either discrete land cover maps, where the majority cover for each pixel is reported, or continuous land cover maps, where the percent cover for each of the classes of interest is derived. The second application for time series of remote sensing data is the detection of land cover change. By examining the spectral characteristics of a pixel through time it is possible to determine a) if the land cover has changed, and b) to infer the change process which has occurred.

This paper will demonstrate the application of data from the Moderate Resolution Imaging Spectroradiometer (MODIS) onboard NASA's Earth Observing System Terra platform to these two problems. This research focuses on the 250m and 500m spectral bands from MODIS.

1.1 Unique Characteristics of the MODIS Instrument

The MODIS instrument provides multiple advantages over the Advanced Very High Resolution Radiometer (AVHRR) heritage instrument [1]. Like AVHRR, MODIS is in a polar orbit which provides for daily global acquisitions for all but the equatorial zone -- where the repeat cycle is approximately 1.2 days. MODIS provides onboard calibration; improved geolocation; narrower spectral bands; improved atmospheric correction; and middle-infrared bands previously available only on Landsat instruments.

But by far the most important characteristic for land cover and land cover change research is the improved spatial resolution. The AVHRR provides data at a minimum resolution of 1.1 km. Landsat ETM+ provides data at 30m and 15m, but with a repeat cycle of 16 days the Landsat temporal resolution is far less than it is for polar orbiting instruments. MODIS provides 250m spatial resolution in the red and near-infrared bands and 500m spatial resolution in the blue, green and middle-infrared spectral bands. For discrete and continuous land cover maps, the combination of improved spatial resolution and daily temporal resolution allows improved representation of landscape heterogeneity and "patchiness." This is particularly germane to conservancy and biodiversity applications.

For land cover change applications, the improved spatial resolution is of even greater importance. The AVHRR rarely provided sufficient detail to monitor land cover change on a per-pixel basis. Land cover change is a process where humans are the primary agent, acting on a local scale through cumulative activities. The nominal spatial resolution of land cover change is usually significantly less than 1km. But at a spatial resolution of 250m the majority of land cover change can be detected [2].

2 Improved Continuous Land Cover Depictions using MODIS

Previous research has demonstrated the ability of spectral data from polar orbiting satellites to provide global discrete land cover classifications [3]. The University of Maryland 1km global land cover classification utilized all AVHRR bands and a one-year time series of monthly composites to produce this result.

However, discrete land cover classifications do not portray the sub-pixel heterogeneity of the landscape. Continuous estimates of the relative proportions of land cover components improve on the discrete classification. At the 8km spatial resolution, AVHRR data were used to provide per-pixel estimates of percent cover for woody, herbaceous and bare land cover [4]. This research employed a linear mixture model to deconvolve the satellite signal. A similar methodology was used to derive percent tree cover at 1 km spatial resolution [5].

New results from the MODIS instrument improve upon this concept through the use of continuous training rather than linear unmixing. Figure 1 illustrates the application of MODIS data to producing continuous fields. Note that the MODIS 250m result (middle) captures the heterogeneity of the Landsat result (left) and is clearly superior to the result from AVHRR (right). A broader spatial extent is shown in Figure 2.

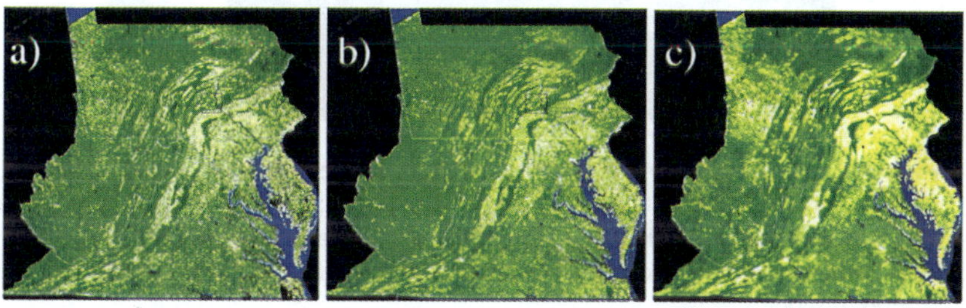

Figure 1. Continuous fields estimate of tree cover for the Mid-Atlantic region at progressive resolutions: a) shows the Multi-Resolution Land Characteristics Consortium (MRLC) result derived from Landsat data; b) shows the MODIS 250m prototype product; and c) shows the 1km product derived from AVHRR data.

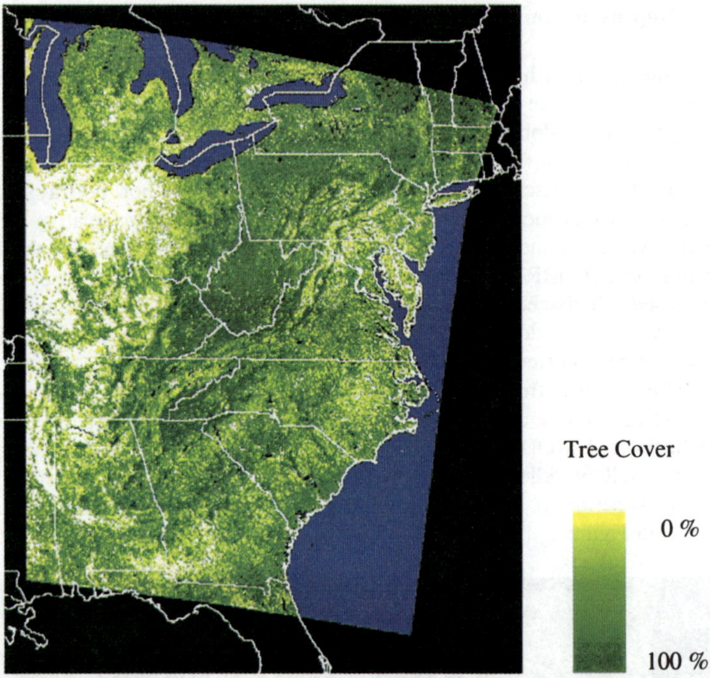

Figure 2. MODIS 250m tree cover estimate for the eastern United States. This type of percent cover data set will be provided annually for MODIS. The product, Vegetation Continuous Fields (VCF), will be valuable for carbon cycle science, land management applications, biodiversity research, and improved parameterization of models.

3 A Global Alarm Product for Land Cover Change using MODIS

During the algorithm development process for MODIS, it was demonstrated that the 250m resolution bands from MODIS could be used to accurately identify a variety of land cover change processes [6].

The MODIS product Vegetative Cover Conversion (VCC) utilizes a suite of three spectral and two spatial methods to determine if the land cover of a pixel has changed [7]. The spectral methods characterize the movement and trajectory of pixels within the red and near-infrared space and are adjusted for seasonal and physiographic variation in the signal. A graphic description of how the spectral space lookup tables are adjusted for seasonal variation is shown in Figure 3.

Two spatial methods detect changes in texture and linear features. The VCC algorithm is intended to detect anthropogenic land cover change (such as deforestation, urbanization, and agricultural expansion) and natural land cover change (such as flooding and burning).

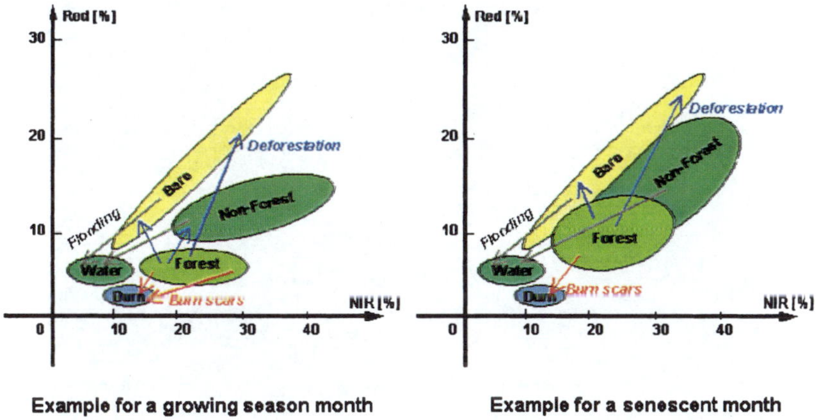

Figure 3. Illustration of how VCC red and near-infrared spectral space lookup tables are adjusted for phenology.

During the summer of 2000, the most significant forest fires in a generation raged across the American West. Figure 4 shows the area burned along the Idaho/Montana border region. Through use of a MODIS time series, VCC is able to discriminate between areas which burned during the sample period and those which had burned during previous events. The total year 2000 burned area shown in the image is 3,391 km^2.

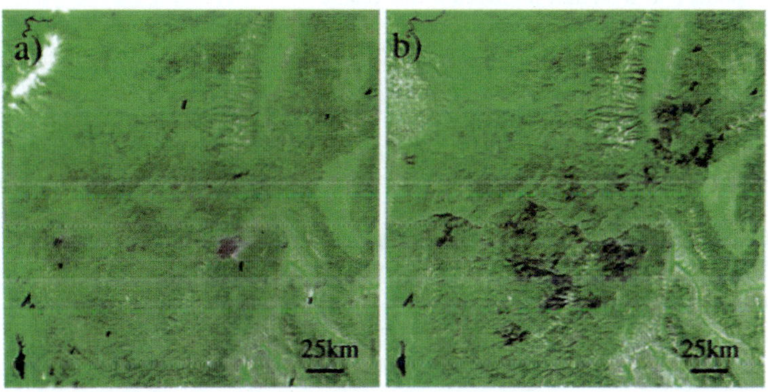

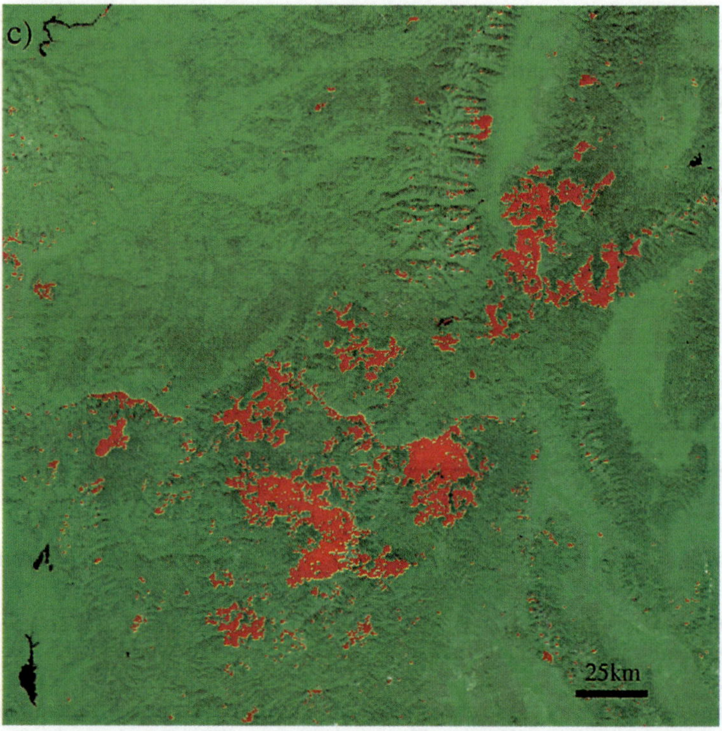

Figure 4. Land cover change resulting from the Clear Creek, Diamond Complex, Wilderness Complex, and Valley Complex fires. Image a) shows the pre-burning condition in June; b) shows the post-burn condition in September; and c) shows the VCC detected burned area in red. The black parallelograms in the June image are data dropouts and were not included in the result.

By combining the results from the five VCC methods it is possible to reduce errors of both omission and commission. The suite of methods also provides the information necessary to label the type of change which has occurred and to assess the accuracy of the result. Results from the five metrics as well as the change label and accuracy estimate are encapsulated within the VCC product file. Multiple Quality Assurance flags are also included.

VCC produces a global "alarm" product for land cover change. The "hot spots" shown in the VCC result can in turn be used to direct acquisition of data with a finer spatial resolution, such as that from Landsat 7 and IKONOS.

4 The USVI Product

In an effort to make MODIS 250m data more accessible to end-users, the University of Maryland Global Land Cover Facility is providing 16-day MODIS 250m composites for the conterminous United States and other selected field experiment locations. After compositing, the USVI is mapped to the Alber's Conic Equal Area projection using standard U.S. Geologic Survey mapping parameters. This data is available for both the full spatial extent and regional subsets.

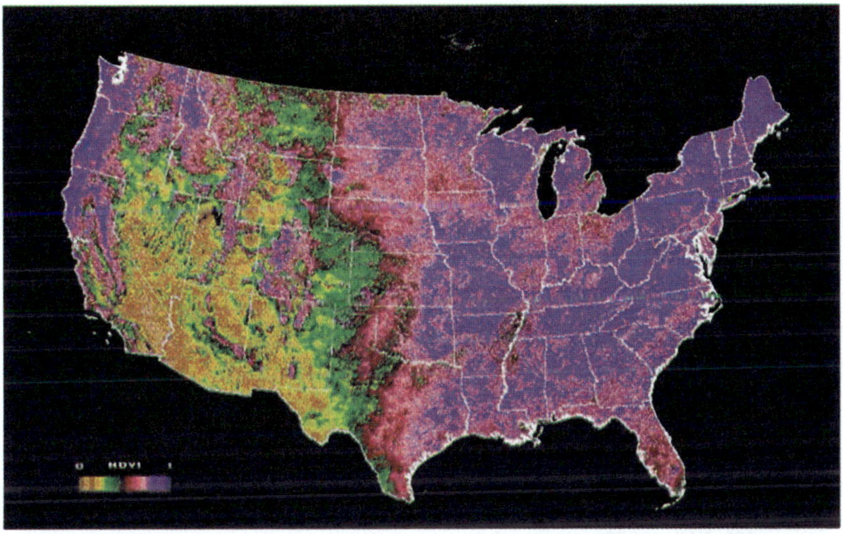

Figure 5. Example of the MODIS 250m USVI product. This is a 16-day maximum NDVI composite. See URL: glcf.umiacs.umd.edu/MODIS/usvi.htm

5 Discussion and conclusions

The MODIS instrument provides several unique advantages for land cover research. The MODIS instrument is well suited for the characterization of land cover and will provide much improved estimates of land cover and land cover change compared with the AVHRR. These improvements arise from its enhanced spatial resolution as well as the better band selection and radiometric performance of the instrument. The resultant products offer the prospect for reliable detection of global land cover change.

6 Acknowledgements

This research was supported by the National Aeronautics and Space Administration under contract NAS596060 and grants NAG59339 and NCC5300. The authors would like to thank Dr. Jacques Descloitres and Dr. Eric Vermote for their assistance with the MODIS 250m input data.

7 References

1. C.O. Justice, E. Vermote, J.R.G. Townshend, R. DeFries, D.P. Roy, D.K. Hall, V.V. Salomonson, J.L. Privette, G. Riggs, A. Strahler, W. Lucht, R.B. Myneni, Y. Knyazikhin, S.W. Running, R.R. Nemani, Z. Wan, A.R. Huete, W. van Leeuwen, R.E. Wolfe, L. Giglio, J.P. Muller, P. Lewis, and M.L. Barnsley, "The Moderate Resolution Imaging Spectroradiometer (MODIS): land remote sensing for global change research," IEEE Trans. Geosci. Remote Sensing, vol. 36, no. 4, pp. 1228-1249, 1998.
2. J.R.G. Townshend and C.O. Justice, "Selecting the spatial resolution of satellite sensors required for global monitoring of land transformations," Int. J. Remote Sensing, vol. 9, no. 2, pp. 187-236, 1988.
3. M.C. Hansen, R.S. DeFries, J.R.G. Townshend, and R. Sohlberg, "Global land cover classification at 1km spatial resolution using a classification tree approach," Int. J. Remote Sensing, vol. 21, no. 6 & 7, pp. 1331-1364, 2000.
4. R.S. DeFries, M.C. Hansen, and J.R.G. Townshend, "Global continuous fields of vegetation characteristics: a linear mixture model applied to multi-year 8km AVHRR Data," Int. J. Remote Sensing, vol. 21, no. 6 & 7, pp. 1389-1414, 2000.
5. R.S. DeFries, M.C. Hansen, J.R.G. Townshend, and A.C. Janetos, "A new global 1-km dataset for percentage tree cover derived from remote sensing," Global Change Biology, vol. 6, pp. 247-254, 2000.
6. X. Zhan, R. DeFries, J.R.G. Townshend, C. DiMiceli, M. Hansen, C. Huang, and R. Sohlberg, "The 250m global land cover change product from the Moderate Resolution Imaging Spectroradiometer of NASA's Earth Observing System," Int. J. Remote Sensing, vol. 21, no. 6 & 7, pp. 1433-1460, 2000.
7. X. Zhan, R. Sohlberg, J.R.G. Townshend, C. DiMiceli, M. Carroll, J.C. Eastman, M. Hansen, and R.S. DeFries, "Detection of land cover changes using MODIS 250m data," Remote Sensing Environment, submitted, 2001.

MULTITEMPORAL PHENOLOGICAL CLASSIFICATION OF ARGENTINA

MARIA LAURA CARRANZA AND ALICIA ACOSTA
Faculty of Mathematics, Physics and Natural Sciences, University of Molise, Via Mazzini 8, 86170 Isernia, Italy
E-mail: {carranza; acosta}@unimol.it

CARLO RICOTTA
Department of Plant Biology, University of Rome "La Sapienza", P.le Aldo Moro 5, 00185 Rome, Italy
E-mail: carlo.ricotta@uniroma1.it

The aim of this paper is to produce a global phenological land cover classification of Argentina using multitemporal NDVI profiles synthesized through multivariate ordination. Decadal AVHRR maximum value composite NDVI time series for the periods 1992-93 and 1995-96 at the kilometer resolution were used to characterize the land cover types of Argentina. The decadal NDVI data were first combined into monthly maximum value composites. Principal component analysis was then applied to the monthly NDVI data set to summarize the seasonal pattern of the NDVI profiles in a lower-dimensional statistical space. The principal components were classified into "greenness classes" using an iterative self-organizing optimization clustering procedure. Following cluster analysis, 19 classes were identified, subdivided into 4 major groups that were mainly distinguished by the intensity and timing of the NDVI peaks. Each greenness class of Argentina is described with reference to its characteristic NDVI profile and then compared with previous phytogeographic classifications. In this sense, multitemporal classification of NDVI time series proved useful to refine previous phytogeographic descriptions and, consequently, to improve the ecological knowledge of the country.

1 Introduction

Phenological studies appeared at the very beginning of the history of modern botany. Phenology has been defined as "the study of the timing of recurring biological events, the causes of their timing with regard to biotic and abiotic forces, and the interrelation among phases of the same or different species" [11]. In this view, since community phenological patterns express broad-scale bioclimatic and ecological diversity, remotely sensed phenological studies represent the appropriate tool for zonal seasonality mapping in areas where traditional ground-based ecological observations and meteorological records are scarce or lacking.

Satellite imagery provides a unique tool for monitoring seasonal dynamics of Earth's vegetation on a global scale. The combination of the normalized difference vegetation index (NDVI) data derived from the Advanced Very High Resolution Radiometer (AVHRR) with a daily repeat cycle and 1-km spatial resolution makes the weather satellites operated by the National Oceanic and Atmospheric

Administration (NOAA) well suited for deriving broad-scale phenological metrics from satellite images. The NDVI is especially suitable for use with NOAA satellites because first, normalization partially compensates for variations in viewing and illumination geometry and, second, the technique of maximum value compositing (MVC) can improve the quality of NDVI imagery [12]. The MVC method considers the maximum value assumed by the NDVI in a time period as a good estimate of the entire period [18]. MVC requires that a series of multitemporal georeferenced satellite data be processed into NDVI images.

For a given region, a spectral-temporal NDVI profile can be obtained by graphing the NDVI MVC values over time. On the basis of these spectral-temporal NDVI profiles, a variety of statistical techniques, such as principal component analysis (PCA), may be applied to analyze broad-scale remotely sensed phenological pattern. PCA is the most widely used multivariate ordination method aimed to reproduce in few orthogonal dimensions the original sample space as accurately as possible [8]. In remote sensing applications, PCA has long been used as a data compression tool by discarding minor components with little explanatory value and minimum loss of total variance.

Some authors used PCA to summarize in few variables (axes) the dimensionality of the spectral-temporal NDVI profiles prior to land-cover mapping [19, 20]. Other studies [1, 6, 10] showed that PCA is an effective tool for detecting seasonal fluctuations of NDVI profiles. In this view, PCA axes can be defined "phenological" insofar as they quantify variations in timing, duration and intensity of the growing season that are determined by seasonal changes in environmental conditions [12].

Argentina encompasses most of the southern half of South America spreading from the Tropic to the cold windswept southern tip of the continent. The latest overall phytogeographic approach dates back to 1976 [2]. The aim of this paper is to produce a new global phenological land cover classification of Argentina using multitemporal NDVI profiles synthesized through multivariate ordination (PCA).

2 Data and methods

Argentina covers an area of approximately 2,766,890 km^2 ranging from 22°S to 55°S latitude. Vegetation types are controlled by a considerable range of bioclimatic conditions supporting extremes as the desert formations, subtropical rain forests and subpolar evergreen forests with different phytogeographic domains (Amazonian, Andean, Patagonian and Antarctic). This variety of phytogeographic types is the result of the vast north-south extension of Argentina, from the tropics to subpolar lands along with the pronounced topographic relief of the Andes Cordillera in the western sector of the country with peaks reaching to over 7000 m associated with pre-Andean ranges and other mountainous areas forming a mosaic of different rainfall, exposure and elevation patches.

Decadal AVHRR maximum value composite NDVI time series for the periods 1992-93 and 1995-96 at the kilometer resolution were used to characterize the land cover types of Argentina. The decadal NDVI data were first combined into monthly MVCs. PCA was then applied to the monthly NDVI data set to summarize the seasonal pattern of the NDVI profiles in a lower-dimensional statistical space.

To estimate the contribution of each principal component to total system's variance, a scree test was used [4]. Following this test, the first five principal components were retained for subsequent unsupervised classification. The principal components were classified into "greenness classes" using an iterative self-organizing optimization clustering procedure (ISOCLUSTER [7]). Next, to identify the hierarchical relation between the greenness classes obtained, a second classification (DENDROGRAM [7]) was performed on the ISOCLUSTER output.

Finally, the output greenness classes were labeled using the average monthly NDVI profiles of each class along with ancillary data such as land cover maps at regional and continental scale [5] and digital elevation models. Each greenness class is described with reference to its characteristic NDVI temporal profile and then compared with the latest overall phytogeographic classification of Argentina by Cabrera [2], which reviewed previous plant-geographical approaches [3, 9, 13, 15]. Where possible, greenness classes were also cross-referenced with more recent regional contributions [14, 16, 17].

3 Results and discussion

Following cluster analysis, 19 classes were identified, subdivided into 4 major groups that were mainly distinguished by the intensity and timing of the NDVI peaks.

The first group has very low NDVI values throughout the year, without any clear seasonality. The second group is characterized by low/medium NDVI values with a short growing season in spring or summer. The third group shows medium/high NDVI values with a minimum in winter and a maximum in spring. The last group is characterized by very high NDVI values throughout the year with slightly higher values during autumn. The resulting 19 classes are described below.

Group 1

1. **High Andean desert:** High Mountain desert with very low vegetative activity. The main vegetation types are grass steppes (desert grasslands) with very low cover. The NDVI profile shows consistently low values throughout the year. Mean altitude: 3000 m.

2. **Andean and Central Patagonia scrub desert:** Scrub desert vegetation on the eastern slopes of the Andes and Central Patagonia. This class encompasses deserts spreading along the western part of the country including the most arid portion of the Patagonia region. The NDVI profile shows consistently low values throughout the year, although slightly higher than in the former class. Mean altitude: 1500-2000 m.

3. **Central Andean and subandean Patagonia steppe:** Scrub vegetation and grasslands on the eastern slopes of the Central Argentine Andes and western Patagonia. This class includes part of the

Central Andes and the subandean Patagonia. The NDVI profile of this class shows seasonality with the lowest values between June and August (austral winter). Mean altitude: 500-1300 m.

4. **Monte and Patagonia steppe:** Open evergreen shrubs along with low steppe and semi-deserts in central Patagonia and the Monte Region. This class includes part of Cabrera's Monte [2] and the less xeric central Patagonia. The NDVI profile is similar to the profiles of class 1 and 2 although with slightly higher values. Mean altitude: about 1000 m.

5. **Atlantic Patagonia steppe:** Bushy steppe consisting of tussock grasses and low thorny shrubs. The NDVI shows low values with a pronounced peak during June and July. The NDVI increment during winter is exclusive of this class and may be related to the influence of the Atlantic Ocean. Mean altitude: about 100 m.

6. **Dry Puna:** This open scrub sub-desert with scattered evergreen species assemblages is locally known as Puna and occurs on the most northeasterly region of Argentina. The NDVI plot shows low values from December to August with an increase during spring. Mean altitude: about 3000 m.

Group 2

7. **Arid Pampa Steppes:** Grass and shrub steppe that occurs in the southern Buenos Aires province and the lower part of Colorado and Negro rivers, including part of Cabrera's Pampa, Espinal and Monte [2]. This class spreads along a gradient from herb steppes to shrub steppes. The NDVI plot shows medium/low values without pronounced seasonality. Mean altitude: about 300 m.

8. **Monte and Arid Chaco:** Thorn open woodlands and thorn shrub-steppe spreading through the western plains of Sierras Pampeanas. This class includes mainly the Monte Region and the transition to the Chaco. Maximum NDVI values in summer, minimum in winter with a slow increase during spring. Mean altitude: about 600 m.

9. **Temperate deciduous forest and Patagonia steppe transition:** Cold temperate deciduous forest of northern and central Andean-Patagonia and Tierra del Fuego, including some Patagonian grass steppes of the southern sector of the Santa Cruz Province and Tierra del Fuego. This class corresponds to the Sub-Antarctic forest of Cabrera [2], which is slightly drier than the Valdivian and Magellanic forests. The NDVI values show a characteristic seasonality with the lowest values in winter (June, July and August), followed by a rapid increase in spring. Mean altitude: about 700 m.

10. **Valdivian evergreen forest:** Cold temperate evergreen forests of the Andean-Patagonia. This class includes very moist formations, such as the *Notofagus dombeyi* forest (Valdivian forest). Rainfall in the Valdivian rainforest is generally between 2000 and 2500 mm, although in some places it may reach 4000 mm. The NDVI plot shows a trend similar to class 9, but with lower values. Mean altitude: about 1200 m.

11. **Subpolar evergreen forest (Magellanic forest):** This class includes the cold evergreen forest of the southern Tierra del Fuego. The Magellanic forest grows in colder and less wet areas than the Valdivian forest. The NDVI plot shows a pattern similar to class 10 with marked seasonality and slightly lower values. Mean altitude: about 400 m.

Group 3

12. **Sierra and Dry Chaco:** This class is characterized by xerophytic forests and savannas, often with thorny species and many halophytic communities. It spreads through the eastern slopes of Sierras Pampeanas, continuing towards south and east including plains and halophytic depressions related to rivers as Salado and Dulce. The NDVI values exhibit the highest values from January to April and the lowest values in July. Mean altitude: about 500 m.

13. **Fringing Pampa:** This class spreads from the northwestern Buenos Aires Province to the north. Class 13, 15 and 16 are part of the Argentina Pampas where potential natural vegetation is

represented by grass steppes or pseudo-steppes with dominant tussock grasses. At present, land use is mainly agriculture and grazing. The NDVI pattern shows the higher values during the growing season and a lower peak during winter. Mean altitude: about 200 m.

14. **Humid Chaco (subtropical savannas and wet grasslands):** This class encompasses grass savannas mainly of eastern Chaco along with wet grasslands and depression related to the Parana, Uruguay and Paraguay rivers. This class is part of a gradual transition from the Gran Chaco of Brazil to the Pampas of Argentina. Here, big tussock grasses are widespread and gallery forests develop along the rivers. Temperature and rainfall both favor biomass production throughout the year. Consequently, the NDVI plot shows generally high values with a minimum in winter. Mean altitude: about 300 m.

15. **Flooding Pampa:** This class includes lowlands related to the basins of the Salado, Quinto and the southernmost part of the Parana and Uruguay rivers. It extends along the Flooding Pampa of Soriano [17], and part of the Flat Pampa, extending also to the southern Argentine "Litoral" (Santa Fe and Entre Rios provinces). The relief is mostly plain. Drainage is sometimes endoreic or areic with extensive marshes and many natural ponds of salty water. The NDVI plot shows high values throughout the year with a slight decrease during winter. Mean altitude: 0 m.

16. **Rolling Pampa:** Mainly located in the Buenos Aires Province, this class includes the rolling Pampa, the southern Pampa and part of the Mesopotamic Pampa [17]. Unlike the previous class, the river network is generally well organized. The lower courses of the Parana-Río de la Plata rivers cut an abrupt terrace with important harbor towns, such as Buenos Aires and Rosario. The NDVI values are generally high during the whole year. Mean altitude: about 100 m.

17. **Western Chaco:** Xerophytic forest corresponding to the Western Chaco of Cabrera [2] with a more arid and continental climate than the Eastern Chaco. Though deciduous species are abundant, trees are mostly semi-deciduous or evergreen. The NDVI plot exhibits maximum values in summer and a minimum in winter. Mean altitude: about 300 m.

Group 4

18. **Subtropical moist forest (Paranaense):** This class is represented by the "selva" vegetation of the Paranaense Region and the uppermost part of Yungas. Although very disturbed, this class was originally composed of subtropical forest. These forests have been intensively exploited for pulp and timber production leaving large areas of degraded forestland with very few remnants of undisturbed vegetation. The NDVI values are very high throughout the whole year with a peak in April. Mean altitude: about 300-400m.

19. **Subtropical rain forest (Yungas):** This class extends along the slopes and spurs of the lower mountains of western Argentina in contact with the Chaco. Main vegetation types are rain forests, corresponding to the Yungas Province of Cabrera [2]. Climate is wet and hot, with summer precipitation and winter frost. The NDVI profile is similar to the profile of the former class but with a more evident minimum during winter. Mean altitude: about 500-600m.

4 Final considerations

In this study, principal component analysis of multitemporal AVHRR NDVI profiles was used to summarize the vegetation seasonality of Argentina. Results show a good correspondence between NDVI annual trends and previous phytogeographic and physiognomic classifications (Figure 1). Despite the extensive agricultural transformations mainly developed in the last 500 years, it is interesting to note that some land cover classes present quite similar spatial distributions to the

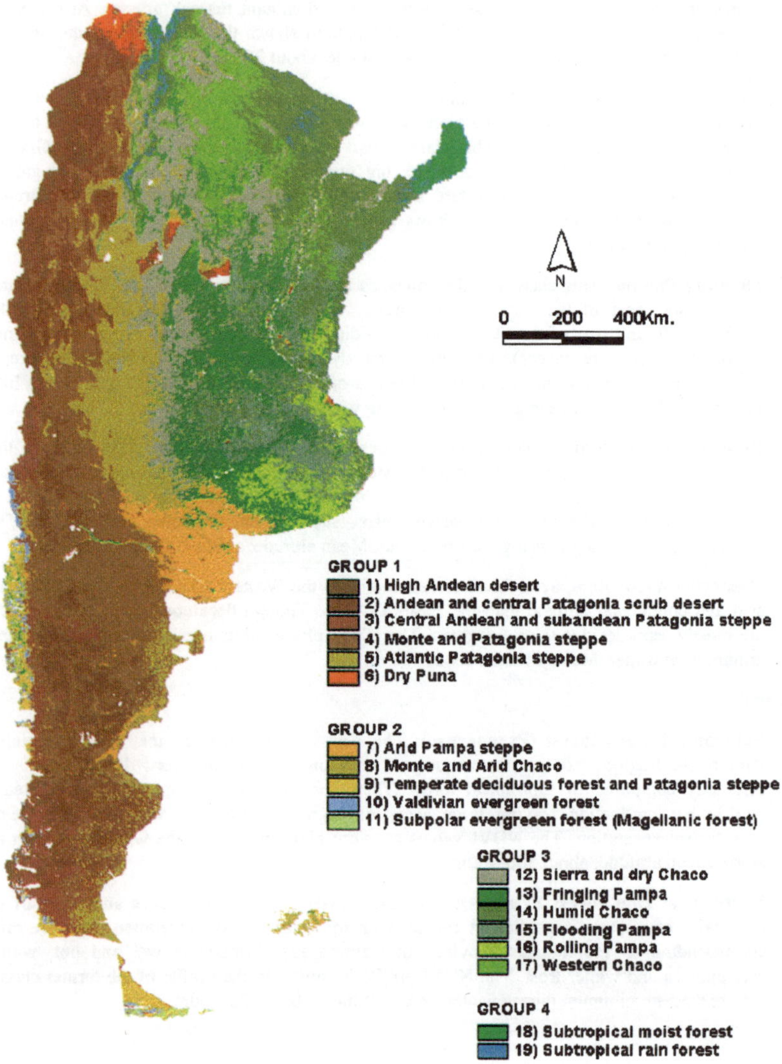

Figure 1. Phenological land cover classification of Argentine landscapes

phytogeographic classes proposed by Cabrera [2]. This could be expected for low disturbed natural areas such as class 1 (High Andean Desert) as Argentina still preserves vast areas with natural vegetation. Although this fact is less obvious for highly disturbed areas such as class 19 (Subtropical moist forest) and it could be explained because in many regions, the extensive agricultural practices are generally coherent with environmental conditions. Consequently, highly disturbed greenness classes, such as the subtropical wet forest maintain certain autonomy as in classical phytogeographic classifications. Other phytogeographic provinces such as the "Espinal" [2, 14] are not identified in our classification. Nevertheless previous authors [2] already questioned its actual phytogeographic and floristic identity. Conversely, the NDVI classification divides the phytogeographic Monte Region into two different classes: a northern one more related to the Chaco (class 8) and a southern one more related to the Patagonia (class 4). Finally, it should be noted that the greenness class 5 (Atlantic Patagonia steppe) was never described in previous phytogeographic approaches, although Paruelo [16] analyzing the region of the Gulf of San Jorge suggested that vegetation is locally controlled by the amount of precipitation and by a local ocean effect that we extended to the whole class. In this sense, since broad scale seasonal NDVI fluctuations are controlled by regional bioclimatic patterns, the outcome of the multitemporal classification of NDVI time series proved useful to refine previous phytogeographic descriptions and, consequently, to improve the ecological knowledge of the country.

References

1. Anyamba A. and Eastman J. R., Interannual variability of NDVI over Africa and its relation to El Niño/Southern Oscillation. *Int. J. Remote Sens.* **17** (1996) pp. 2533-2548.
2. Cabrera A. L., Regiones fitogeograficas argentinas. In *Enciclopedia Argentina de Agricultura y Jardineria, Vol. 2* (Ed. ACME, Buenos Aires, 1976).
3. Castellanos A. and Pérez Moureau, R. A., Los tipos de vegetacion de la República Argentina. *Monografías del Instituto de Estudios Geográficos, Vol. 4* (Universidad Nacional de Tucumán, Tucumán, 1944).
4. Cattell R. B., The scree test for the number of factors. *Multivar. Behav. Res.* 1 (1966) pp. 245-276.
5. Dinerstein E., Olson D. M., Graham D. G., Webster A. L., Primm S. A., Bookbinder M. P. and Ledec G., *A conservation assesment of the terrestrial Ecoregions of Latin America and the Caribbean* (WWF and The World Bank, Washington D.C., 1995).
6. Eastman J. R. and Fulk M., Long sequence time series evaluation using standardized principal components. *Photogramm. Eng. Remote Sens.* **59** (1993) 991-996.

7. ESRI, *ARC/IFO User's Manual* (Environmental System Research Institute INC., Redlands, California, 1995).
8. Feoli E. and Orlóci L., The properties and interpretation of observations in vegetation study. In *Handbook of Vegetation Science, Vol. 11: Computer Assisted Vegetation Analysis,* ed. by E. Feoli and L. Orlóci (Kluver Academic Publishers, Dordrecht, The Netherlands, 1991) pp. 3-13.
9. Hauman-Merck L., Burkart, L. R., Parodi, L. R. and Cabrera, L. A., La vegetación de la Argentina. In *Geografía de la República Argentina, Vol. 8* (1947).
10. Hirosawa Y., Marsh S. E. and Kliman D. H., Application of standardized principal component analysis to land-cover characterization using multitemporal AVHRR data. *Remote Sens. Environ.* **58** (1996) pp. 267-281.
11. Lieth H., *Phenology and Seasonality Modeling* (Springer Verlag, Berlin, 1974).
12. Lloyd D., A phenological classification of terrestrial vegetation cover using shortwave vegetation index imagery. *Int. J. Remote Sens.* **11** (1990) pp. 2269-2279.
13. Lorentz P. G., Cuadro de la vegetación de la República Argentina, ed. by R. Nap (La República Argentina, Buenos Aires, 1876) pp. 77-136.
14. Morello J. and Adamoli J., Las grandes unidades del vegetacion y ambiente del Chaco Argentino. Segunda parte: Vegetacion y ambiente de la Provincia del Chaco. In *Serie Fitogeografica, Vol. 13* (INTA, Buenos Aires, 1974).
15. Parodi L. R., Las regiones fitogeograficas argentinas. In *Enciclopedia Argentina de Agricultura y Jardineria, Vol. 3* (Ed. ACME, Buenos Aires, 1964).
16. Paruelo J. M., Aguiar M., Golluscio R., Leon, R. J. and Pujol G., Environmental controls of NDVI in Patagonia based on NOAA-AVHRR satellite data. *J. Veg. Sci.* **4** (1993) pp. 425-428.
17. Soriano A., Río de la Plata Grasslands. In *Ecosystems of the World,: Natural Grasslands, Vol. 8A* (Elsevier, Amsterdam, 1992) pp. 367-407.
18. Taddei R., Maximum value interpolated (MVI): a maximum value composite method improvement in vegetation index profile analysis. *Int. J. Remote Sens.* **18** (1997) pp. 2365-2370.
19. Townshend J. R. G., Goff T. E. and Tucker C. J., Multitemporal dimensionality of images of normalized difference vegetation index at continental scales. *IEEE Trans. Geosci. Remote Sens.* **23** (1985) pp. 888-895.
20. Tucker C. J., Townshend J. R. G. and Goff T. E., African land-cover classification using satellite data. *Science* **227** (1985) pp. 369-375.

MONITORING NATURAL DISASTERS AND 'HOT SPOTS' OF LAND COVER CHANGE WITH SPOTVEGETATION DATA TO ASSESS REGIONS AT RISKS

FRÉDÉRICK LUPO, ISABELLE REGINSTER, ERIC LAMBIN

Lab of Remote Sensing, Dpt. of Geography
University of Louvain-la-Neuve (UCL),
Bat.Mercator Place L. Pasteur, 3,
1348 Louvain-la-Neuve Belguim,
tel +32 10/47.28.70, fax +32 10/47.28.77,
e-mail: lupo@geog.ucl.ac.be, reginster@geog.ucl.ac.be, lambin@geog.ucl.ac.be

Abstract

A multi temporal change vector algorithm has been applied to SPOT VEGETATION data over Europe and Africa in their entirety. The change map represents the impact of natural disasters and rapid land-cover changes (such as droughts, floods, etc...) which occurred in the different regions (according to the White map) during the growing seasons between May 1998 and April 2000. This change map displaying the impact on surface attributes of natural disasters was based on the Normalised Difference Vegetation Index with a Monthly compositing periods. The validation and interpretation of this SPOT VEGETATION-based map of natural disasters and extreme land-cover changes were done with collateral data on natural disasters and 'hot spots' of land-cover change. For each important change, information on the process of change and their environmental significance is presented.

1 Introduction

All recent scientific evidences clearly point to the fact that the impact of global change on land surface attributes will not be uniformly distributed geographically. Assessing the regions "at risk" of rapid land-cover changes and/or natural disasters is therefore a priority for global change research and for policies aimed at mitigating the impact of these changes. The objectives of this project are to: (1) Use SPOT VEGETATION data to monitor over large regions the impact on ecosystems of natural disasters such as droughts, fires, floods and vegetation diseases, as well as land-cover change 'hot spots'; (2) Validate and interpret SPOT VEGETATION-based maps of natural disasters and extreme land-cover changes with collateral data on natural disasters and 'hot spots' of land-cover change; (3) Integrate this validated product in the current efforts of the global change scientific community, sponsored by the International Geosphere-Biosphere Programme (IGBP) and International Human Dimensions Programme on Global Environmental Change (IHDP), (4) to assess regions 'at risk' of rapid environmental change in order to focus research on most vulnerable areas and support the design of appropriate mitigation policies.

2 Method

The land-cover change detection approach is based on a comparison of the seasonal development curve for successive years of a remotely-sensed land-cover indicator - e.g. a vegetation index or a measure of spatial heterogeneity. When the time trajectory of the indicator over a particular pixel departs from the pixel's reference time trajectory, a change in land-cover is detected. This multitemporal approach is very sensitive to changes in seasonality and ecosystem dynamics, in addition to more abrupt landscape disturbances. It is quite insensitive to atmospheric and sensor noises that only affect isolated periods. The seasonal dynamic of a land-cover indicator can be represented by a multitemporal vector p (i, y) for pixel i and year y. Any change in accumulated value and/or in seasonal dynamic of the indicator between the current and reference years can be measured as: c (i) = p (i, ref) - p (i,y) where c (i) is the change vector for pixel i between the reference year ref and the year y. The magnitude of the change vector |c| measures the intensity of the change in land cover. It is calculated by the Euclidean distance between the pixel position, for the current and reference years, in the multitemporal space of the observations (i.e. each dimension of that space represents the land-cover indicator for one observation period; the coordinates of a pixel position are the values for that pixels, for that year, of the land-cover indicator for the observation periods). Note that this arithmetic calculation assumes that the land-cover indicator I is a quantitative measure which is linearly related to some land-cover attributes. Through this vector difference, all the input images (decadal or monthly composites for two years) are reduced to a single land-cover change magnitude map. Tests and validations of this methodology have already been published: (see reference [1], [2]), [3], [4]), [5]), [6], [7], [8]).

The methodology has been applied to the SPOT VEGETATION data. The change image based on monthly composites is developed by temporal aggregation (maximum value compositing method) of the decadal images.

3 Results

3.1 *Description of data sets*

3.1.1 Remote sensing data

We used the ten-day synthesis product (VGT-S10, see reference [15]) which corresponds to the highest value of top-of-atmosphere Normalized Difference Vegetation Index (NDVI) for every pixel during a ten days period. The radiance data were atmospherically corrected with the Simplified Method for the Atmospheric Correction (SMAC) procedure. The VGT-S10 products were delivered in the Plate carrée 1km map projection. The data cover growing seasons of different agro-ecological regions (from May 1998 to April 2000). Over this period, one expects to detect mostly the influence of interannual climatic variability and natural hazards.

An additional cloud screening was necessary as some clouds remained, particularly along the Gulf of Guinea. The cloud mask provided with the data was expanded by a neighbourhood analysis. Firstly, all pixels in a 5x5 moving window were added to the cloud mask when one pixel of the window was a cloud. Then, a 10x10 majority filter was applied. Thirdly, the cloud mask was widened to include the pixels with a high change magnitude which were neighbours of cloudy pixels.

3.1.2 Reference data

A database on natural disasters was assembled to develop an independent estimate of change. Decadal rainfall data were extracted from the Africa Data Dissemination Service (see reference [10]). Daily data on fires detected from NOAA AVHRR were taken from the World Fire Web (see reference [11]). Information on the major natural disasters in the region was assembled from FAO's Global Information and Early Warning System (see reference [12]), the Famine Early Warning System programme (see reference [13]) and the Contributions for Natural Disasters (see reference [14]).

3.1.3 Vegetation indices

Three vegetation indices were tested in a previous study over a test site west Africa (see reference [8]): the NDVI, the Soil-Adjusted Vegetation Index (SAVI) and the Normalised Difference Infrared Index (NDII) computed by replacing the red band by the short-wave infrared band (centred on 1.65 µm) in the NDVI formula. In this band, reflectance is related to water content of the canopy. It is also less sensitive to aerosols than the red band. Only the NDVI based map was retained and validated.

3.2 Presentation of the results

The results are:

1. a digital map representing the impact of the natural disasters and rapid land-cover changes which occurred during the growing seasons in the different agro-ecological regions of Africa and Europe between May 1998 and April 2000 (see figure 1);
2. a detailed validation of this map, with collateral data on natural disasters (droughts, floods), rainfall, and information on the process of change and their environmental significance.

In Africa, the regions were defined from the Ecosystem Map of White (1983),. The period corresponding to the growing season for each region was defined from rainfall and spectral vegetation index profiles. In Europe, the limits of the agro-ecological regions and their growing season are based on rainfall and air temperature distributions. The limits of the northern Europe region are based on the Agro-ecological Zones of the FAO (see reference [9]).
A region of severe moisture constraints is also defined in Asia. It is also based on the FAO maps.
Following the figure 1, there is two illustrations of severe floods in Mozambique and northern Ghana, all the red circle on the map mean that a validation has been done (see reference [16])

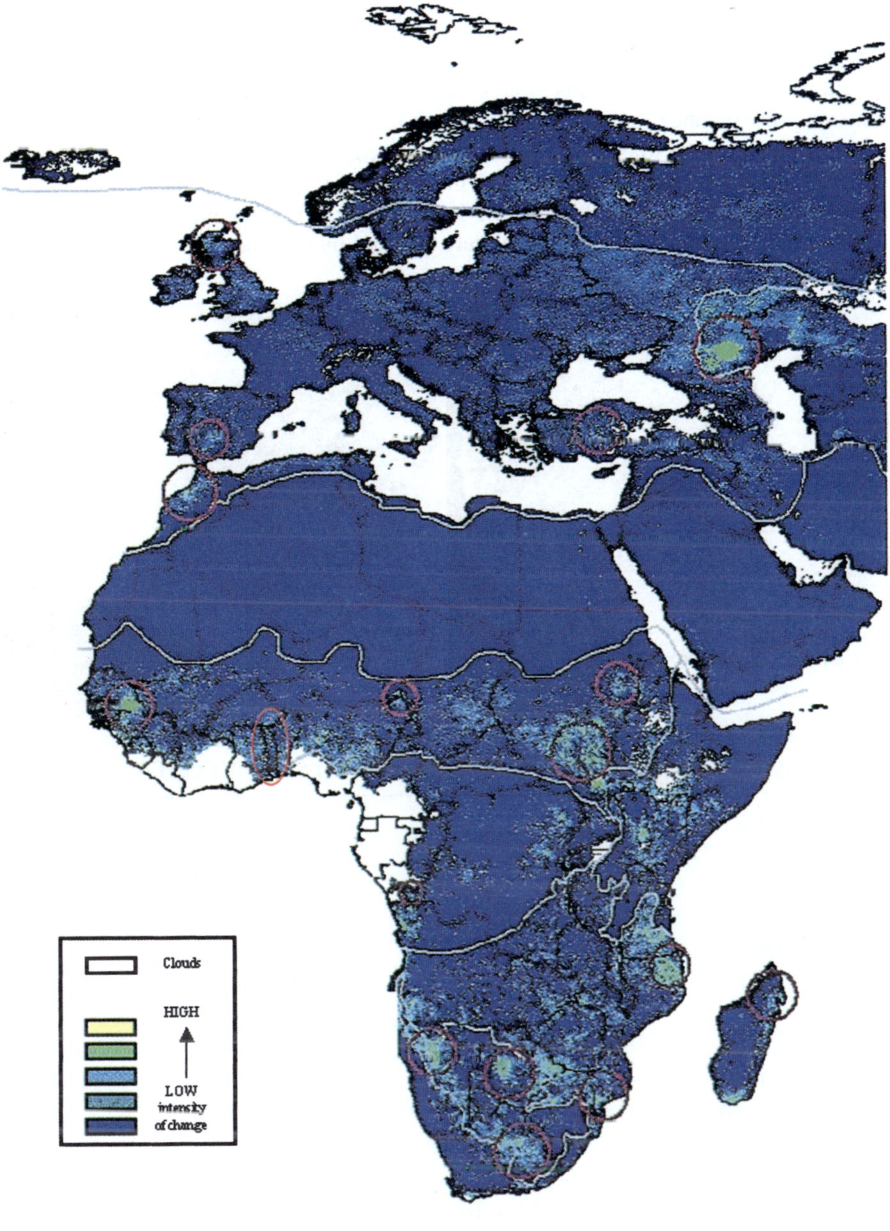

Figure 1 : Land cover change map from SPOT VEGETATION; Impact of Natural disasters from May 1998 to April 2000

3.2.1 February 2000- Southern Mozambique - Swaziland : severe floods

Torrential rains since the beginning of February in the south of the country have resulted in the worst floods for decades, causing a number of deaths and considerable damage to infrastructure and housing. Then, the cyclone-Eline hit Mozambique on Tuesday 22 February, ripping roofs off houses, washing out roads and stranding thousands of residents. The center of Eline skirted the main towns of Mozambique leaving vast stretches of land submerged under muddy brown water. Some 1 million people have been severely affected by the floods.

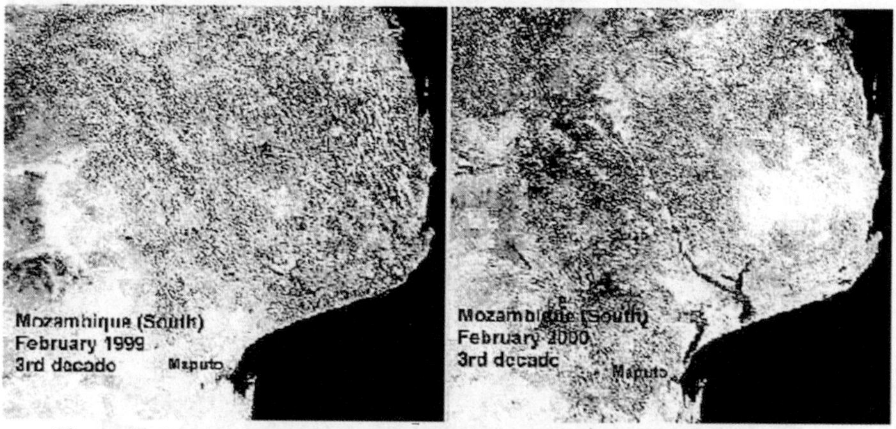

Figure 2 : Decadal composite of NDVI images, Southern Mozambique and Limpopo River in February 1999 and 2000 from SPOT VEGETATION

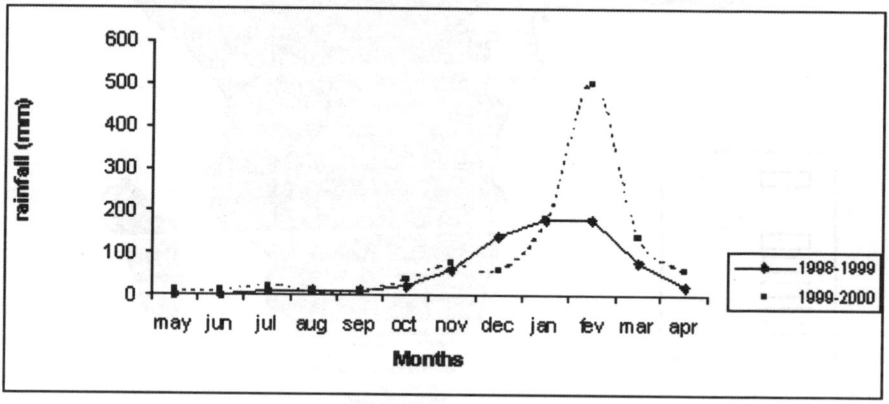

Figure 3 : Rainfall for May 1998 to April 2000 in mm/month:

Mozambique 22°N/ 32°E

3.2.2 October 1999 - Ghana, Togo, Burkina Faso, Benin and Nigeria: Severe floods

A large area has been affected with 117,000 acres of farmland destroyed and with 290,000 people homeless (followed by cholera outbreaks) (United Nations, Contributions for natural disasters http://www.reliefweb.int/). Significant anomalous rainfalls are shown in the figure 4a (source: FEWS Famine Early Warning System).

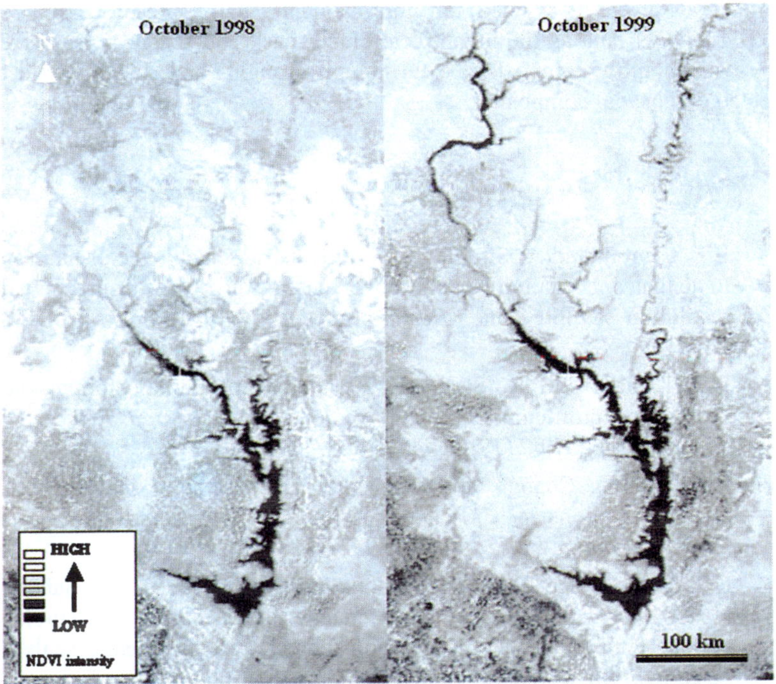

Figure 4 : Monthly composites of NDVI images, Ghana, Lake Volta, in October 1998 and 1999 from SPOT VEGETATION.

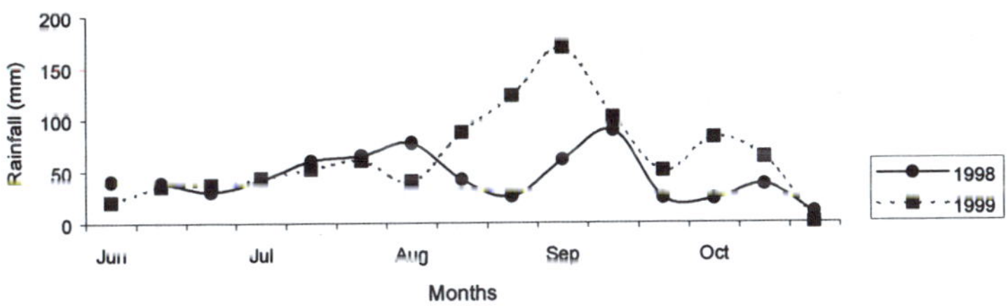

Figure 5 Seasonal distribution of rainfall in northern Ghana in mm/month.

4 Discussion and conclusions

This new land surface product represents the impact on surface attributes of the major natural disasters which are mainly caused by rainfall anomalies. Most of the validation information is related to severe floods, heavy rains, cyclones or important difference in rainfall distribution between two growing seasons. High change magnitudes detected over known deforestation 'hot spots' are too close from clouds to be validated with confidence. The validation highlights the good performance of the method. There is a need for improvements in the pre-processing of SPOT VEGETATION data as the incomplete cloud mask and sensor noises forced us to work with monthly composites, even though droughts and floods would be better detected on decadal composites.

5 References

[1] E.F. Lambin and A. Strahler,. "Multitemporal change-vector analysis: A tool to detect and categorise land-cover change processes using high temporal resolution satellite data," *Remote Sensing of Environment*, vol.48, pp.231-244, 1994.

[2] E.F. Lambin and A. Strahler, "Remotely-sensed indicators of land-cover change for multitemporal change-vector analysis," *Int. J. Remote Sens.*, vol. 15,no.10, pp.2099-2119, 1994.

[3] E.F. Lambin, Change detection at multiple temporal scales: Seasonal and annual variations in landscape variables, "*Photogrammetric Engineering & Remote Sensing*, vol.62, no.8, pp.931-938, 1996.

[4] E.F. Lambin and D. Ehrlich, "Land-cover changes in sub-Saharan Africa (1982-1991): Application of a change index based on remotely-sensed surface temperature and vegetation indices at a continental scale," *Remote Sensing of Environment*, vol.61, no.2, pp.181-200, 1997.

[5] E.F. Lambin and D. Ehrlich, "Identification of tropical deforestation fronts at broad spatial scales, " *Int. J. Remote Sens.*, vol.18, no.17, pp. 3551-3568, 1997.

[6] E.F. Lambin, "Modeling and monitoring land-cover change processes in tropical regions," Progress in Physical Geography, vol.21, no.3, pp.375-393, 1997.

[7] J.S. Borak, E.F. Lambin and A. Strahler, "Use of temporal metrics for land-cover change detection at coarse spatial scales, " *Int. J. Remote Sens.*, vol.21, no 6/7, pp1415-1432, 2000.

[8] F. Lupo, I. Reginster and E.F. Lambin, "Monitoring land-cover changes in West Africa with SPOT VEGETATION: Impact of natural disasters in 1998-1999, " *Int. J. Remote Sens.*, in press, 2001.

[9] Agro-ecological Zones of the FAO: http://www.iiasa.ac.at/Research/LUC/GAEZ/index.htm/.

[10] Africa Data Dissemination Service, USGS: http://edcintl.cr.usgs.gov/adds/adds.html

[11] World FIRE Web, JRC: http://ptah.gvm.jrc.it

[12] FAO, Global Information and early warning System: http://geoweb.fao.org

[13] FEWS Famine Early Warning System: http://www.info.usaid.gov/fews

[14] United Nations, Contributions for Natural Disasters: http://www.reliefweb.int/fts/

[15] Vegetation Program: http://sirius-ci.cst.cnes.fr:8080/index.html

[16] UCL, http://www.geo.ucl.ac.be/Disasters.htm.

MULTITEMPORAL REMOTELY SENSED INDICES AND A PROPOSED INTEGRATED FUNCTIONAL VEGETATION INDEX (IFVI) TO MONITOR HOLM-OAK WOODS ALONG A SPATIAL GRADIENT

FAUSTO MANES, MONICA GIANNINI, SILVIA ANSELMI, ALESSIA ALLEGRINI AND SUSANNA MELINI

Department of Plant Biology, University of Rome «La Sapienza», P.le A. Moro 5, 00185 Rome, Italy

E-mail: fausto.manes@uniroma1.it

This work forms part of the studies on change detection, with special reference to the study of the functional character of plant ecosystems in relation to the anthropic impact and to the assessment of sensitivity to global changes.

The study has been conducted in the Rome metropolitan area on populations of *Quercus ilex* L., an evergreen species typical of the Mediterranean environment.

Remote sensing vegetation indices, NDVI, MSI, PRI-TM (Landsat 5 TM images) have been calculated, and through these a synthetic index called the Integrated Functional Vegetation Index (IFVI) has been developed. Temporal variations in IFVI in the spectral response of the holm-oak wood have also been analyzed in natural and urban sites in order to assess the response to natural stresses and to atmospheric pollution.

1 Introduction

Human activities cause land cover changes and consequently landscape and microclimate variations. In particular around metropolitan areas the anthropic impact needs to be analyzed for land use planning and environmental monitoring. [1,2]. Remotely sensed vegetation indices are useful for assessing the impacts of natural or anthropic origin on ecosystems and, in particular, for quantifying and monitoring the health conditions of vegetation over space and time [3]. The application of remote sensing is becoming a widespread tool in change detection analysis of landscape systems. It is possible to develop an adequate change detection methodology based on spectral images to identify areas in which significant changes occur in the landscape, in natural and urban areas. Early detection of changes in tree physiological conditions allows actions to be taken that may prevent a long-term decline in biodiversity or losses in productivity of forest ecosystems [4].

The aim of the present work is to develop a synthetic index which, considering the satellite remotely sensed indices NDVI (Normalized Difference Vegetation Index) [5], MSI (Moisture Stress Index) [6], and PRI-TM (Photochemical Reflectance Index – Thematic Mapper) [7,8,9], can describe the state of the tree vegetation from the functional point of view. The vegetation type studied is the holm-oak wood, as a typical ecosystem of the Mediterranean region, sampled in urban and rural sites of the Rome metropolitan area, along a spatial gradient from the coastal area towards the inner hills characterized by different intensity of anthropic impact. Remotely sensed data have been integrated with ecophysiological field data carried out on the holm-oak (*Quercus ilex* L.). Multitemporal variations in the spectral response of the holm-oak wood have also been analyzed in order to assess the response to natural stresses and to atmospheric pollution, to provide a tool for the studies on forecasting climatic effects.

2 Methods

2.1 Study sites

The metropolitan area of Rome has a vast historic centre (16 km^2) and an extensive surrounding area; large urban parks, historic villas and large seminatural areas characterize its landscape.

Figure 1 shows the band 5 of the Landsat 5 TM image acquired on the 21 July 1999 in which the different sites of the study area are indicated. In this study some vegetation indices have been analyzed using Landsat 5 TM images of the city of Rome and of some of its outskirts. Field measurements of chlorophyll content have been carried out in some of the historic villas in the city, Villa Pamphili (VP), Villa Borghese (VB), and Villa Ada (VA), in a seminatural area, Veio (VE), and in natural protected areas, Macchia Grande (MG) and Castel Porziano Estate (CP). In this last area we have sampled two sites: Lo Scopone (CP-ls) and Grotta di Piastra (CP-gp).

2.2 Remote sensing analysis

The satellite data used in this study are Landsat 5 TM images, taken on 3 August 1998 and on 21 July 1999 (Frame 19131) (provided by Telespazio SpA), geocoded by the nearest-neighbour method which yielded a mean square error of less than 1 pixel. All processing was performed by ILWIS 2.2 (Integrated Land Water and Information System) for WINDOWS (ITC - ILWIS Group, Enschede, The Netherlands).

Remotely sensed data were processed to obtain multitemporal vegetation indices: NDVI (TM4-TM3)/(TM4+TM3), MSI (TM5/TM4), and PRI-TM (TM3-TM2)/(TM3+TM2) which provided information on the green biomass, on water stress conditions of vegetation, and on the functionality of the plant communities, respectively.

The proposed integrated functional vegetation index (IFVI) has been calculated by assessing the field of variation, in the absence of anomalous data, of the three remotely sensed indices analyzed and constructing 5 range classes for each index. The five classes have been assigned a score (1 to 5), in accordance with a progressive improvement in the state of the vegetation (high values of NDVI, and low values both of MSI and PRI-TM), in relation to the significance of each remotely sensed index. The scores obtained for each of the three indices were then summated for each site.

2.3 *Pigment extraction and analysis*

Concentrations of total chlorophyll have been measured on leaf extract prepared with Dimethylformamide. The leaf disk is removed with a 9-mm-diameter cork borer and each disk is immersed in 3 ml of Dimethylformamide for 48 hrs. The chlorophyll content is assayed spectrophotometrically according to Wellburn [10]; the absorbance of each solution is determined respectively as 480, 646.8 and 663.8 nm with a spectrophotometer (Jasco UV-VIS, Japan).

2.4 *Statistical analysis*

Analysis of variance (ANOVA), the Student-Neuman-Keuls' test (performed at a significance level of 0.05) and the regression and correlation analysis were carried out by SPSS statistical software package (SPSS Inc. - USA).

3 Results

It results from the analysis of the NDVI data that both in 1998 and in 1999 (Table 1) the highest values recorded were those for the holm-oak populations in the Castel Porziano and Macchia Grande sites, while in Villa Borghese in both years the lowest values were observed. In particular, these values were 23% lower in 1998 and 27% lower in 1999 than in the two protected areas.

Regarding the MSI in the Veio site, in both years the highest values were observed (on average 1.032 in 1998 and 0.880 in 1999), followed by Villa Borghese, while the Macchia Grande holm-oak wood had the lowest values both in 1998 (averaging 0.720) and in 1999 (0.620 on average).

For the PRI-TM data, too, once more the highest values for holm-oaks occurred in Villa Borghese.

An analysis of the temporal variations showed that between 1998 and 1999 there was a statistically significant increase ($p<0.05$) in the NDVI and a statistically significant decrease ($p<0.05$) in the MSI and in the PRI-TM for the holm-oak stands in the sites examined.

For the synthetic index (IFVI) (Fig. 2) it is seen that both in 1998 and in 1999 the differences among the sites tend to remain unchanged, with the highest values in 1999.

The highest values of the IFVI in both years were observed in the protected areas of Castel Porziano, at Grotta di Piastra, and of Macchia Grande (11 in 1998 and 14 in 1999). Regarding the historic villas, in Villa Ada, the index, calculated on holm-oaks , in both years (10 in 1998 and 13 in 1999), is comparable with that of Castel Porziano. Villa Borghese, instead, had the lowest value of the index among all the sites in both years (6 in 1998 and 7 in 1999).

Figure 3 shows the leaf content in total chlorophyll for the trees sampled in the sites examined in 1999. It is observed that the values for the Villa Borghese site (averaging 8.7 μg/ml) are lower than those for the protected areas of Castel Porziano (on average 11.0 μg/ml) and Macchia Grande (on average 11.3 μg/ml). These data are in negative correlation with the PRI-TM values calculated in the corresponding sites (y = -28.617x+8.759, r = –0.859, p = 0.013).

Table 1. Mean values and relative standard deviation of vegetation indices, calculated on 3 August 1998 and on 21 July 1999 from Landsat 5 TM images. Different letters refer to significant differences (significance level ≤0.05) between sites.

Sites	*NDVI*		*MSI*	
	1998	***1999***	***1998***	***1999***
CP-ls	0.474±0.017 c	0.562±0.012 cd	0.788±0.047 a	0.702±0.039 b
CP-gp	0.483±0.023 c	0.570±0.011 d	0.783±0.050 a	0.704±0.0260 b
MG	0.486±0.012 c	0.563±0.015 cd	0.720±0.036 a	0.620±0.041 a
VP	0.441±0.036 b	0.519±0.021 b	0.815±0.054 a	0.726±0.037 b
VB	0.370±0.030 a	0.410±0.040 a	0.907±0.064 b	0.860±0.078 c
VA	0.429±0.021 b	0.549±0.010 c	0.775±0.064 a	0.713±0.027 b
VE	0.398±0.044 a	0.510±0.013 b	1.032±0.158 c	0.880±0.114 c

Sites	*PRI-TM*	
	1998	***1999***
CP-ls	-0.0527±0.0127 a	-0.0783±0.0202 b
CP-gp	-0.0521±0.0129 a	-0.0963±0.0108 a
MG	-0.0527±0.0260 a	-0.0781±0.0187 b
VP	-0.0180±0.0208 b	-0.0490±0.0127 cd
VB	-0.0001±0.0192 b	-0.0133+0.0203 e
VA	-0.0376±0.0157 a	-0.0723±0.0152 bc
VE	-0.0057±0.0098 b	-0.0330±0.0121 de

4 Discussion

The vegetation indices considered supply information on the green biomass (NDVI), on the leaf moisture content (MSI) and on the photosynthetic functionality (PRI-TM). However the developed Integrated Functional Vegetation Index highlights better the vigour of the holm-oak woods. This index has made it possible to compare, in different sites, the same vegetational type on the basis of functional characters, obtaining a synoptic vision. The gradient observed for vegetation indices along the transect performed from natural sites toward urbanized sites seems more evident in the light of IFVI data. However the three satellite indices contribute towards hypothesizing the principal factor that contributes towards causing the altered conditions. At the Veio site water stress conditions are the principal factor that influence the functionality of holm-oaks as supported by MSI data. At the Villa Borghese site the main stress factor is due to air pollutants which affect the chlorophyll content and thus the PRI-TM, as reported in the results.

Moreover the results show an overall improvement in the vigour of the holm-oak woods in 1999 with respect to 1998. This improvement was not attributable to climatic conditions but, probably, related to a decrease of atmospheric pollutants concentrations, as shown in Figure 4. The natural sites are in a better state than sites inside the Rome urban area where the historic villas can be classified according to a gradient of alteration in relation to the different impact to which they are liable. Villa Ada, characterized by better preserved conditions, has woody species over extensive areas having a higher level of cover than in the other villas. Villa Borghese, instead, situated in the centre of the city, surrounded by streets where the traffic is very intense, and with vast asphalt paved internal zones, had the highest level of impact.

Numerous studies have been conducted on the relations between remotely sensed indices and field data: the NDVI, for example, has been correlated with numerous biophysical parameters of the canopy, including the Plant Area Index (PAI) [11,12,13]. In the present work the relation has been analyzed between the PRI-TM index and the total chlorophyll content, and a good correlation has been obtained. The PRI-TM index was obtained by using the visible reflectance which proves to be highly sensitive to the photosynthetic pigment content [14,15]. The field analyses run on the leaf chlorophyll content bear out the functional differences among the holm-oaks in the various sites, which differences are evidenced by means of IFVI.

In conclusion, integrating field data and data remotely sensed by satellite and that relating to the IFVI synthetic index makes it possible to monitor and assess on a multi-temporal basis the variations in the state of the vegetation attributable to natural and anthropic stress factors, as also in response to stress factors in relation to global change. Further studies are in progress to evaluate this index in other study areas in the Mediterranean environment and over a longer time span.

References

1. Fung T. and Siu W., Environmental quality and its changes, an analysis using NDVI, *Int. J. Remote Sensing* **21** no. 5 (2000) pp.1011-1024.
2. Ridd M.K. and Liu J., A comparison of four algorithms for change detection in an urban environment, *Remote Sens. Environ.* **63** (1998) pp. 95-100.
3. Luther J.E. and Carroll A.L., Development of an index of balsam fir vigor by foliar spectral reflectance, *Remote Sens. Environ.* **69** (1999) pp. 241-252.
4. Mohammed G.H., Sampson P.H., Colombo S.J., Noland T.L. and Miller R.J., Bioindicators of forest sustainability: development of a forest condition rating system for Ontario, *Project strategy, Forest Research Information*, Paper N°137, Ministry of Natural Resources, Ontario Forest Research Institute, Ontario (1997) pp. 22.
5. Rouse J.W.,. Haas J.A, Schell J.A., and Deering D.W., Monitoring vegetation systems in the Great Plains with ERTS, Proceedings Third Earth Resource Technology Satellite-1 Symposium, Greenbelt: NASA SP-**351** (1974) pp. 309-317.
6. Rock B.N., Vogelmann J.E., Williams D.L., Vogelmann A.F., and Hoshizaki T., Remote detection of forest damage, *Bio Science* **36** (1986) pp. 439.
7. Gamon J.A., Peñuelas J., and Field C.B., A narrow-waveband spectral index that tracks diurnal changes in photosynthetic efficiency, *Remote Sens. Environ.* **41** (1992) pp. 35-44.
8. Peñuelas J., Llusia J., Pinol J., and Filella I., Photochemical reflectance index and leaf photosyntetic radiation-use-efficiency assessment in Mediterranean trees, *Int. J. Remote Sens*. **18** no. 13 (1997) pp. 2863-2868.
9. Manes F., Anselmi S., Giannini M., Melini S., Remote sensing and field analysis of urban green in Rome. *Remote Sensing for Agriculture, Ecosystems and Hydrology II.* Barcelona, September 25-29 2000. EUROPTO Series, Proceedings of SPIE **4171** (2001) pp. 320-327.
10. .Wellburn A.R, The Spectral determination of Chlorophylls *a* and *b*, as well as Total Carotenoids, Using Various Solvents with Spectrophotometers of Different Resolution, *J. Plant Physiol.* **144** (1994) pp. 307-313.
11. White J.D., Running S.W., Nemani R., Keane R.E., Ryan K.C.,. Measurement and remote sensing of LAI in Rocky Mountain montane ecosystems, *Canadian Journal of Forest Research* **27** (1997) pp.1714-1727.
12. Manes F., Anselmi S., Giannini M., Melini S., Relationships between Leaf Area Index (LAI) and Vegetation Indices to analyze and monitor Mediterranean ecosystems, *Remote Sens. for Agriculture, Ecosystems and Hydrology* II. EUROPTO Series, Proceedings of SPIE **4171** (2001) pp. 328-335.
13. Manes F., Blasi C., Anselmi S., Giannini M., A Multiscale Study to Analyse the Response of Vegetation to Climatic Conditions In: Visconti G., Beniston M., Iannorelli E.D. and Barba D., *Global Change and Protected Areas.*

Advances in Global Change Research. Kluwer Academic Publishers. **9** (2001) pp.271-280.
14. Lichtenthaler H.K., Vegetation stress: an introduction to the stress concept in plants, *J. Plant Physiol.* **148** (1996) pp. 4-14.
15. Chappelle E.W., Kim M.S. and McMurtrey J.E., Ratio analysis of reflectance spectra (RARS): An algorithm for the remote estimation of the concentrations of chlorophyll a, chlorophyll b, and carotenoids in soybean leaves, *Remote Sens.Environ.* **39** (1992) pp. 239-247.

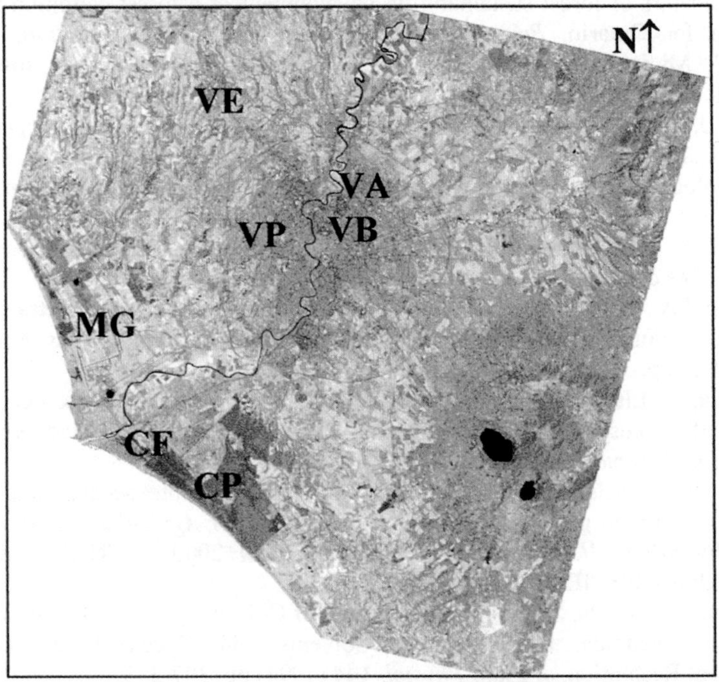

Figure 1. Band 5 of the Landsat 5 TM image (taken on 21 July 1999) of the metropolitan area of Rome in which the different sites are located. The abbreviation for each site is reported in the text. Scale 1: 600000

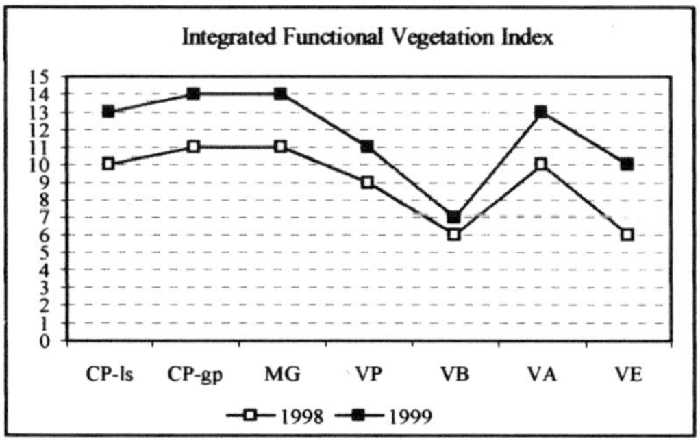

Figure 2. Values of Integrated Functional Vegetation Index (IFVI) for the holm-oak woods sampled in the different sites of the study area in 1998 and 1999.

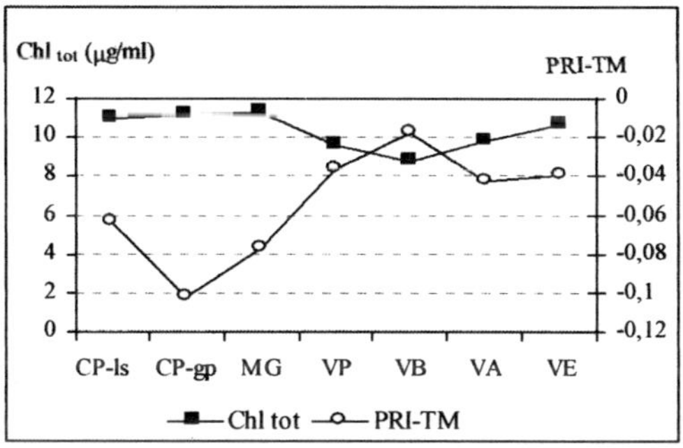

Figure 3. Mean values of leaf content in total chlorophyll and PRI-TM for the holm-oak woods sampled in the different sites of the study area.

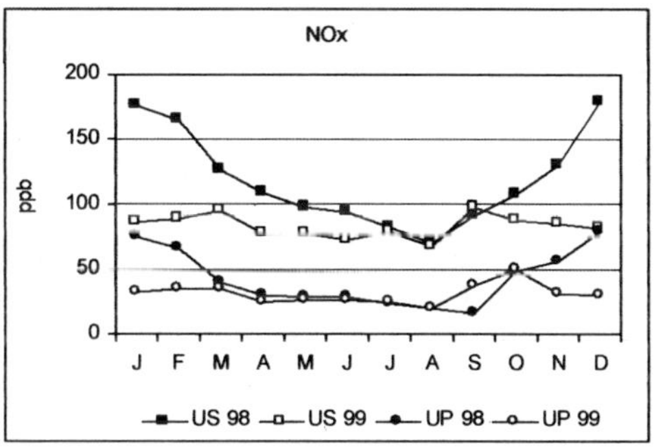

Figure 4. Trend of monthly mean concentrations of nitrogen oxides, for the years 1998 and 1999. UP=urban park (Villa Ada); US=urban street, characterized by dense motor vehicle traffic.

MULTITEMPORAL EVALUATION OF MIVIS DATA FOR WASTE DISPOSAL AND LANDSLIDE APPLICATIONS

C.M. MARINO, G. DEL PERO, M. LONGO, R. COLOMBO

Dip. Scienze Ambientali e Territoriali, Univ. Degli Studi di Milano Bicocca, Piazza delle Scienze, 1 – 20124 MILANO - ITALY

E-mail: carlo.marino@unimib.it

In this paper multitemporal hyperspectral data, acquired by airborne MIVIS are used as source data for determining spatial and temporal evolution of processes related to biogas emission in waste disposal and to biophysical indicators for landslide activity.

Waste disposal is located in Cavenago Brianza (Mi) in a flat alluvial area, while landslide is referred to the Corniglio landslide sited in Northern Apennines.

Image processing technique coupled with expert-based image interpretation showed as MIVIS data can be satisfactorily employed in mapping biogenic emissions and in recognition of potential areas suitable for local landslide.

1 Introduction

Remote sensing techniques are useful to describe a wide portion of landscape, in synoptically way, obtaining objective information of the investigated surfaces.

Airborne MIVIS (Multispectral Infrared and Visible Imaging System) data has showed in the last decade a series of potential application in environmental sciences (Del Pero et al., 1987; Bianchi and Marino, 1994).

MIVIS data are collected by an hyperspectral sensor, that allows to have surface's behavior all around visible, infrared and thermal region of the electromagnetic spectrum, with high spatial and radiometric resolution. MIVIS data are collected sensor with a spatial resolution of 2 milliradiants, for example when the airplane is flying at 1.000 meters of height the spatial resolution is 2 meters resolution.

Optical band usually involved in researches for biogas emissions are bands 18 (0.773 - 0.793 μm), 14 (0.693 - 0.713 μm) and 7 (0.553 - 0573 μm), while in thermal domain MIVIS data are directly recorded as brightness temperature in band 93 (8.200 -8.500 μm), (Del Pero et al., 1997). For landslide applications also middle infrared region were employed (1.200 - 1.450 μ), (Cavazzini et al., 1999).

Remotely sensed in researches for solid waste monitoring are usually used in qualitative way and methods based on image interpretation by expert are mainly proposed in literature (Del Pero et al., 1985; Del Pero et al., 1986, Marino et al., 2000).

Both, optical and thermal bands of MIVIS data are used in this study Such a paper is focused to the monitoring of temporal evolution of natural phenomena concerning biogas emission and landslide by using an expert visual approach.

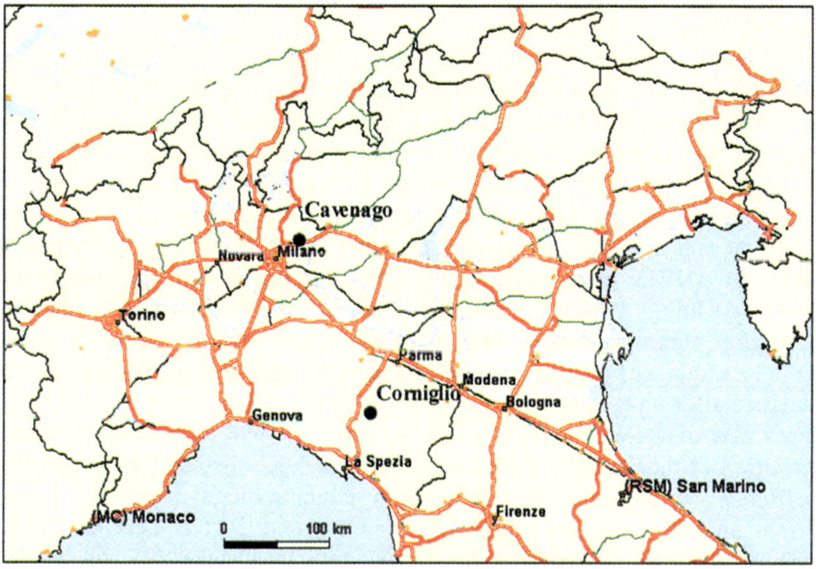

Fig 1 Study areas: Cavenago waste disposal and Corniglio landslide.

2 Waste disposal application

Waste disposal is located in Cavenago Brianza, near Milan, (fig.1), and it is characterized by environmental problems related to the proximity of urban areas and, although it is located in agricultural area, many people live in the neighborhood.
Spatial and spectral resolution of MIVIS data were used for deriving information about the spatial and temporal distribution of the waste emitted biogas. The knowledge of areas with different behavior of biogas emission is necessary to develop an intervention technique to control and reduce such emissions and, in this context the experiment was developed in collaboration with the Public Authority managing the solid waste disposal
Hyperspectral airborne remote sensing has been provided to be an useful tool for mapping thermal anomalies related to the subsoil movement of the biogas.

Thermal anomalies are generally derived as the effects of the esothermical reaction produced by the firecoat methane oxidation and by the indirect effects on the vegetation canopy.
Radiometric apparent temperature, appears as the most appropriate spectral region for monitoring biogas emissions, because biogas effects produces thermal variation that can be read on the surface as a consequence of the thermal diffusion in the underground.
Multitemporal remotely sensed data were initially georeferred by using an image to image procedure. Nearest neighbor resampling technique was applied to retain radiometric integrity and an accuracy of one pixel was determined in geometric correction procedure. Any atmospheric correction in thermal band were applied, while in the visible domain an image normalization based on the assumption of constant reflectors in time was performed.
Initially, it was mapped the spatial distribution of brightness temperature by analysis of MIVIS data acquired in 1996. Slicing and image interpretation allowed defining thermal anomalies useful to establish a criterion of intervention planning in according with local authority. The actions to knock down the effects of biogas escapes by a well field were executed by local authorities after image analysis.
Using a new data set of MIVIS image acquired in 1998 and 2000 a new spatial distribution of thermal anomalies was obtained. Such distribution maps showed the efficiency of the proposed structures in reducing biogas emissions.
Thermal anomaly in each image (fig.2-a, 2-b and 2-c) is here deducted by means of stretching technique of digital data in channel 93, as a relative difference in surface temperature from different homogeneous areas. A classified map in areas with temperature ranges in seven subjective levels is then produced.
The study of the variations (persistency) of the thermal anomalies in conducted by comparing the thermal interpreted images by a slicing technique and mapping the multitemporal variations in order to describe the evolution of the biogas emission (fig. 2-d). Such a map represent the effect of the variations in term of spatial distribution, without including a quantitative evolution of the surface temperature.
Moreover a procedure mapping of thermal anomalies is in realization by considering visible and reflected infrared bands in addition to the temperature map.

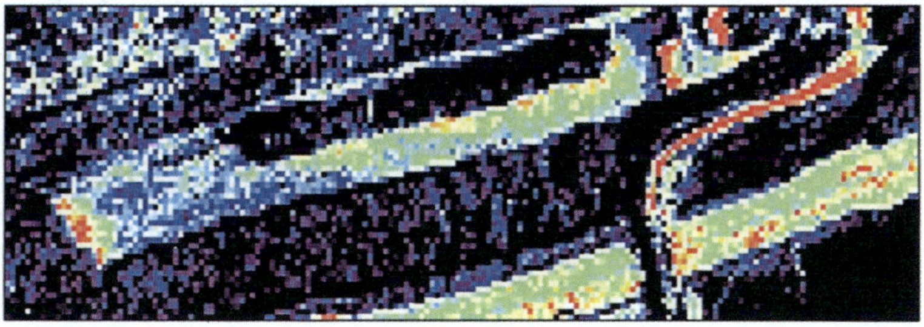

Fig. 2-a Thermal anomalies in 1996

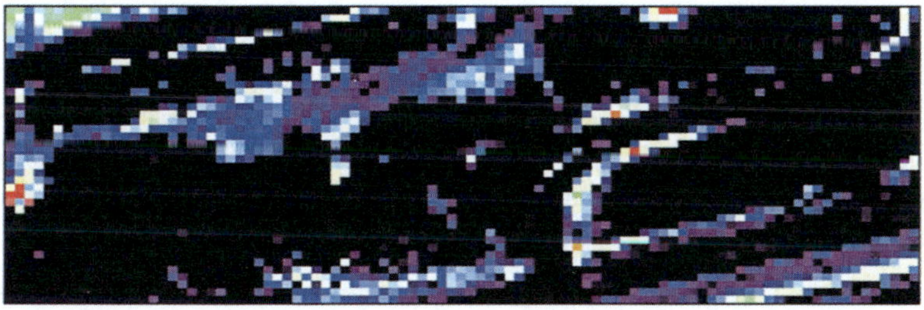

Fig. 2-b Thermal anomalies in 1998

Fig. 2-c Thermal anomalies in 2000

LEGEND

High thermal level

Middle thermal level

Low thermal level

Fig. 2-d Evolution of biogas emission.

• Persistence of biogas escapes. • Reduction of biogas escapes.

A combined approach between anomalies maps and false color composite images can be exploited keeping into account the rule of the vegetation to fix in the time the effects of the emission by the solid waste disposal (fig. 2 e). A multicriteria evaluation, based on a heuristic approach that keep into account the biogas effects on the combined soil-vegetation system and their thermal variation, appear as a complementary procedure in defining thermal anomalies.

Fig. 2-e RGB false color composite showing general quality of vegetation useful as indicator of vegetation "status".

• Good "status". • Bad "status "/ absence of vegetation.

3 Landslide application

The same general simple method of analysis showed in the previous section was adopted for studying the area of Corniglio landslide: In addition to MIVIS data, airborne ortho-photos obtained in different times was used to produce two Digital Elevation Model (DEM) in order to evaluate the temporal change in elevation related with the landslide activity.

DEM and MIVIS data was coupled for understanding change in morphology and in land use conditions.

Channel 18 (near infrared), and channels 14 and 7 (red and green), has been used for evaluating the vegetation status by false color composite analyses (fig 3b) and by using channels 23, 25, 26 (middle IR) and 93 (thermal infrared) it was obtained a soil moisture indicator. Such a indicator was based on the product integration of reflectance data with thermal data (fig 3c). Both elaboration were useful to investigate the landscape characteristics. A series of consideration can be done by the expert in order to understand the spatial and temporal distribution of the vegetation and to obtain qualitative information about the vegetation health. Information regarding land movements and water presence under the surface were necessary to obtain a good comprehension of the landslides activity in the test area. Multitemporal MIVIS data, acquired before the landslide and after the process, were used to produce two classification maps by means of a maximum likelihood algorithm and to establish the spatial vegetation change in relation to the geomorphic process. Natural vegetation reduction in corresponding of the landslide was result of 60%.

Moisture distribution obtained by the hyperspectral data has been also superposed on DEM, to map the water presence in the soil. In flat areas, water presence was related to stagnation while in the slope it was related to spring water from water table intersection with the new generated topography.

Fig. 3-a Corniglio's landslide visible bands sintesys draped on DEM.

Fig 3-b False color composite draped on DEM.

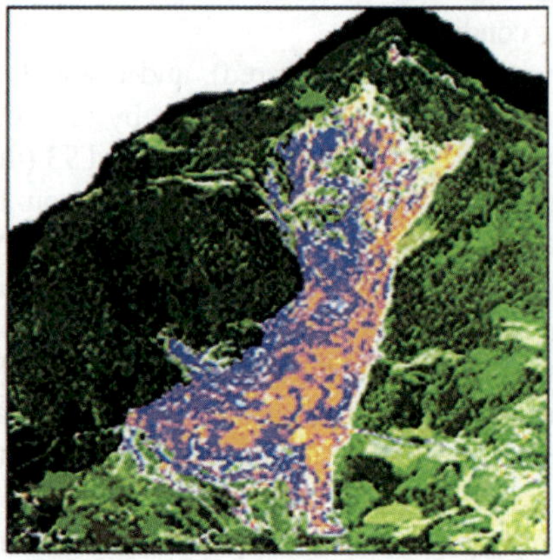

Fig 3-c Soil moisture indicator draped on DEM.

4 Discussion

In this study a simple approach for mapping thermal anomalies and to understand some indicators of landslide activity was applied in the Cavenago and Corniglio areas. The proposed approach is conducted using only few spectral bands of the hyperspectral MIVIS airborne sensor by an expert-based interpretation of visible and thermal temporal images. A series of indicators can be detected from MIVIS data and used to map biogas emission in waste disposal and to analyse landslide areas.

The waste disposal study has showed how a correct multitemporal airborne MIVIS image interpretation allow to address a series of practical intervention about the positioning of well-point batteries in order to reduce the gas emissions.

MIVIS data, acquired before and after the landslide process, were used to investigate change in moisture conditions and vegetation distribution as useful indicators in landslide processes.

Future planning is focused to the retrieval of quantitative exploitation of MIVIS data for determining an automatic and objective technique to describe the principal parameters related to the presence of biogas emission and for researching moisture indicators interesting in landslide process.

In this context we believe to obtain a really objective classification for waste disposal analysis, and to give additional elements for studies in landslides activities.

References

1. Del Pero G., Pizzaferri G. & Tonelli A.M. (1986) - "L'impiego delle tecniche termografiche nel controllo dei suoli adibiti a discarica di rifiuti solidi urbani". I° Conv.Naz."Telerilevamento per la gestione del suolo". Parma - Settembre 1986;
2. Del Pero G., Marino C. M. & Pizzaferri G. (1987) - "L'impiego delle tecniche di telerilevamento in Italia: esempi di applicazioni ambientali in contesto geologico". Atti del IV Congresso nazionale dell'ordine dei geologi. Venezia - Settembre 1987pp. 95-99;
3. Bianchi, R; Marino, C.M. (1994) – C.N.R. Lara Project, Italy: M.I.V.I.S./M.I.D.A.S. Environmental Airborne Hyperspectral Remote Sensing System – Proceedings of the First International Airborne Remote Sensing Conference, Strasburg, France, Vol. III pp.613-619.
4. Del Pero G., Ferroni P, Marino C. M. & Ronzoni M. (1995) " Utilizzo dello scanner iperspettrale MIVIS per il monitoraggio di ambiti a discarica". VII° Convegno Nazionale Ass. It. di Telerilevamento - Chieri (To) – Ottobre 1995.
5. Del Pero G., Ferroni P, Marino C. M. & Zucca F. (1997) - "Nuove applicazioni dello scanner iperspettrale MIVIS per valutazione della compatibilità ambientale di siti di discarica di R.S.U.". 1a Conferenza Nazionale delle Associazioni Scientifiche per le Informazioni Territoriali e Ambientali (A.S.I.T.A.) - Parma, Ottobre 1997.
6. Cavazzini A; Del Pero G., Dacomo A., Marino C. M. & Pizzaferri G. (1999) - "Integrazione di telerilevamento iperspettrale e aerofotogrammetria sulla frana di Corniglio (PR) " 3a Conferenza Nazionale ASITA - Napoli, Ottobre 1999.
7. Marino C. M, Del Pero G., Longo M.. & Panigada C. (2000) - "Analisi multitemporale di un sito adibito a discarica di R.S.U.". 4a Conferenza Nazionale ASITA, - Genova, Ottobre 2000.

MULTITEMPORAL THERMOGRAPHY AIMED AT MONITORING VOLCANIC AREAS

ARNALDO M. TONELLI

Via Miramonti, 4 38068 Rovereto (TN)

Italy

Key words: volcanoes, multitemporal monitoring, fumarolic activity, thermal surveys

The Monitoring of volcanoes characterized by fumarolic activity can be usefully helped by thermal multitemporal surveys. Aimed at anticipating possible paroxysmic events, this kind of survey has been experimentally tested since the early '70s. In order to monitor the thermal activity of the fumaroles field over time (maximum value of temperature, total power, migration of the thermal baricentres, etc.) a very careful registration of the multitemporal thermal data is required. This paper presents a further approach to the problem based on the analysis of the symptoms accompanying the increasing or decreasing thermal transitory.
At the surface of a volcanic area showing fumarolic activity, the global thermal effect is given by the sum of the contribution of each single fumarole. Temperature near the fumarole duct decreases with the radial distance, following roughly a negative logarithmic law. Any variation of thermal power in the volcanic area modifies the radial profile of temperature around the fumaroles, till a new condition of stability has been reached. The ratio between the module of the radial temperature gradient $|G(r)|$ and the temperature itself $T(r)$, that is the normalized gradient $|G(r)|/T(r)$, changes in such a way that by an increase of thermal power the gradient rises faster than the temperature and vice versa. The histogram of $|G(r)|/T(r)$, called "characteristic function", related to surveys carried out at different dates, allows to understand if a given thermal distribution is in a phase of transitory compared to a previous one. In case of increasing thermal transitory the histogram of $|G(r)|/T(r)$ "includes" the one referred to a previous date.
The mechanical equivalent model of a fumarole and its thermal environment is given by an elastic membrane fastened to a ring with a very large radius and forced at the center by a thin prop. If the prop is forced up, a transitory is induced and the radial slope of the elastic membrane changes faster than its height. If the prop reduces its force, coming a little back, radial slope varies slower than the height.
To assess the properties of the characteristic function, a series of thermograms have been collected on a physical model simulating the behaviour of a single fumarole. The model was formed by a conductive cylinder filled with sand and heated at the centre by a thin electrical resistor. A digital thermovision system was mounted so that it pointed vertically the centre of the model as close as to fill up its view field. The recording of thermograms has started with the start of heating of the resistor, and many thermograms have been collected during the warming up phase till reaching a steady state condition. Then the heating power has been reduced, in order to produce a decreasing thermal transitory.
Data have been processed to obtain the "normalized gradients" function and the corresponding histograms, the "characteristic function". Approaching the steady state configuration, the characteristic function tends to "include" the ones related to the increasing thermal transitory.

The method was applied to two thermal surveys on the crater of Vulcano island, collected by night by an airborne line scanner respectively in 1993 and 1994. The characteristic function has shown that in the year 1994 the volcano had slightly reduced its fumarolic activity in comparison with the previous year 1993.
The comparison of the characteristic functions of multitemporal thermal surveys allows to show the stability or, on the contrary, a trend towards an increase or a decrease of the volcanic activity, with a very good sensitivity and using just a comparison among corresponding areas, without any accurate registration among the set of multitemporal thermal images.

1 Introduction

The monitoring of volcanic areas, in particular those characterized just by fumarolic activity, can achieve a useful contribution from thermal multitemporal surveys. This topic has been experimentally tested since the early '70s.
The evaluation of the thermal activity over time (maximum value of temperature, total power, migration of the thermal baricentres, etc.), aimed at anticipating possible paroxysmic events, has presented a serie of difficulties, like the requirement of a very careful registration of multitemporal thermal data and the low sensitivity concerning the temperature variation.
This paper presents a different way to monitor the status of activity of volcanic areas, based on the comparison between the thermal symptoms bound up with the situation of increasing or decreasing thermal transitory. When a system is heated by an inner source, as it happens in volcanic areas, a "characteristic function" can be associated to the distribution of temperature on the surface The evolution of the characteristic function over time shows if a system is in a situation of stability or of transitory.

2 The characteristic function

On the surface of a volcanic area showing fumarolic activity, the global thermal effect is the result of the contribution of each single fumarole. The temperature around the fumarole duct decreases with the radial distance following roughly a negative logarithmic law. Let us consider a single fumarole like a cylindrical duct, supplying warm gases and vapors, surrounded by materials with low thermal conductivity (fig.1, above left). Any variation of the thermal power produces a variation of the temperature on the fumarole mouth which, in turn, modifies the radial thermal profile as far as a new condition of stability has been reached. The ratio between the module of the radial temperature gradient $|G(r)|$ and the temperature itself $T(r)$, that is the normalized gradient $|G(r)|/T(r)$, changes in such a way that by an increase of thermal power the gradient rises faster than the temperature and vice versa. The histogram of $|G(r)|/T(r)$ is called "characteristic function". The comparison of the characteristic functions obtained by surveys carried out at different dates allows to understand if a given thermal distribution is in a phase of transitory and permits to state the trend (increasing, decreasing, not changing). The histogram of $|G(r)|/T(r)$ in case of an increasing thermal transitory "includes" that obtained in a previous date. The gradient to temperature ratio shows higher values in this transitional situation, and vice versa. The choice of analysing the histogram, and not just the maximum value of $|G(r)|/T(r)$, is based on the need of having a statistical coherent response on a large number of pixels.

The mechanical equivalent model is that shown in fig.1, above right. An elastic membrane, fastened to a ring with a very large radius, is forced at the center by a thin prop. If the prop is forced up, a transitory is induced and the radial slope of the elastic membrane changes faster than the height. The contrary happens if the prop reduces the force coming a little back: radial slope varies less than height.

3 Experimental measures on a model

Recently, to assess the properties of the characteristic function, a physical model has been created to simulate the behaviour of a single fumarole. The model consists in a conductive cylinder, with inner diameter of 12 cm, filled with sand and heated, in the central zone, by a thin cylindrical electrical resistor (fig. 1, bottom). A digital thermovision system – operating within the range 8÷12 μm - was fixed so that it pointed vertically the centre of the model as close as to fill up its view field. The recording of thermograms has started as soon as the resistor began to be heated and a large number of thermograms have been collected during the warming up phase till reaching a steady state condition. Then the heating power has been reduced, in order to create the conditions of a decreasing thermal transitory.

The thermal frame has the dimensions of 320 x 240 pixels, like the 2D matrix of the microbolometric system used (an AVIO TVS 610). A selection of four situations is shown in fig. 2. The module of thermal gradient $|G(r)|$ has been computed using a flying sub-matrix of 3x3 pixels on which maximum *max* and minimum *min* values of temperature have been individuated. Then, on the flying submatrix the normalized gradient $|G(r)|/T(r)$ has been computed as 2(*max*-*min*)/(*max*+*min*).

On the bottom of figure 2 are reported the characteristic functions associated to the thermal situations at time "1", "2" and "3". It can be easily seen that as the system runs toward a steady state, the histogram of $|G(r)|/T(r)$, referred to time t2, "includes" the one of the previous time t1, and so on.

4 Application to an actual case

An example of application of the characteristic function is shown in figure 3, where two thermal images of the crater of Vulcano island (Aeolian archipelago, Southern Italy) are presented. They have been collected by night, respectively in the late Spring of 1993 and 1994. The surveys have been carried out by an average altitude of 1500 m, using an airborne line scanner (MIVIS hyperspectral Daedalous, property of the National Research Council of Italy), operating in band 8÷12 μm. The characteristic function indicates that in 1994 the volcano had slightly reduced its fumarolic activity in comparison with to the situation of the previous year, 1993.

5 Conclusions

The comparison of multitemporal thermal surveying on areas interested by fumarolic activity performed by the histogram of the normalized gradient $|G(r)|/T(r)$ (that is the characteristic function) offers the following advantages:

- It shows with good sensitivity the stability (histograms overlapping each other) or, on the contrary, a trend towards an increase (histogram of $|G(r)|/T(r)$ of time t2 including that of the previous time t1) or a decrease of the activity (histogram of $|G(r)|/T(r)$ of time t1 including the one of the subsequent time t2)
- It does not require an accurate registration, but just a comparison among corresponding areas

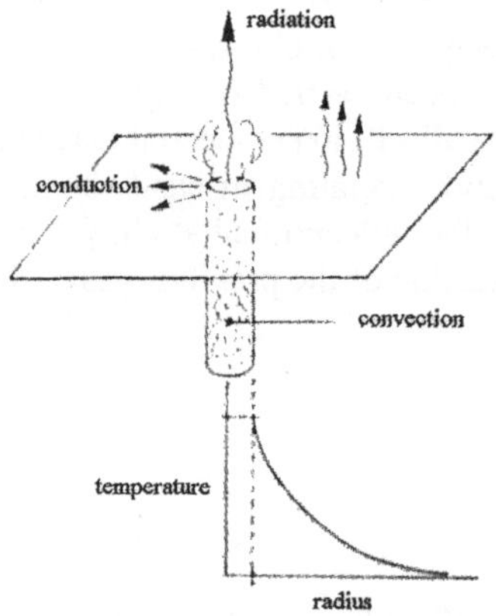

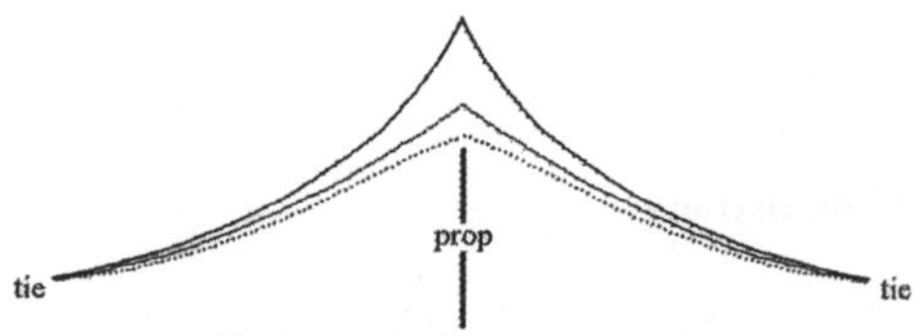

equivalent analogic model of the temperature profile: the deformation of an elastic membrane by means of a prop pushing up or releasing down

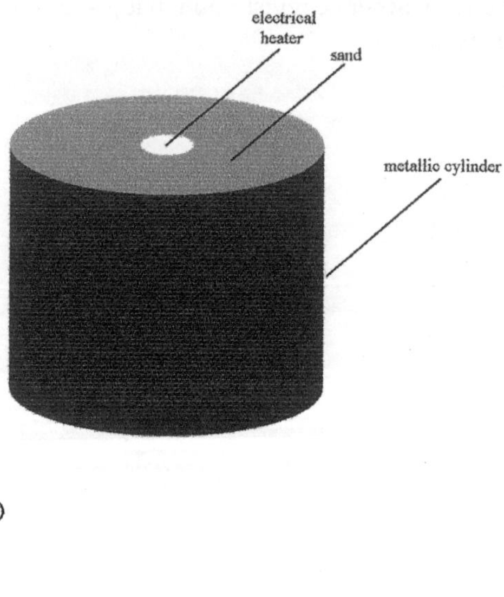

Figure 1 (see text)

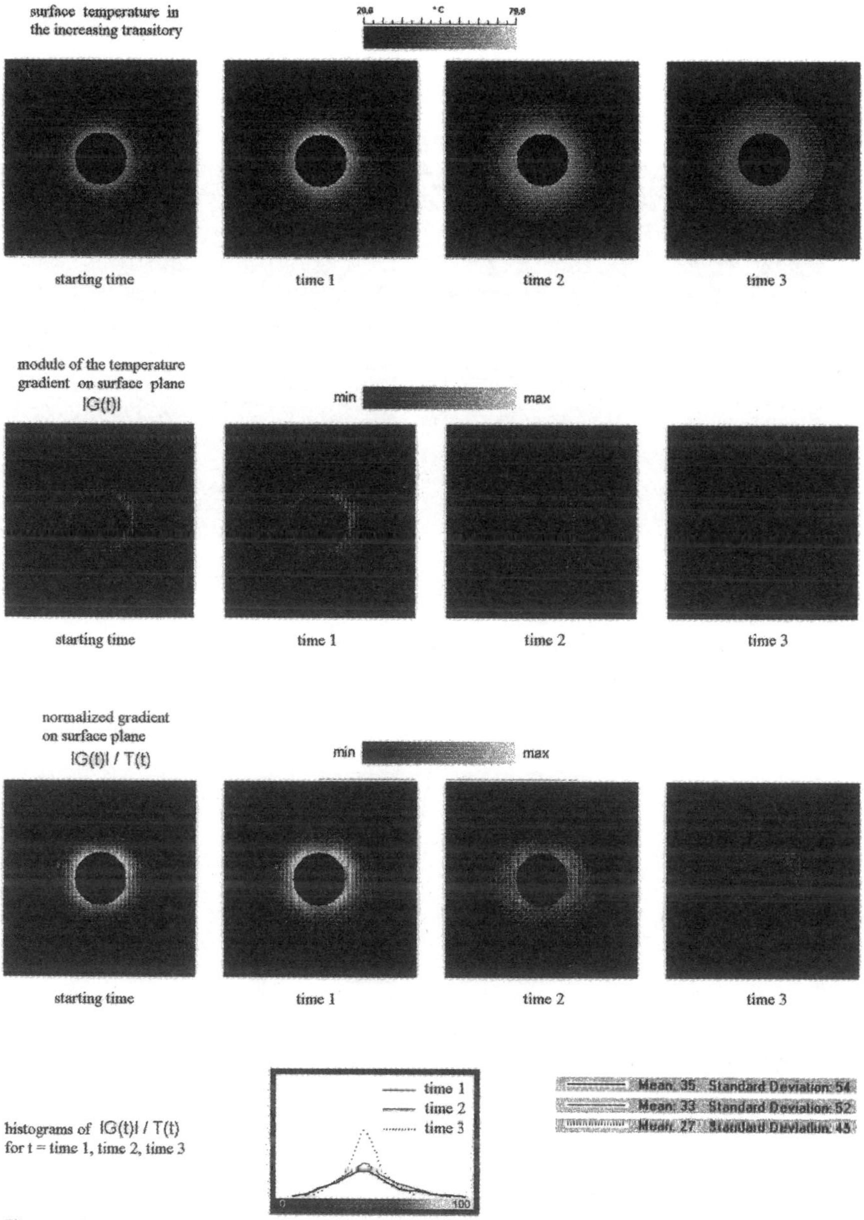

Figure 2 (see text)

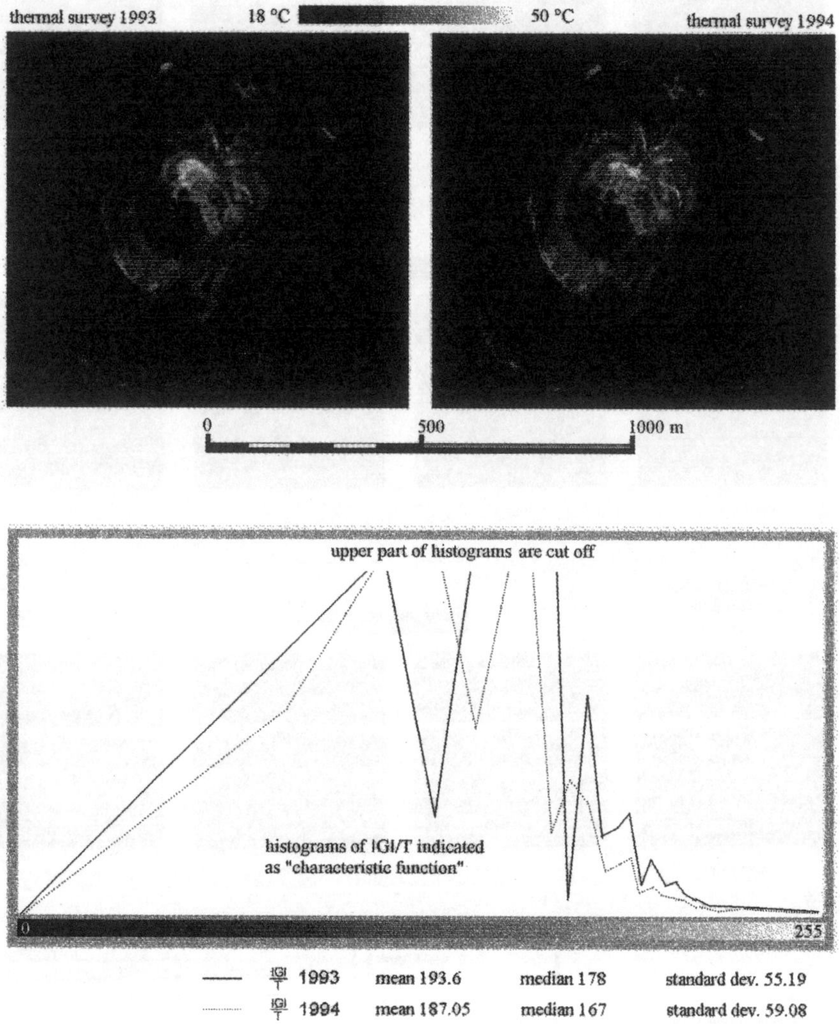

Figure 3 (see text)

6 References

CASSINIS R., MARINO C.M., TONELLI A. M. (1970) "Ground and airborne thermal imagery on italian volcanic areas". Paper presented at the U. N. Symposium on the Development and Utilization of Geothermal Resources - Pisa 1970. Published on the review Geothermics, 1970 - Vol. 2, Part I

CASSINIS R., MARINO C.M., TONELLI A. M. (1971) "Evaluation of thermal I.R. imagery on italian volcanic areas: ground and airborne surveys". Proceedings of 7th International Symposium on Remote Sensing of Environment - Ann Arbor Michigan U.S.A. - 17÷21 May 1971

LECHI G.M., MARINO C.M., TONELLI A. M. (1971) "Geological applications of Remote Sensing surveys in Italy with special reference to their utilization on volcanic areas". Proceedings of 24th International Congress of Geology - Section 9 - Montreal, Canada, 1971

CASSINIS R., MARINO C.M., TONELLI A. M. (1974) "Remote Sensing techniques applied to the study of italian volcanic areas: the results of the repetition of the airborne I.R. survey compared with the previous data" Proceedings of the Ninth Symposium on Remote Sensing of Environment Ann Arbor, Michigan, U.S.A., April 1974

TONELLI A. M. (1975) "Application of the gradient/radiance characteristic function in the study of the thermal trend of volcanic areas" Proceedings of the Tenth International Symposium on Remote Sensing of Environment Ann Arbor, Michigan, U.S.A., October 1975

KAMINSKY H. TONELLI A.M. (1976) "The use of satellite data for monitoring zones of possible geothermal interest" Proceedings of International Congress on Thermal Waters, Geothermal Energy and Volcanism of the Mediterranean Area, pagg. 286÷299 - Athens, Greece, October 1976

LO GIUDICE E., TONELLI A. M. (1976) "Computing methodologies for monitoring the thermal trend of volcanic areas" Atti del "International congress on thermal waters, geothermal energy and vulcanism of the mediterranean areas" volume 1 pagg. 354÷364 Atene, ottobre 1976

TONELLI A. M. (2000) “Some operative applications of remote sensing” Annali di Geofisica, Vol. 43 n° 6, 2000 Editrice Compositori srl – Bologna (Italy)

MONITORING OF AIR POLLUTION INDUCED VEGETATION DAMAGE IN PASVIK-PECHENGA-NIKEL USING HYBRID CLASSIFICATION AND NDVI BASED CHANGE DETECTION TECHNIQUES

HANS TØMMERVIK

Norwegian Institute for Nature Research, N-9296 Tromsø, Norway
E-mail: hans.tommervik@ninatos.ninaniku.no

KJELL ARILD HØGDA

Norwegian Institute for Nature Research, N-9296 Tromsø, Norway
E-mail: kjell-arild.hogda@itek.norut.no

Monitoring based on remote sensing has been applied in an air pollution contaminated area (mainly SO_2 and heavy metals) in the border areas between Norway and Russia. A hybrid classification method and change detection methods based on a modified vegetation index (MNDVI) and principal component analysis (PCA) have been utilised in order to extract information about vegetation changes. The different methods gave important information about the environment, and one of the main changes was the reduction of lichen dominated vegetation communities in the area. Lichen *(Cladina* spp.) dominated forests and heaths had a significant relationship (r^2=0.9) with the differences in the SO_2 emissions during the whole period 1973-1999, meaning that the variations of the area of these vegetation cover types reflected the large differences in the emissions of air pollutants during the period. The overall accuracy (PCC) of the vegetation- and land cover maps from 1973 to 1999 were estimated to be in the interval between 75% and 83% at a confidence level of 95%. The change detection methods based on the PCA analysis and the MNDVI-indices gave satisfactory associations between the increase and decrease of biomass and the differences of SO_2 emissions during the period 1973-1994 (r^2= 0.8 and r^2=0.7, respectively), while use of the hybrid classification method gave satisfactory results concerning vegetation cover changes, including lichen dominated vegetation formations.

1 Introduction

The north-eastern regions of Norway and Finland are influenced by the effects of air pollution emanating from the Kola industrial complex in Russia [1]. Especially lichen dominated areas have been affected because the lichens are sensitive to air pollution [1,2]. Air pollution from the nickel-processing industry in Nikel, Zapolyarnij, and Monchegorsk in the Kola region accounts for a large part of these environmental problems. The Pechenganikel Company in Nikel, Russia, was established in 1933 by Finnish and Canadian interests. The factory was taken over by the Soviet-Union in 1944, and after a period of reconstruction the production was started again in 1945. Emissions reached a maximum level of 100 000 tons of SO_2 during the first thirty years of production. Until 1974, indigenous nickel with only 6.5% sulphur content was processed in the smelters at Nikel and Zapolyarnij. After

1974, nickel ores from the Norilsk mines in Siberia were also processed in the smelters, and the emissions of SO2 increased rapidly from a level of 204,000 tons in 1973 up to nearly 400,000 tons at the most in 1979 [2]. This was followed by a decrease to 354 000 tons in 1985, about 273 000 tons in 1990, 230 000 tons in 1993 [2], and a further reduction to c. 200 000 tons in 1999. In the period 1988-2001 we performed a research project on the effects of this air pollution on the terrestrial ecosystems in the Varanger (Norway) and Nikel-Pechenga (Russia) areas (Fig. 1). The study area is approximately 7 100 km^2. Important vegetation entities in the area are birch and pine forests, bogs and mires, dwarf shrubs, lichen heaths, and sparsely vegetated areas in the mountain regions [1,4,5].

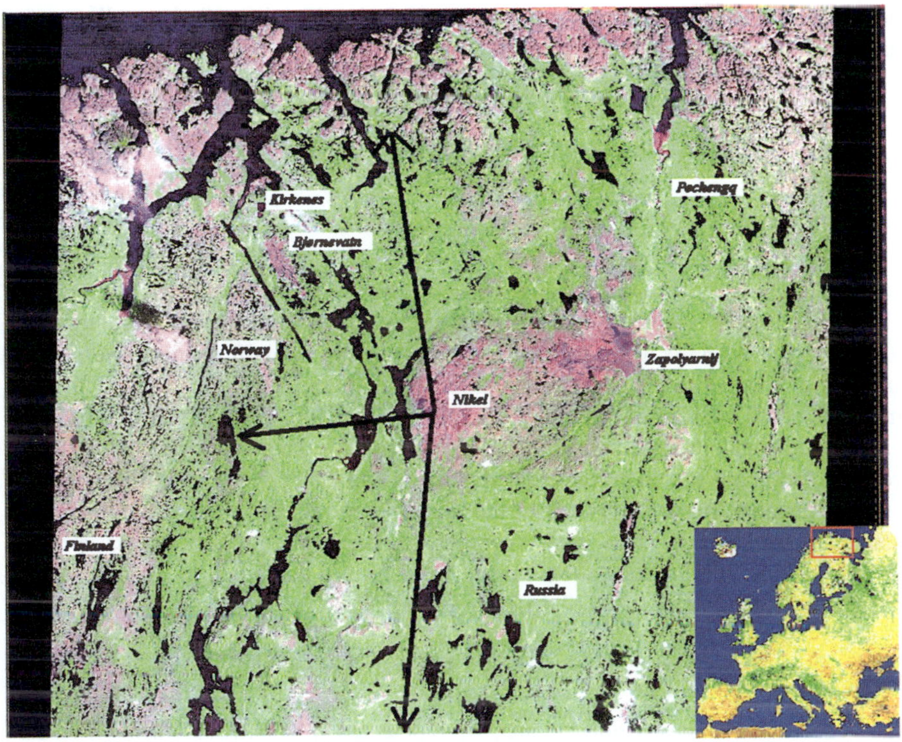

Figure 1. Location of the study area presented as a Landsat 5 TM image from 1994. Fieldwork done in 2000 was conducted along transects presented as arrows on the image. The vegetation damage and mining areas in the vicinities of Nikel and Zapolyarnij is shown in pink-violet colours.

The main objectives of the study have been:

- ✓ to document the present status of the environment,
- ✓ to monitor long term vegetation changes due to the pollution impact
- ✓ to determine the present rate of change in the area

1.1 Satellite data applied

Landsat MSS data from 1973, 1979, 1985, and 1988 together with Landsat TM data from 1985, 1988, 1992, 1994, and 1999 have been the remote sensing data sources for this study. Selection of images was based on coverage of the study area before and after the presumed dates of vegetation degradation (before 1974 and after 1985), similarity in annual precipitation patterns, and coincidence with the peak vegetation growth period of the area. Five sets of field inventory data collected in the period 1970 to 1999 have also been used for verification of the satellite data output results, i.e. the vegetation maps.

2 Methods

2.1 Land cover maps

All satellite data sets were registered to a common UTM format with a resolution of 30x30 meters. With the MSS data set the green, red and near-IR channels were used for hybrid unsupervised/supervised classification method [1,2,3,6], which means that the method consists of a pre-classification stage before ground truth control and classification and re-classification phases utilising the knowledge and the results from the field controls to produce the vegetation and land cover maps. With the TM data a 6 channel image composed of the blue, green, red, two near-IR, and mid-IR channels was applied. The initial number of (spectral) vegetation classes was selected to 36. The 36 vegetation classes were further combined to 12 land cover classes for all the satellite data (7 different years).

2.2 Vegetation change

The vegetation changes within the area were detected simultaneously using a visual interpretation approach and one approach applying vegetation indexes or principal component analysis. In the first approach the land cover maps were compared by visual inspection and changes within each vegetation cover unit during the period were detected. In the second approach the Modified Normalised Vegetation Index (MNDVI) was computed from the TM images to detect areas with changes that were not due to normal seasonal fluctuations in the vegetation. This vegetation index correct the image for atmospheric and soil noise [7,8]. This approach could not be used on the MSS scenes because the MNDVI incorporates the blue channel not available in the MSS data. Instead, a multi dimensional principal component analysis (PCA) was used. The PCA is an orthogonal transformation of n-dimensional image data in order to reduce the number of image bands by de-correlating the original bands [9]. During this process, each of the new bands or principal components portrays certain proportion of the information contained in the original data set. The second principal component generally of Landsat MSS images highlights green vegetation pattern with positive loadings on the near infrared bands and negative loadings on the visible bands [9].

3 Results and discussion

3.1 Land cover maps

In figure 2 land cover maps from 1973 (Landsat MSS) and 1994 (Landsat TM) are presented. Observe the significant difference in overall colour appearance. This is due to the transition in the vegetation cover from lichen heaths and lichen forests coloured in yellow/light green to bilberry and meadow forests coloured in darker green. Also to be noted is the increase in areas with industrial barrens (coloured in red) around the smelters in Nikel and Zapolyarnij.

The land cover map from 1988 was compared with the sulphur dioxide-concentration data proved by NILU [5], and the measurements of nickel (Ni) and sulphur (S) in reindeer lichens *(Cladina* spp.). The land cover types with the greatest stress and damage had the largest modelled SO_2 concentration levels in the ambient air ($r=0.87$, $p=0.006$, $F=11.43$), while the vegetation cover types with the lowest degree of stress (lichen dominated vegetation cover types) had the lowest levels of SO_2 in the ambient air ($r=-0.72$, $p=0.07$, $F=3.83$). The conclusion from this analysis was that the in situ data support the observations of changed, damaged vegetation and industrial barrens imaged by the Landsat satellites, especially in the surroundings of Nikel and Zapolyarnij [5].

3.2 Classification accuracy

The classification accuracy for land cover maps was assessed, and the barrens and industrial barrens and partly damaged vegetation was better than 80% during the

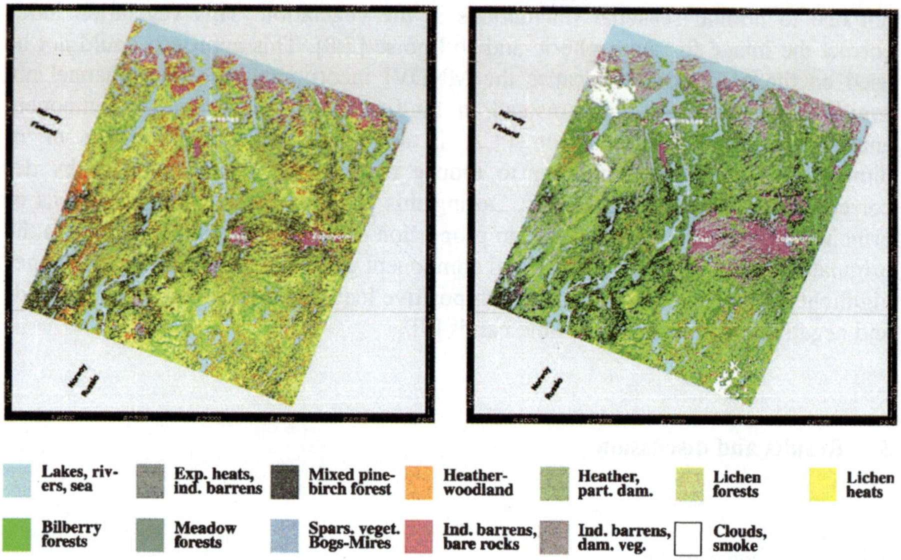

Figure 2. Land cover maps from the study area.

period 1973-2000. Bilberry dominated vegetation was found to have more variable accuracy (60-86%) during the same period possibly due to the transitional changes in the species composition for this type, and air pollution induced transformation of Empetrum-lichen vegetation formations into a bilberry dominated vegetation formation [1,10]. For the lichen dominated types we found a remarkable stable classification accuracy varying from 78-89%. Finally, the feasibility of using optical remote sensing (Landsat MSS and TM data) for mapping of vegetation cover changes due to air pollution impact was successful with an overall significant classification (hybrid classification) accuracy (PCC) varying from 75% - 85 % during 1973-1999 [4].

3.3 *Vegetation changes based on the hybrid classification method*

One of the main changes in the vegetation and land cover maps produced for the different years was that the area with lichen heath and lichen forest vegetation types decreased from c. 37% of the area in 1973 to 10% of the area in 1992 (Fig. 2). Especially the lichen heaths and lichen forests situated in the higher altitudes, in areas exposed towards Nikel and Zapolyarnij, suffered during the period 1973 to 1999. The lichen cover in these areas has been removed and bare soils, gravels, and bare rocks have been exposed. The lichen dominated areas increased in the period 1992 to 1999 and in 1999 they covered 17%. This last increase is interpreted as a possible response to reduced emissions. Lichen *(Cladina* spp.) dominated forests and heaths showed a significant relationship (r^2=0.9, p=0.01, F=29.6) with the differences in the SO_2 emissions during the whole period 1973-1999, meaning that the variations of the area of these vegetation cover types reflected the large differences in the emissions of air pollutants during the period. Bilberry and meadow forests increased from c. 7.5% in 1973 to 32% in 1999 and a possible explanation is that basic compounds (pH greater than 7.0) are spread as dust in the surroundings of the smelters as well as with the smoke and fertilised the soils. Barrens, contaminated and partly damaged areas increased from c. 7.5% to c. 19% in the period 1973 to 1994, with an subsequently decrease to 16% in 1999.

By analysis of the land cover maps, the rate of the changes in the vegetation have been estimated. In terms of km from the emission sources the annual increase of affected area was about 5 km in the period 1973-79, 4 km in the period 1979-85, and 1 km in the period 1985-88, respectively [1]. During the period from 1973 to 1988 the total area of affected environment increased in average by 300 km^2 each year. After 1988 the situation has been more stable due to the reduced emissions of SO_2 during this period [1,4]

3.4 Vegetation changes based on the modified vegetation index and PCA

Table 1 shows the percentage of pixels in each change image that had an increase or decrease in biomass based on the principal component (PCA) and the MNDVI change detection scheme.

Table 1. Percentage of pixels in each change image that had an increase or decrease in biomass based on the principal component (PCA) and the MNDVI-indices.

	PCA	
Scene	**up (%)**	**down (%)**
mss73mss79	13.2	13.5
mss73mss85	11.9	13.0
mss79mss85	10.6	12.2
tm85tm92	0.4	3.1
tm85tm94	0.9	2.4
	MNDVI	
tm85tm92	0.4	6.3
tm85tm94	1.0	4.5

Following main changes have occurred during the period 1973 to 1994:

1. Distinct negative changes in biomass have occurred in the Nikel-Zapolyarnij area in the period 1973 to 1994.These areas coincide with the industrial barrens, exposed/eroded heaths, and partly damaged areas with high concentrations of SO2 in the ground air layer.
2. Negative changes are also observed in the Jarfjordfjell/Korpfjell areas in northern direction from the Cu-Ni smelters. This was most obvious in the period 1979 to 1985.
3. Positive changes in biomass have occurred in the Jarfjord and Kirkenes areas where Lichen and Empetrum-dominated vegetation formations were transformed into a more or less bilberry *(Vaccinium myrtillus)* dominated formations with increased biomass. This coincides with the increase of the areas of classes 8 and 9 (Bilberry forests/heaths and meadow forests) from c. 7.5% in 1973 to c. 32% in 1985. The positive changes have additionally occurred in areas with reduced crown density in the birch and pine and increased contribution from the field layer to the radiances from these areas. This was most obvious in the period 1973 to 1979.
4. The changes were more significant (both positive and negative) in the period 1973 to 1985 than in the period 1985 to 1994.
5. The changes in and the transformation of the lichen heaths into a more barren like environment or dwarf shrub heath in the mountain areas were not reflected in the PCA-analysis or the MNDVI-indices.
6. The association between the increase (up) and decrease (down) of biomass for the whole study area and the differences in SO_2 emissions during the period 1973 to 1994 showed a high degree of co-variation. This association between the increase (up) of biomass and the differences in SO_2 emissions during the period 1973 to 1994 was significant on a 95% level (r^2= 0.8, F=10.5, p=0.05), while the association between the decrease (down) of biomass and the differences in SO_2 emissions during the period 1973 to 1994 was significant on a 90% level (r^2=0.7, F=8.2, p=0.06). The coefficients of determination (r^2 adj.) were 0.7 for increase and 0.6 for decrease, respectively.

4 Acknowledgements

We thank The Norwegian Research Council and The State Pollution Control Authority in Norway for financial support during the project.

References

1. Tømmervik, H., Johansen, B.E., and Pedersen, J.P., Monitoring the effects of air pollution on terrestrial ecosystems in Varanger (Norway) and Nikel-Pechenga (Russia) using remote sensing. The Science of the Total Environment **160/161** (1995) pp. 753-767.
2. Tømmervik, H., Johansen, B., and Lauknes, I., Use of airborne *casi* spatial mode data for mapping of Sub-Arctic mountain heaths in Pasvik, Northern Norway. Canadian Journal of Remote Sensing **23** (1997) pp. 230-242.
3. Price, K.P., Pyke, D.A., and Mendes, L. Shrub Dieback in an Semiarid Ecosystem: The Integration of Remote Sensing and Geographic Information Systems for Detecting Vegetation Change. Photogrammetric Engineering & Remote Sensing **58** No. 4 (1992) pp. 455-463.
4. Tømmervik, H., Høgda, K.A., Solheim, I., and Marhaug, Ø., Monitoring of Vegetation Decline in Kola Peninsula. In *Proceedings for the 27th International Symposium on Remote Sensing of Environment. Information for Sustainability, June 8-12, 1998, Tromsø, Norway* pp. 630-633.
5. Tømmervik, H., Johansen, M.E., Pedersen, J.P., and Guneriussen, T., Integration of remote sensed and in-situ data in an analysis of the air pollution effects on terrestrial ecosystems in border areas between Norway and Russia (Russia). Environmental Monitoring and Assessment **49** (1998) pp. 51-85.
6. Tømmervik, H. 2000. Vegetation damage studies in the Jarfjordfjell area, Northern Norway, by use of airborne *casi* spatial mode data. Remote Sensing Reviews **18** (2000) pp. 19-51.
7. Liu, H.Q., and Huete, A., A Systems Based Modification of the NDVI to minimize Soil and Atmospheric Noise. In *Proceedings from IGARSS-94, Pasadena, Ca.* U.S.A. 2pp.
8. Huete, A. R., and Liu, H. Q., An error and sensitivity analysis of the atmospheric- and soil-correcting variants of the NDVI for the MODIS-EOS. IEEE Trans. Geosci. Remote Sens. **32** No. 4 (1994) pp. 897-905.
9. Fung, T., and LeDrew, E., Application of principal component analysis to change detection. Photogrammetric Engineering and Remote Sensing **53** No. 12 (1987) pp. 1649-1658.
10. Odasz-Albrigtsen, A.M., Tømmervik, H., and Murphy, P., Decreased photosynthetic efficiency in plant species exposed to multiple airborne pollutants along the Russian-Norwegian Border. Canadian Journal of Botany, **78** (2000) pp. 1021-1033.

MULTITEMPORAL CLASSIFICATION OF AGRICULTURAL CROPS USING THE SPECTRAL-TEMPORAL RESPONSE SURFACE

C. A. O. VIEIRA

Universidade Federal de Viçosa,Departamento de Engenharia Civil, Campus da UFV-DEC
Viçosa MG 36.571-000, Brazil
E-mail: carlos.vieira@ufv.br

P. M. MATHER AND P. APLIN

The University of Nottingham, The School of Geography, University Park, Nottingham
NG7 2RD, England
E-mail: Paul.Mather@nottingham.ac.uk and Paul.Aplin@nottingham.ac.uk

A method for classifying agricultural crops using multitemporal, multi-spectral and multi-source remotely sensed data is described. The procedure characterizes all the pixels in a scene by considering their intensity values as a function of time of imaging and spectral waveband. An analytical surface is interpolated through these data points, which may be irregularly spaced. For purposes of comparison, two interpolation methods were used to generate and parameterize the analytical surfaces. The surface coefficients were then input to different supervised classifiers (Maximum Likelihood and Artificial Neural Network algorithms). Results show that classification accuracy is significantly improved in comparison with the use of any single-date image. Classification accuracies in excess of 87% were achieved. The advantages of the methodology described in this paper are that it takes account of the reflectance spectra at different points in the growing season, and that neither the time periods between images nor the wavebands used need be the same at each date. Thus, the procedure can handle data from sensors such as SPOT HRV and Landsat TM. In addition, the use of coefficients to represent the analytical surfaces significantly reduces the amount of data processing, whilst maintaining information reliability.

1 Introduction

Efficient crop management at province and national level requires accurate and timely information about the spatial distributions of different crops. Commonly, multispectral remotely sensed images are used to distinguish crop types on the basis of their spectral properties [1]. However, such analyses involving single-date images have the drawback that, since maximum discrimination between different crop types occurs at different stages in the growth cycle, not all differences are incorporated in the procedure. Moreover, different crop types represented in the area under study may be at different stages of growth. In addition, the temporal 'profile' of the spectral reflectance curve of each crop is not taken into account. Such profiles may be of considerable value in discriminating between crop types, which may be difficult to distinguish at certain points in the growth cycle. Furthermore, results derived from data obtained by different sensors may not be comparable due to differences in spectral and spatial characteristics. Finally, since agricultural crops are dynamic phenomena, it is often useful to observe their development over time (e.g., for crop yield estimation purposes). A solution is to use multitemporal images for crop monitoring [2]. For most current multitemporal classification techniques, a correspondence of time to growth state is established for each possible crop category that minimises the smallest difference between the given multispectral-multitemporal vector and the category mean vector indexed by growth state [3]. These techniques, however, are fairly inaccurate since only relatively few static spectral and temporal 'snapshots' contribute to crop identification. That is, images with specific spectral wavebands acquired on specific dates are used, rather the full spectral and temporal continuum. Using the full multitemporal data set may increase crop classification accuracy, since it contains more information than a selection of bands [4].

This paper demonstrates new methods of classifying agricultural crops using the spectral-temporal signatures of remotely sensed images. Per-pixel classifications are performed using multispectral, multitemporal and multisource data, whereby analytical surfaces representing the spectral and temporal continua of each pixel are computed. The surface coefficients are then used as discriminating variables (features).

2 Study Area and Data

The study area is located near the town of Littleport in Cambridgeshire, eastern England. This area was approximately at mean sea level with gently undulating topography. The agriculture of the region was characterized by rotational crop plantation techniques.

Eight remotely sensed images acquired during the 1994 summer growing season were used for analysis. The data set included four Landsat TM images (11 June, 27 June, 20 July, 14 August 1994) and four SPOT HRV images (13 May, 28 June, 30 July, 14 August 1994). Only six spectral wavebands of Landsat TM imagery were used as the thermal infrared band (band 6) was omitted from the analysis. In addition, local farmers' Field Data Printouts for 1994 were collected and used to generate a ground reference data set. All images were geometrically registered to the Ordnance Survey of Great Britain's National Grid. For each image, registration was performed using 17 ground control points and nearest neighbour re-sampling, since this technique maintained the original pixel values [5]. In each case, the root-mean-square error associated with registration was less than 0.5 pixels.

Atmospheric correction was performed to account for atmospheric differences between multitemporal images. Initially, image digital numbers were corrected to radiance using information supplied with the image data files [6]. Radiance was then converted to apparent reflectance (recorded at the sensor) and finally to surface reflectance. The final step used an inversion of the 5S (Simulation of the Satellite Signal in the Solar Spectrum) model [7].

3 The Spectral-Temporal Response Surface (STRS) Model

Badhwar [2], Haralick [3], and Lambin and Strahler [4] consider the problem of characterising the temporal dimension in remote sensing classification but none utilizes the method proposed by Vieira *et al.* [8], involving the use of the spectral-temporal response surface (STRS), which provides for the characterisation of the spectral reflectance properties of individual pixels over time.

The STRS approach is based on a view of multiband and multitemporal imagery from different sources represented in a three-dimensional space, the axes of which are time (x), spectral waveband (y) and reflectance (z). Measurement from a number of different sensors in the optical wavebands can be plotted in this space. A bivariate polynomial of the form: $z = F(x, y)$, where $F()$ indicates a polynomial function of some order, is generated for each of the crop types in the area of study. Two methods were used in order to generate the fitted surfaces: polynomial trend surface analysis (PTS) and collocation (COL), since fitted function interpolation can impose a prescribed general behaviour on the surface to override aberrant, anomalous, or noisy data. Watson [9] gives a comprehensive reviews on these interpolations methods and Mather [10] reviews polynomial trend surfaces.

These analytical functions are then parameterised and their coefficients, rather than the pixel values in each spectral band, are used as input features in the image classification process.

4 Methodology

4.1 Sampling Techniques and Classification Phase

From the co-registered and radiometrically corrected image set, two independent sample sets (comprising 1440 pixels in total) were selected using a stratified random sampling technique. These pixels represented the six most common cover types in the study area: *Potatoes, Sugar Beet, Wheat, Fallow, Onions, and Peas.* Each sample has 120 pixels (patterns) per class (total 720 pixels). One sample (selected at random) was used to train the classifier and the other was reserved for validating the results

The image acquisition dates were expressed in the form of Julian days (x-axis) and the spectral dimensions (y-axis) were characterized by their medial waveband values computed in the form of wavelengths. Thus, the spectral bands were labelled using the medial wavelength values of 0.458, 0.56, 0.66, 0.83, 1.645, 2.215 µm – given to the six available Landsat TM channels (except the thermal infrared TM band 6) - and 0.545, 0.645, 0.84 µm – given to the three SPOT HRV channels respectively. The radiometric properties are expressed in the form of reflectance values along the z-axis.

36 three-dimensional control points were generated for each pixel, using the four TM images (6 bands each) and the four SPOT HRV images (3 bands each), as described above. It is important to mention that the values along the x, y, and z-axes are scaled onto the interval 0-1 before the interpolation phase; this is sometimes referred to as normalization.

Initially, the control points were used to fit a second-degree polynomial surface to estimate the spectral-temporal response surface (STRS) using the PTS method, as described earlier. Although a surface of order 7 (with 36 coefficients) accounted for over 99% of the sum of squares, the use of a surface of order 3 (with 10 coefficients) proved to be sufficient to characterise the pattern of variation in the pixel's reflectance over time and across the available wavebands. The advantage of using a lower-order surface is that it takes into account the general characteristics of the pixel. A high-order surface fits the detailed pattern of the spectral-temporal response of the pixel, and this detail may well be specific to the individual pixel.

The same control points were used to fit a second surface using the Collocation Interpolator. As the interpolated coefficients show different magnitudes on their values, they were again scaled collectively to the interval between 0 and 1 before the training and test phases. Examples of the PTS and Collocation surfaces for one pixel are shown in Figure 1 (a to f) for each crop.

According to Vieira *et al.* [8] the Maximum Likelihood (ML) classifier is the algorithm that best combines classification accuracy and computational economy when the STRS coefficients are used as input to the classification process. Therefore, a supervised classification was performed using the Maximum Likelihood (ML) algorithm developed by Mather [1] and adapted to classify 3D surface coefficients.

For purposes of comparison, a single-date image (Landsat TM, acquired on 27th June 1994) was used to perform a standard classification in order to compare the results of the multitemporal and multisource method against a classification based on a single-date image. For each pixel, the six reflectance values are considered together and, therefore, generate a six-dimensional vector, to be also used as input into the supervised classifiers: Maximum Likelihood (ML), Artificial Neural Network (ANN) and two variants of an Artificial Neural Network (ANN and ANNT).

Both artificial neural network architectures chosen are multilayer perceptrons using the backpropagation algorithm [11]. The only difference between the models is in the input layer. The first ANN model was uses one pixel per spectral band in the input layer. Therefore, this neural network had 6 nodes in the first layer. The input nodes in the ANNT model represent a 3 x 3 window of pixel data centred on the pixel of interest from each band of the image as the input (giving a total of 54 nodes in the input layer) [12]. This input modification takes local textural information into account.

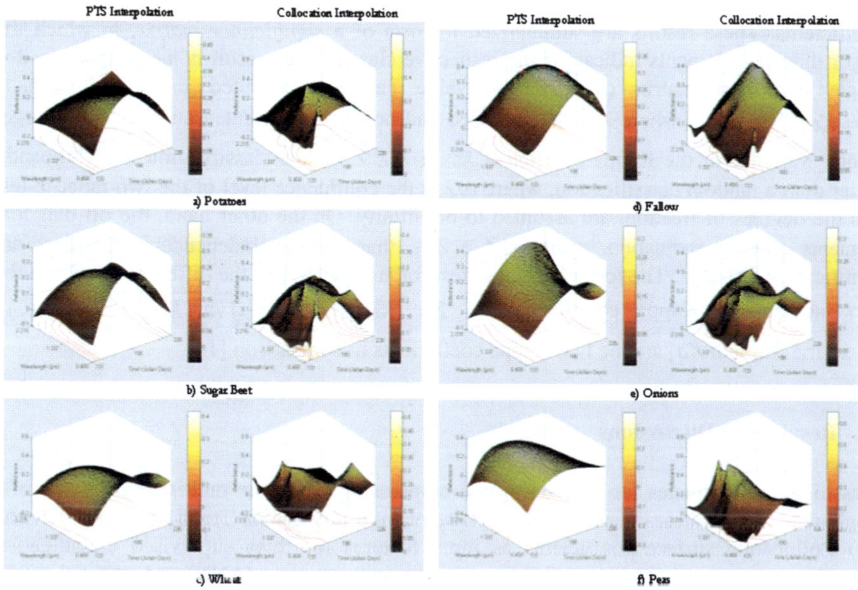

Figure 1. Analytical surfaces and contours for several crops.

All neural networks configurations tested had an output layer with 6 nodes, corresponding to the 6 general crop classes. The number of hidden layers (1) and the number of hidden nodes (10) were found using the Hirose *et al.* [13] building up procedure. The learning rate and momentum were set initially at 0.2 and 0.9 respectively. The learning rate was reduced during training to 0.1 after 1000 epochs.

For the second experiment, two sample sets were selected using stratified random sampling based on the reference image ("ground truth"), which was generated at the same scale and with the same projection system as the remotely sensed data. Each sample has also 120 patterns per class, giving a total of 720 patterns. One sample set was used to train the classifiers and the other independent sample set was reserved for the assessment of the accuracy of the classification.

4.2 *Accuracy Assessment*

In order to perform a systematic investigation of the relative cost of incorporating the temporal dimension into the crop classification process, standard accuracy measures derived from a confusion matrix were computed, using an independent test data set based on the Field Data Printouts collected from individual farmers in the study area. The measures used were: overall accuracy, individual class accuracy, producer's accuracy and user's accuracy. The calculations associated with these measures are described in standard textbooks (e.g., [1]). The Kappa coefficient, conditional Kappa for each class, and test Z statistics, all of them widely used statistics derived from the contingency matrix, were also computed [14].

In addition, a pairwise test statistic for evaluating the significance of the classifiers (represented here by their respective confusion matrices), was calculated utilizing the Kappa coefficients. These results are summarized in form of a *significance matrix*, in which the major diagonal elements indicate if the respective classification result is meaningful. In this single confusion matrix case, the Z value can be computed using the formula $Z = Ka/\sqrt{\mathrm{var}(Ka)}$, where Z is standardized and normally distributed and *var* is the large-sample variance of the Kappa coefficient *K*. If $Z \geq Z_{\alpha/2,}$ the classification is significantly better than a random classification, where $\alpha/2$ is the confidence level of the two-tailed *Z* test and the degrees of freedom are assumed to be infinity. On the other hand, the off-diagonal elements give an indication, again if $Z \geq Z_{\alpha/2}$, that the two independent classifiers are significantly different. The formula used to test for significance of the difference between the two independent Kappa coefficients is: $Z = |Ka_1 - Ka_2| \big/ \sqrt{\mathrm{var}(Ka_1) + \mathrm{var}(Ka_2)}$, where the Ka_1 and Ka_2 are the two Kappa coefficients in comparison [14].

5 Results and Discussions

Classification accuracies for six agricultural crops using (i) the six multispectral bands of a single-date TM Landsat image, (ii) the coefficients of a Polynomial Trend Surface (PTS) and (iii) Collocation surface coefficients as input features for three supervised classification algorithms - maximum likelihood (ML), artificial neural networks (ANN) and artificial neural network texture (ANNT) - are presented in Table 1. Individual classification accuracy for each crop (Conditional Kappa * 100), overall accuracy, the value of the Kappa coefficients and their variances, and test Z statistic are reported in this table. These accuracies were calculated from an independent dataset (720 patterns). The pixels received the label of the output class having the highest probability.

As the absolute value of the test Z statistic is greater than the critical value of 1.96, all the classification results are significant better than a random classification at the 95% confidence level. The accuracy levels for the single date image increased from 72.9% (NL) to

Table 1. Classification accuracies for six agricultural crops using Single-Date LANDSAT Image, Polynomial Trend Surface (PTS) and Collocation (COL)) and three classification algorithms - maximum likelihood (ML), artificial neural networks (ANN) and artificial neural network texture (ANNT). The table shows individual classification accuracy for each crop (Conditional Kappa * 100), overall accuracy, the value of the Kappa coefficients and their variances, and test Z statistic. If the absolute value of the test Z statistic is greater than 1.96, the result is significant better than a random classification at the 95% confidence level. These accuracies were calculated from an independent dataset test (720 patterns).

INTERPO.	LANDSAT(27/06/94)			STRS	
CLAS/MET	ML	ANN	ANNT	PTS-ML	COL-ML
Potatoes	64.5	66.8	71.9	95.9	94.9
Sugar Beet	53.8	57.9	58.6	73.8	75.3
Wheat	70.9	75.6	95.5	89.3	92.8
Fallow	80.4	81.8	79.7	70.8	63.3
Onions	84.9	89.7	88.0	95.9	97.8
Peas	53.5	67.9	80.8	93.0	100.0
OVERALL(%)	72.9	77.6	81.7	87.4	87.2
Kappa	0.675	0.732	0.780	0.848	0.847
Variance	0.000394	0.000347	0.000299	0.000219	0.000222
Z	33.99	39.28	45.09	57.27	56.88

77.6% (ANN) and to 81.7 %(ANNT), showing that ANN-based procedures perform better than statistical methods in this case, and also demonstrating that increased classification accuracy resulted from the incorporation of texture information. However, the use of the STRS coefficients as inputs to a ML classifier gave a 5.7% increase in classification accuracy, compared to the single-date image. Oddly, fallow is the only individual category for which accuracy decreased using the STRS approach.

The lower performance achieved with ML classifier using only the TM multispectral bands is believed to be due in part to a non-linear separability of the classes under study and also to the fact that the size of the training data set was inconsistent with the design properties and assumptions of the supervised maximum likelihood algorithm. Moreover, for some of the crops (e.g., sugar beet and potatoes, or onions and peas) the multispectral profiles for the chosen date are not very well separated. Even so, the neural models produce a satisfactory performance on the same data set. Furthermore, the separability of the classes is considerable improved when the local spatial variance of individual pixels is implicitly taken as input into the neural network model by employing a window as implemented in the ANNT algorithm and described earlier.

Table 2 provides the computed Z values for a pairwise statistical test in order to check how significant are the improvements in classification accuracy. The classification accuracy obtained using the STRS approach (PTS and Collocation coefficients and the ML algorithm) was found to be significantly better than the accuracy achieved by any of the individual classifiers (ML, ANN and ANNT) using a single-date image (see yellow pair, $Z > 1.96$ at the 95% confidence level). This demonstrates the need to utilise the STRS approach in order to achieve the highest possible accuracy in crop classification. It should be observed that there is no significant difference between the performance of the ML procedure using either the PTS or the Collocation coefficient as input features (see blue pair, $Z = 0.05 < 1.96$). It can be concluded that, for this data set, these two sets of features are equivalent. If two different techniques or algorithms were being tested and if they were shown to be not significantly

Table 2. Results of Kappa Analysis for comparison among the classifiers. The table also presents the Kappa coefficients and variance for each classifier. The Z values (in major diagonal and off diagonal elements) were computed using formula as describe in subsection 4.2.

CLASSIF	ML	ANN	ANNT	TSA	COL
KAPPA	0.675	0.732	0.78	0.848	0.847
VAR	0.000394	0.000347	0.000299	0.000219	0.000222
ML	34.01				
ANN	2.09	39.30			
ANNT	3.99	1.89	45.11		
TSA	6.99	4.88	2.99	57.30	
COL	6.93	4.82	2.94	0.05	56.85

different, then it would be appropriate to use the cheaper, quicker, or more efficient approach.

As expected, the performance of the neural network classifiers exceeds that of the ML classifier using a single date Landsat TM image. However, our results indicate that there are no significant differences in performance between the ANN and ANNT algorithms ($Z = 1.89 < 1.96$) at the same confidence level.

6 Conclusions

A methodology for classifying agricultural crops combining multitemporal, multi-spectral and multisource remotely sensed data has been shown to be effective in identifying general agricultural crop classes over an area in East Anglia (UK). Classification accuracies in excess of 87% were achieved, even though parts of some of the images are covered by clouds. The basic assumption of the method, that different crops have different spectral-temporal trajectories, has been used in earlier studies. However, the methods used to characterise the spectral reflectance changes over a growing season using a spectral-temporal surface represents a promising new approach, for several reasons. First, the method can deal with multisensor data, as the spectral bands measured at each date do not need to be the same. Second, data points obscured by clouds can be filtered out throughout the interpolation and parameterisation procedures of the analytical surfaces. Third, the overall spectral variation of a given crop class over the growing season is captured by a set of coefficients, which are fewer in number than the training data pixels and hence produce computationally more efficient classifiers.

7 Acknowledgements

Preliminary research by Dr. Vieira was supported by the Brazilian Research Council (CAPES). The later stages of this study were conducted as part of the FET-ENVIS project, a European Commission-funded RTD (Proposal Number IST-1999-29005) performed in collaboration with Nansen Environmental and Remote Sensing Centre, Norway and Ruhr-Universität Bochum, Germany. We are grateful to Logica PLC and SPOT Image for permission to use their images. Computing facilities were provided by the School of Geography, The University of Nottingham.

References

1. Mather, P.M., Computer Processing of Remotely-Sensed Images: An Introduction (John-Wiley and Sons, Chichester, Second edition, 1999).
2. Badhwar, G.D., Autin, W.W., Carnes, J.G., A semi-automatic technique for multitemporal classification of a given crop within a Landsat scene. *Pattern Recognition* **3** (1982) pp. 217-230.
3. Haralick, R.M., Hlavka, C.A., Yokoyama, R., Carlyle, S. M., Spectral-temporal classification using vegetation phenology. *IEEE Trans. on Geoscience and Remote Sensing* (1980) **GE-18/2** pp. 167-174.
4. Lambin, E.F., and Strahler, A.H., Indicators of land-cover change for change-vector analysis in multitemporal space at coarse spatial scales, *International Journal of Remote Sensing* (1994) **10** pp. 2099-2119.
5. Jensen, J.R., Introductory Digital Image Processing: A Remote Sensing Perspective (Prentice-Hall, Englewood Cliffs, NJ, 1986).
6. Teillet, P.M. and Fedosejevs, G., On the dark target approach to atmospheric correction of remotely sensed data, *Canadian Journal of Remote Sensing* **21**(1995) pp. 374-387.
7. Tanré, D., Deroo, C., Duhaut, P., Herman, M., Morcrette, J.J., Perbos, J. and Deschamps, P.Y., Description of a computer code to simulate the satellite signal in the solar spectrum: the 5S code, *International Journal of Remote Sensing* **11** (1990) pp. 659-668.
8. Vieira, C.A.O., Mather, P.M., and McCullagh, M.J., The Spectral-Temporal Response Surface and its use in the multi-sensor, multitemporal classification of agricultural crops. *In ISPRS: IAPRS* **33** (2000), Amsterdam, The Netherlands, Part B2 (Amsterdam, International Society for Photogrammetry and Remote Sensing) pp. 582-589.
9. Watson, D.F., Contouring: A guide to the analysis and display of spatial data. (Pergamon Press, Oxford, 1999) pp. 101-161.
10. Mather, P.M., Computational Methods of Multivariate Analysis in Physical Geography, (John Wiley and Son, Chichesters, 1976).
11. Benediktsson, J.A.; Swain, P.H. and Ersoy, O.K., Neural network approaches versus statistical methods in classification of multisource remote sensing data. *IEEE Trans. on Geoscience and Remote Sensing*, **GE-28** (1990) pp. 540-552.
12. Paola, J.D., and Schowengerdt, R.A. A review and analysis of backpropagation neural networks for classification of remotely-sensed multispectral imagery, *International Journal of Remote Sensing* **16** (1995) pp. 3033-3058.
13. Hirose, Y., Yamashita, K. and Hijiya, S. Back-propagation algorithm which varies the number of hidden units, *Neural Networks* **4** (1991) pp. 61-66.
14. Congalton, R.G., and Green, K., Assessing the Accuracy of Remotely Sensed Data: Principles and Practices (Lewis Publishers, New York, 1999).

MULTI-TEMPORAL, MULTI-SENSOR SAR DATA FOR OPERATIONAL RICE FIELD MAPPING

F. HOLECZ , E. DWYER , S. MONACO

sarmap, Cascine di Barico, 6989 Purasca, Switzerland
E-mail francesco.holecz@sarmap.ch,

B. SCHMID

geoconcept, Dammstrasse 54, 8037 Zurich, Switzerland
E-mail: barbla.schmid@geoconcept.ch,

R. FISCHER [3], AND U. FREI [3]

Remote Sensing Laboratories, Winterthurerstrasse 190, 8057 Zurich, Switzerland
E-mail: frei@rsl.geo.unizh.ch

This paper presents the main achievements of a project that was carried out in the framework of ESA's Data User Programme, and which was dedicated to rice field mapping in tropical areas. The major result of the project is *SARscape*®, a tailor-made software for land applications, which is now commercially available. In this project it was used for rice mapping and rice growth monitoring utilising a time series of ERS and Radarsat-1 Synthetic Aperture Radar data as input. The modular SAR processing chain is embedded in the commonly used *ArcView*® (Environmental Systems Research Institute, Inc.) Geographic Information System environment and allows novice users, even if they are unfamiliar with SAR technology, to generate thematic products such as rice maps and to monitor rice growth in an automatic way. Co-registration, speckle filtering and other algorithms have been adapted to exploit the information in time series data. The performance of this system and the resulting products are shown for a test site located in Sri Lanka.

1. Introduction

Over 90% of the world's production and consumption of rice is in Asia. The acerage of irrigated rice is expected to increase in coming years, but given the conflicting demand for limited water resources, combined with concern about the adverse affects of irrigation, the increase in production may be limited. There is therefore a need for efficient monitoring of rice fields in order to optimise use of resources and forecast rice yield. Conventional methods for rice monitoring are based on ground collected statistics, which have proved time consuming, inaccurate and expensive. Spaceborne remote sensing offers an effective alternative to these traditional methods through operational, weather independent systems utilizing Synthetic Aperture Radar technology.

To address the limited availability of commercial software systems, designed with the SAR user in mind, sarmap has developed *SARscape*®, a complete SAR processing and analysis system, embedded in the widely used GIS software package *ArcView*®. The use of *SARscape*® for mapping and monitoring of rice fields is illustrated for an important rice growing area located in eastern Sri Lanka, where multi-temporal ERS SAR acquisitions combined with Radarsat-1 SAR data were used.

2. Test site and data

The test area of Polonnaruwa (about 30 km by 40 km) is located in the flat lowlands of eastern Sri Lanka. Agricultural activities are mainly rice cultivation and homesteads. The cultivated land is irrigated with water from tanks and rainfall during the monsoon from November to March. There are normally two rice seasons: the Maha season from November to March, and the Yala season from May to September. Within this project, the rice growing during the Maha season 1999/2000 was monitored.

An ERS SAR time-series of 9 images was acquired in order to monitor a complete rice growing cycle. Six Radarsat-1 SAR images were also used to evaluate ENVISAT ASAR characteristics with respect to temporal resolution, variable incidence angle, and alternate polarization mode. An overview of all acquisitions is given in Table 1.

Table 1: ERS and Radarsat-1 data acquired for the Polonnaruwa test site

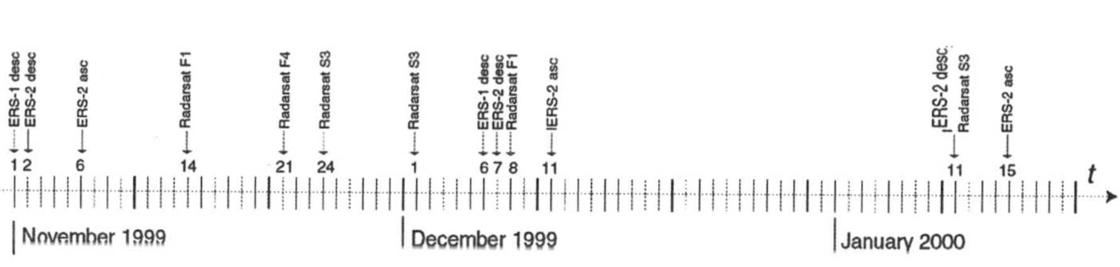

3. Data processing

The basic idea behind the generation of rice maps using remote sensing SAR techniques is the analysis of changes in the acquired data over time. Measurement of temporal changes in reflectivity of the plants relates to the phenological status of the rice: an increase in the radar backscatter corresponds to a growth in the rice plants and therefore to an increase in rice biomass. The radar response to rice fields at different growing stages during the crop cycle can be distinguished in three main growing stages, namely sowing-transplanting (surface scattering), growing (surface-volume scattering) and flowering (volume scattering) [1].

SARscape® was used to carry out all data processing and analysis steps. Figure 1 shows the main functionailty of this system. It handles both ERS and Radarsat SAR data, from which multilooked and interferometric products can be generated. Landsat TM and IKONOS data may also be ingested to facilitate fusion of SAR and optical data. The system is currently being updated to be ENVISAT compatible.

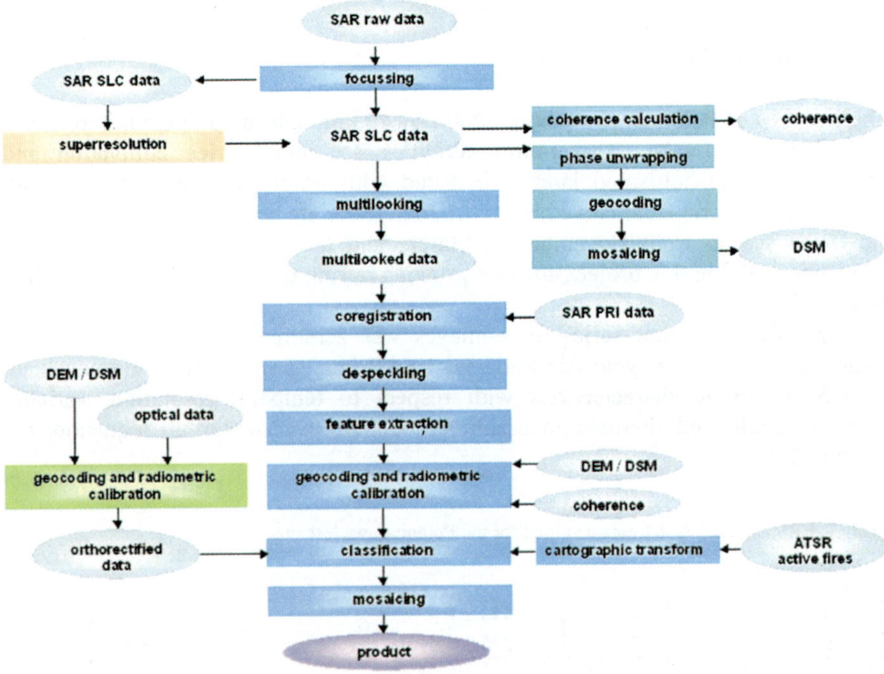

Figure 1. Block diagram showing the main functionality of SARscape®, which allows processing to interferometric and multilooked SAR data products.

Firstly, SAR raw data were focused using an omega-k processor and subsequently multi-looked to amplitude data. Images having the same orbit and frame were automatically coregistered by maximising the correlation in a number of subwindows in the image. Speckle reduction was carried out using a recently developed method for multi-temporal images which involves a combination of a modified version of Lee's spatial filter with an averaging in time. The full spatial resolution of all the imagery is retained while there is a significant increase in the signal-to-noise ratio [2]. This method gives superior results when compared with other methods, in particular if more than four images are used. For example in a deforestation mapping project carried out by us using a time-series of 15 ERS images, which had been multilooked to 5 azimuth and 1 range looks, the speckle filtered images had an average of 25 equivalent number of looks at a ground spatial resolution of 25 metres.

Since the various data sets were acquired from different orbits and systems, the SAR data were terrain geocoded into the local cartographic reference system and radiometrically calibrated [3]. Only in this way is it possible to combine and compare σ^o data. The appropriate part of the GTOPO30, a Digital Elevation Model with a grid spacing of 30 arc seconds (roughly corresponds to 1 km), was resampled and used [4]. Supervised classification based on the maximum likelihood approach was used for the rice map generation.

4. Results

4.1 Multitemporal data sets

Backscattering coefficients from ERS-2 SAR data show changes of the order of +8 dB during the rice cultivation. This increase in the radar backscatter is shown in Figure 2 (a), where SAR data have been acquired during the descending orbits of November 2 1999 (red), December 7 1999 (green), and January 11 2000 (blue). Rice fields can be clearly distinguished as red areas, while forest and other agricultural extents, whose variations over time are constant or different, appear in green. The SAR data have been speckle filtered using the multi-temporal approach and subsequently terrain geocoded and radiometrically calibrated, using ground control points and the resampled DEM data .

Due to the coarse resolution of the GTOPO30, actual relief is poorly approximated , especially in areas where topography is highly varying. The height inaccuracy of those areas has an impact on the location of the pixel and on the scattering area correction, where its effect on the SAR data can be only partially compensated (see bottom left in Figure 2 (a)). Nevertheless, the use of GTOPO30 leads to far better results, in regions where relief variations are present, than an ellipsoidal geocoding.

The possibility of combining data acquired from different orbits and SAR systems is important for rice mapping purposes, mainly because of time synchronisation problems between rice growth (in particular at sowing-transplanting stage) and orbit repeat cycle. A failure of the system or delay of the acquisition at the earliest stages (due to windy conditions, etc.) can compromise the mapping accuracy of the final product. This is illustrated in Figure 2 (b), where ERS data have been acquired during the ascending orbit of December 11 1999 (red), and descending orbits of November 1 1999 (green) and January 11 2000 (blue). In

Figure 2 (c) ERS-1 data from the descending orbit of November 1 1999 (red), and the ascending orbit December 11 1999 (green) and Radarsat-1 S3 data from the descending orbit of December 1 1999 (blue) have been combined. Note, the high similarity between the two images (b) and (c): rice fields can be clearly identified in red tones while forest appears in grey-green.

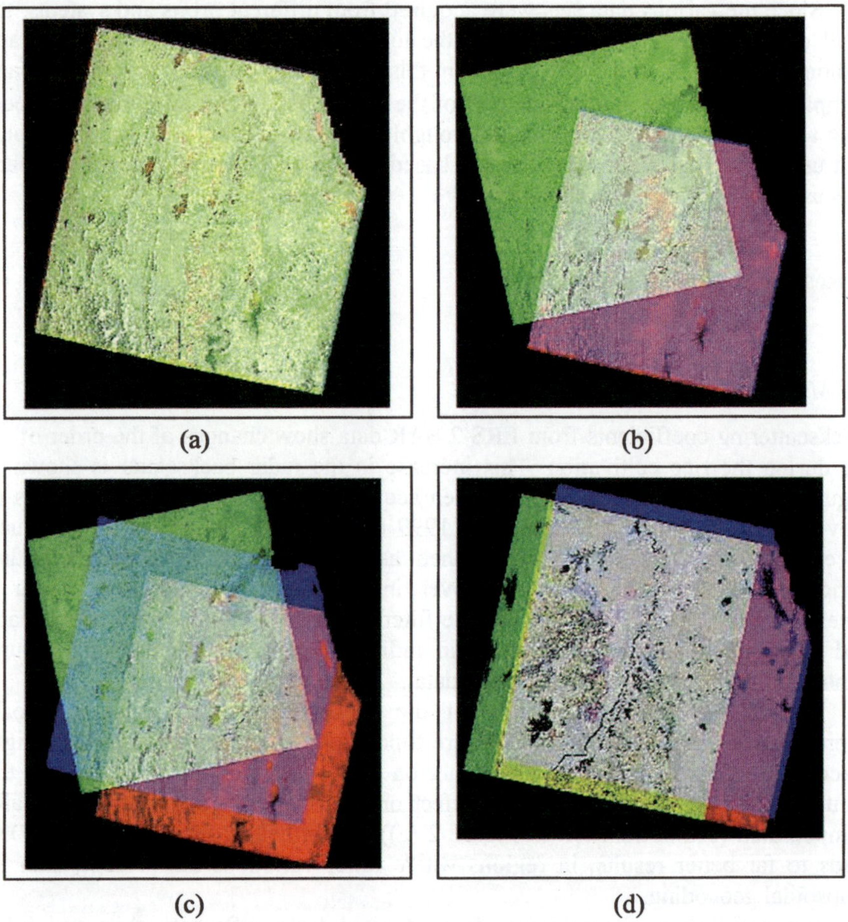

Figure 2. Terrain geocoded ERS-2 multi-temporal backscattering coefficients. (a) descending (b) descending and ascending orbits. (c) ERS-2 descending and ascending orbits combined with Radarsat-1 S3 descending orbit. (d) Three terrain geocoded backscattering coefficients of Radarsat-1 fine beam acquisitions.

The spatial resolution of the SAR system plays an important role for precise rice mapping, especially in those regions like Polonnaruwa, where the dimensions of rice parcels are often under half a hectare. Furthermore, the possibility of merging SAR data from different acquisition modes improves the data availability.

Three Radarsat-1 Fine Beam mode acquisitions were acquired and combined as shown in Figure 2 (d). Note, that due to its higher spatial resolution (10 m), rice fields (green-yellow) are more easy to recognise than in Figure 2 (a) with ERS at 25m.

4.2 Classification results

An example of a semi-automated supervised classification, based on the maximum likelihood approach, considering the backscattering coefficients from ERS-2 descending orbits over the whole rice season, is shown in Figure. 3 (left). Since *SARscape®* products are already in a GIS environment, they can be easily integrated with other data for further kinds of analysis and/or planning tasks. Figure. 3 (right) shows the rice map generated overlaid with the ERS backscattering coefficient, hydrological data (light blue), road transport data (grey), and some rice field boundaries (red), which were mapped during a field campaign undertaken in November and December 1999.

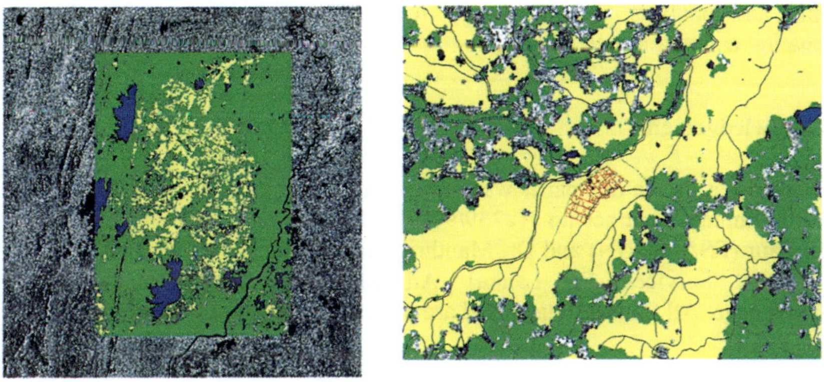

Figure 3. Rice map derived from multi-temporal ERS-2 backscattering coefficients (left) and enlarged area overlaid with available GIS data (right) . Rice: yellow, water bodies: blue, non-rice land: green.

5. Conclusions

The results of this project show that it is feasible to map and monitor rice in an operational way but that no standard procedures can be applied. Each area has to be considered individually with respect to crop calendar, plant varieties, rice field dimensions, and data availability (topographic maps, DEM data, etc.). The temporal and spatial resolution of the SAR data to be used must be selected accordingly.

However, satellite overpass and flooding of the paddy fields should coincide temporally, and the weather and surface conditions at the beginning of the rice growing cycle should be favourable, i.e. without wind. In addition expert knowledge with respect to rice is required to generate properly the desired products.

The utility of time series SAR data for environmental monitoring (e.g. crop development, deforestation, burnt area mapping, etc.) is now proven and with the imminent launch of Envisat, the continuing availability of SAR data is assured. Sarmap has developed its own commercial, tailor-made software package *SARscape*®, for land applications. This package is embedded in the commonly used *ArcView*® GIS environment and handles different SAR and optical systems [5]. It is particularly suitable for the processing and analysis of time series data, as it has optimised co-registration, speckle filtering and parameter extraction algorithms built in. In addition many of the tasks can be carried out in batch mode.

In coming years with the vast quantities of EO data that are becoming available, the use of time series data will become more commonplace as software systems and analysis techniques are further developed to exploit the large information content in such data sets.

6. Acknowledgements

The work described in this paper was carried out in the framework of the ESA Data User Programme, contract no. 12239/97/I-HGE. The authors would like to thank Dr. H. Laur (ESA/ESRIN) and Dr. Manthrithillake and his team (Natural Resources Management Services, Sri Lanka). All ERS data used are copyright of ESA/Eurimage. All Radarsat data used are copyright of Radarsat International.

References

1 Le Toan T. et al., *Rice crop mapping and monitoring using ERS-1 data based on experiment and modelling results*, IEEE Transactions on Geoscience and Remote Sensing, **35**, **1**,(1997).

2 De Grandi, G.F. et al., *Radar reflectivity estimation using multiple SAR scenes of the same target: techniques and applications*, Proceedings of IGARRS conference, Singapore, (1997).

3 Holecz F. et. al., *Calibration of airborne and spaceborne SAR data*, Proceedings of the SAR Calibration Workshop, CEOS Calibration/Validation Working Group, Ann Arbor, (1994).

4 http://edcdaac.usgs.gov/gtopo30/gtopo30.html.

5 sarmap homepages: http://www.sarmap.ch, 2001.

SEASONAL LANDSAT TM DATA TOPOGRAPHIC DEPENDENCE IN RUGGED DECIDUOUS FOREST AREAS

MASSIMO VINCINI*, DAVID REEDER** AND ERMES FRAZZI*

**Centro Ricerca Analisi Spaziale e Telerilevamento, Università Cattolica del Sacro Cuore, via Emilia Parmense 84, 29100 Piacenza, Italy, E-mail: mvincini@pc.unicatt.it*
***Department of Earth Sciences, Dartmouth College, HB6105, Hanover, New Hampshire, 03755 USA, E-mail:David.H.Reeder@Dartmouth.edu*

ABSTRACT: The topographic dependence of multi-temporal Landsat TM data (band 3, 4 and 5) collected on different periods of the growing season (April–August) was studied for two contiguous deciduous forest areas in northern Apennines (beech, 366 ha, and mixed deciduous forest, 343 ha). Minnaert topographic correction constants (*K*s) and *c* correction factors were derived for the two forest types and temporal significance levels of the models were examinated showing a consistent behavior of the models from image to image. A robust linear relationship between log *c* and Minnaert *K*s was evident. Minnaert normalization constants as well as *c*-correction factors value were dependent on sun zenith angle for band 3 and 5.

1 Introduction

As shown in Figure 1 where a Landsat ETM image of the study site is reported, remote sensing of vegetation in rugged areas is severely affected by the topographic effect. The solar irradiance incident on a slope varies strongly with slope azimuth relative to the sun and reflectance of the slope varies with the incidence angle (i) between the solar vector and the slope normal and the excitance angle (e) between the slope normal and the observer.

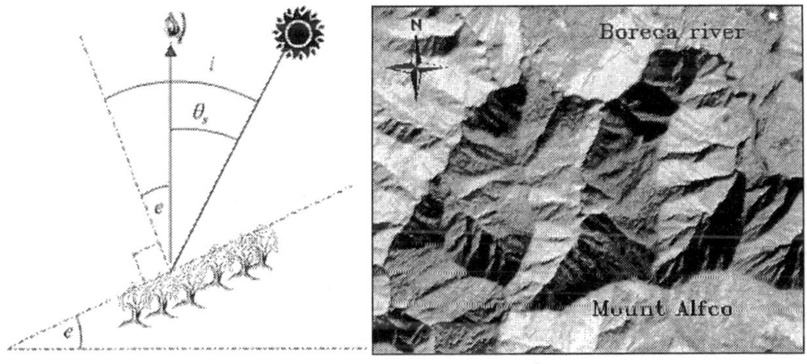

Figure 1. Geometrical presentation of reflection angles for a sensor viewing at nadir and Landsat ETM image of the study site for 29 Sept. 1999; i incidence angle, θ_s solar zenith angle, e excitance angle.

Although many *in situ* measurements have been carried out [1,2], complete bidirectional reflection distribution functions (BRDFs) for forest canopies are not available. A consistent topographic correction method is therefore required for regional and local forest mapping and inventorying using high resolution images such as obtained by Landsat TM. The model should be as simple (i.e. generally applicable) as possible while accounting for most topographic effect.

Among classical parametric models using DEMs for estimating the topographic normalized radiance (L_n, for i=e=0) from slopes observed radiance (L) are:

- The *Lambertian model* (cosine correction method) assuming for all wavelengths (bands) that the bidirectional reflectance factor is a constant independent of incidence and excitance angles,
$$L = L_n \cdot \cos i$$
being fallacious in its assumption it usually underestimates the reflectance of sun-facing slopes and overestimates the reflectance of slopes facing away;
- The *Minnaert Model* [3] deriving for each band K constants, empirical coefficients describing the bidirectional reflection distribution of the surface where K=1 represents the Lambertian model and decreasing K values indicate increasing anisotropic behaviours; Minnaert constants are estimated calculating the slope of the regression line:
$$\log(L_\lambda \cdot \cos e) = \log L_{\lambda n} + K_\lambda \cdot \log(\cos i \cdot \cos e)$$
obtained by linearizing the Minnaert empirical photometric function [4]:
$$L_\lambda = L_{\lambda n} \cdot \cos^{k_\lambda} i \cdot \cos^{k_\lambda - 1} e$$
if K=1 the function reduces to the Lambertian model;
- The *c-correction method* [5], based on the observed empirical linear correlation between L and cos i, estimates for each band a correction factor c by taking the ratio of the regression offset a to the regression slope m:
$$c_\lambda = \frac{a_\lambda}{m_\lambda} \qquad L_\lambda = L_{\lambda n} \cdot (\cos i + c_\lambda)$$
if c=0 the function reduces to the Lambertian model.

While the cosine method is basically an application of the Lambert photometric function, the Minnaert and the c-correction are empirical methods estimating the correction factors for each band from the data (image) to be corrected. A relevant difference between the two is that the c-correction method doesn't take in consideration the excitance angle which, according to the Helmoltz reciprocity principle, should act in the same way as the incidence angle.

Both methods produce fewer topographic residuals than the Lambert correction and have been used, with comparable results, for reducing the topographic effect on spectral signatures of several forest types in a wide range of geographic locations [3,5,6,7]. However, the Minnaert model, as well as the c-correction method, have been reported to perform inconsistently from image to image [8] because of the variability of the correction constants values for the same vegetation cover type.

Forest-canopy BRDF vary with species [9], stand age and the season of the year [10], as a consequence of variations of the biophysical properties governing the vegetation optical behaviour. Therefore, an analogous variability of correction factors values should be expected for images collected on different phenological stages of the same vegetation cover. Previous results of the present study [11] have shown the dependence of Minnaert *K* values on the period of the growing season.

This study addresses the evaluation of the consistency of the Minnaert and c-correction models from image to image for two deciduous forest types in the northern Apennines. The work has been conducted as a part of the project 'X-SAR SRTM Mission Italian Activities' under the ASI contract n° ARS-B.0/99-08.

2 Methods

Surfaces of the study area in the northern Apennines (Italy) are characterized by a wide range of slopes and aspect angles and, as a consequence, by a wide range of values of the parameter regulating the topographic effect, the incidence angle (*i*). Elevation ranges in the study area from about 650 m for the Boreca riverbed to 1620 m for the Alfeo mountain. Two contiguous areas where chosen in the study site for topographic correction, the former (366 ha) covered by beech (*Fagus Sylvatica*) forest, roughly above 900 m of elevation, and the latter (343 ha), at lower elevations, by mixed deciduous forest.

The Landsat 5 TM data, selected to represent typical spring and summer forest phenological stages and imaging conditions, were for 21 April 96, 5 May 95, 8 June 96, 13 July 97 and 3 August 87 with a solar zenith angle of respectively 42.5, 39.0 33.5, 32.6 and 36.7 degrees. No direct shadowing was present over the study area for all images. The 30x30 m spatial resolution DEM used was derived by linear interpolation of the contours from the 1:25.000 scale regional digital topographic map. The resulting DEM was smoothed by the application of a mean (low pass) 3x3 filter. The images of the study site and the DEM were co-registered by nearest neighbor resampling with twelve control points and a root mean square error of less than 0.5 pixels. Minnaert constants (*K*) and *c* correction factors were derived, as done by most authors, from uncalibrated radiance data (DNs), using the IDL© procedure developed at Dartmouth College. The atmospheric correction of TM data by the "dark object" method was not considered because of its assumption of an isotropic skylight distribution and a lambertian surface both of which are fallacious [12].

3 Results and Discussion

In Figure 2 are reported calculated Minnaert *K* values, for all dates and both forest types, and topographically corrected NDVI mean values [11]. In Table 1 are reported the squared regression correlation coefficient values of the *K* and *c* estimates, indicating the variance fraction of the observed radiance explained by the topographic effect according to the two models.

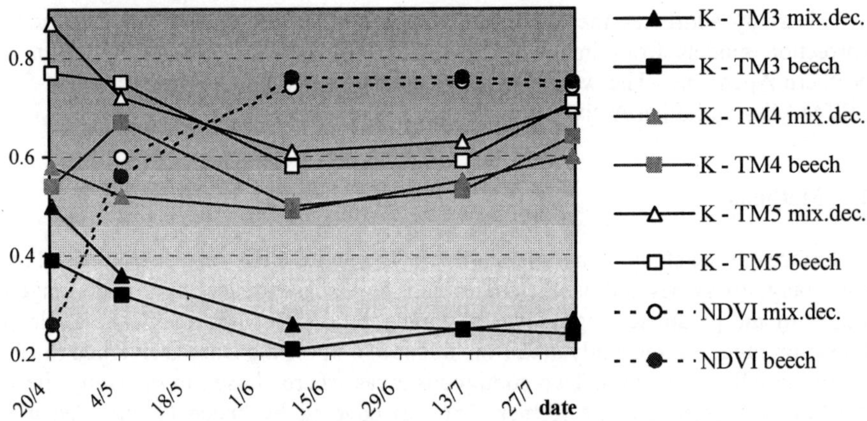

Figure 2. Minnaert K constants and mean NDVI values for all dates and both forest types.

As visible in Figure 2 Minnaert *K* values for the three bands and both forest types seem to show a similar seasonal change pattern with the exception of the band 4 *K* value for the May image which was however characterized by the lowest significance level of the K estimates for band 4 (Table 1) as a consequence of the unevenness of early foliage development, evident from the NDVI data, over the large elevation range of the study area.

Table 1. Squared correlation coefficients (r^2) of the *K* and *c* estimates.

		TM band 3		*TM band 4*		*TM band 5*	
		mix.dec.	*beech*	*mix. dec.*	*beech*	*mix.dec.*	*beech*
Apr.	*K*	0.83	0.64	0.83	0.76	0.87	0.79
96	*c*	0.79	0.58	0.83	0.74	0.87	0.79
May	*K*	0.74	0.55	0.71	0.64	0.83	0.72
95	*c*	0.71	0.46	0.69	0.62	0.83	0.74
June	*K*	0.66	0.59	0.77	0.69	0.83	0.74
96.	*c*	0.59	0.53	0.74	0.67	0.83	0.71
July	*K*	0.66	0.62	0.81	0.69	0.83	0.72
97	*c*	0.61	0.49	0.77	0.67	0.81	0.71
Aug.	*K*	0.72	0.66	0.85	0.76	0.88	0.79
87.	*c*	0.66	0.59	0.83	0.72	0.88	0.77

In Figure 3 the scatterplots of Minnaert-corrected and uncorrected TM band 3 and 4 vs. cos *i* are reported for the mixed deciduous forest July image. Figures 3 display the strong topographic control of TM data in the study area as well as the effectiveness of the Minnaert model even for band 3 in summer when canopy absorbs the majority of light.

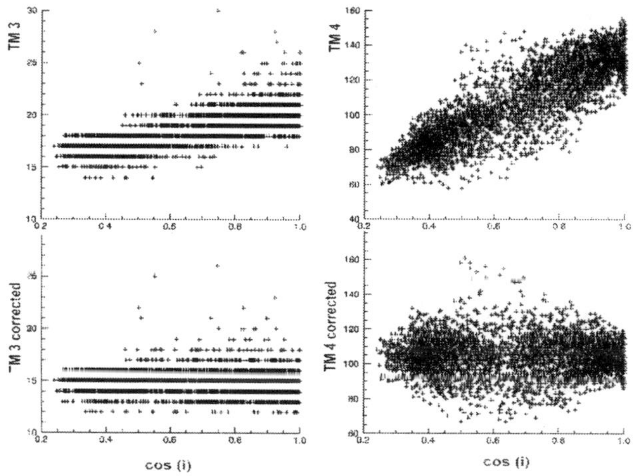

Figure 3. Scatterplots of corrected and uncorrected TM bands 3 and 4 vs. cos *i* for mixed deciduous forest, July 1997.

Slight inaccuracies of image and DEM co-registration, affecting the significance level of the estimates, can mask the seasonal variation of the significance level of the models. Nevertheless, four main factors contributing to the significance level of the estimates can be figured out from the data reported in Table 1:

- *Forest type dependence*, for all dates and bands r^2 values were smaller for beech forest than for mixed deciduous forest, possibly as a consequence of a smaller evenness level of the vegetation cover;
- *Wavelength dependence;* for all summer dates, when both forest types has reached maximum NDVI, regression significance increased from band 3 to band 5 as a consequence of the canopy absorption of the majority of the light in the red band (Figure 3) allowing the atmosphere path radiance, maximum for band 3 and minimum for band 4 and 5, to dominate [12,13];
- *Phenological stage dependence;* likewise, the low vegetation coverage in the early season and the consequent low absorption in the red band, led for band 3 to the maximum significance of the estimates for the lowest NDVI (0.25) April image;
- *Solar elevation dependence;* the seasonal variation of the significance levels of the estimates for bands 4 and 5, not affected by marked seasonal differences of the vegetation absorption, reflected the strength of the topographic control, i.e.

the solar zenith angle, with higher significance levels far from the summer solstice; again the only exception was for the low significance level of the May image due to the unevenness of early foliage development.

The difference existing between the significance levels of two models (Table 1), both reducing to the lambertian model for lambertian reflectors, was proportional to BRDF anisotropy, with significance levels of the Minnaert model appreciably higher for small K values (band 3 especially in summer), slightly higher for band 4 and practically equivalent for band 5 characterized by a near lambertian behavior. As shown in Figure 3 where the regression line between the decimal logarithm of the *c* correction factors and the *K* values for all dates and both forest types is reported, a robust relationship between the correction factors derived by the two models was evident.

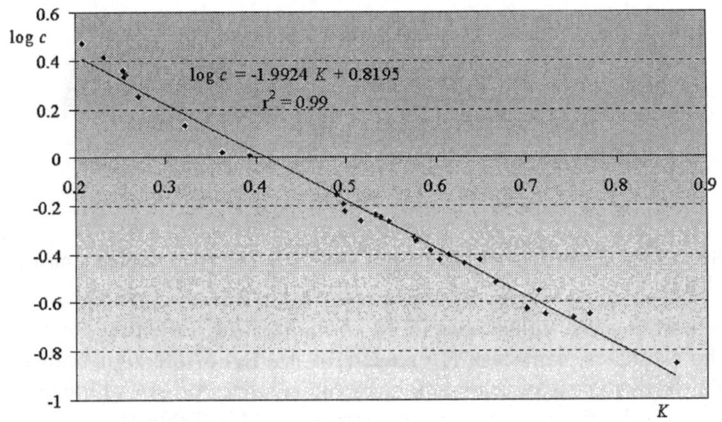

Figure 4. Regression line between log c and K for all dates and both forest types.

Both models involve an empirical estimate of the shift of the vegetation cover from the lambertian behavior and a certain correlation between the estimated correction factors should therefore be expected, but the remarkably neat fitting reported in Figure 4 seems rather to indicate that the *c* factor, obtained without taking in consideration the excitance angle, is as good as the Minnaert constant at describing forest BRDFs and modeling the topographic effect in TM data.

As visible in Figure 2 for Minnaert *K* values, the two models seem to show, regardless of the forest type, a seasonal change pattern of the correction constants similar for the three bands, with reflection approaching the lambertian behaviour moving far from the summer solstice. In Figure 5 the regression lines between the *c* factors for TM band 3 and 5 and the cosine of the solar zenith angle are reported for mixed deciduous forest, characterized by a higher topographic normalization significance level.

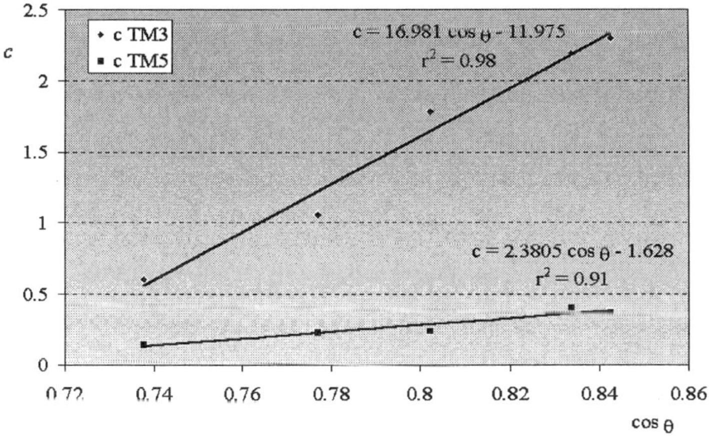

Figure 5. Regression lines between *c* and the cosine of the solar zenith angle for mixed deciduous forest.

Such relationships, evident also for the same bands in beech forest, but not significant for band 4 data of both forest types possibly as a consequence of the phenological stage (i.e., foliage development affecting NIR reflection behavior), suggest the dependence of topographic normalization factors on the image solar zenith angle. For a sensor viewing at nadir the solar zenith angle can be considered as the phase angle, the angle between the illumination source and the observer, which, although disregarded in literature for topographic normalization purposes, do affect the topographic normalization factors [3,4]. Further work is required to support the existence of this dependence on solar zenith angle that would explain most of the seasonal variability of normalization factors for the same vegetation cover.

4 Conclusions

Provided that the Minnaert model and the c-correction method are applied to a homogeneous forest vegetation cover they have consistent behavior and converging results from date to date with significance levels depending on wavelength, on the strength of topographic control (i.e. the solar zenith angle) and on the phenological stage of the vegetation cover.

For the studied deciduous forests the *c* correction factor, disregarding the excitance angle, seems to be as good as the Minnaert constant at describing forest BRDF and modeling the topographic effect in TM data.

Most of the variability of topographic normalization factors from image to image for the same vegetation cover, reported as an inconsistency affecting both models, seems to be explained by the image solar zenith angle (i.e. the phase angle for sensor viewing at nadir).

References

1. Deering D.W., Middleton E.M. and Eck T.F., Reflectance anisotropy for a spruce-hemlock forest canopy. *Remote Sens. Environ.* **47** (1994) pp. 242-260.
2. Ranson K.J., Irons J.R. and Williams D.L., Multispectral bidirectional reflectance of northern forest canopies with the Advanced Solid-State Array Spectroradiometer (ASAS). *Remote Sens. Environ.* **47** (1994) pp. 276-289.
3. Smith J.A., Tzeu Lie Lin and Ranson K.J., The Lambertian Assumption and Landsat Data. *Photogramm. Eng. Remote Sens.* **46** (1980) pp. 1183-1189.
4. Minnaert M., The reciprocity principle in lunar photometry. *Astrophys. J.* **93** (1941) pp. 403-410.
5. Teillet P.M., Guindon B. and Goodenough D.G., On the slope-aspect correction of multispectral scanner data. *Can. J. Remote Sens.* **8** (1982) pp. 84-106.
6. Justice C.O., Wharton S.W. and Holben B.N., Application of digital terrain data to quantify and reduce the topographic effect on Landsat data. *Int. J. Remote Sens.* **2** (1981) pp. 213-230.
7. Colby J.D., Topographic normalization in rugged terrain. *Photogramm. Eng. Remote Sens* **57** (1991) pp. 531-537.
8. Gu D. and Gillespie A., Topographic normalization of Landsat TM images of forest based on subpixel sun-canopy-sensor geometry. *Remote Sens. Environ.* **64** (1998) pp. 166-175.
9. Abuelgasim A.A. and Strahler A.H, Modeling of bidirectional radiance measurements collected by the Advanced Solid-State Array Spectroradiometer (ASAS) over Oregon transect conifer forests. *Remote Sens. Environ.* **47** (1994) pp. 261-275.
10. Kleman J., Directional reflectance factor distribution of two forest canopies. *Remote Sens. Environ.* **23** (1987) pp. 83-86.
11. Vincini M. and Reeder D., Seasonal changes of Minnaert topographic normalization constants (*Ks*) in rugged deciduous forest areas. Proceedings *IGARSS 2000* Congress; Honolulu, USA, July 24-28 2000, pp. 1603-1606.
12. Holben B.N. and Justice C.O., An examination of spectral band ratioing to reduce the topographic effect on remotely sensed data. *Int. J. Remote Sens.* **2** (1981) pp. 115-133.
13. Justice C. O. Wharton S. W. and Holben B.N., Application of digital terrain data to quantify and reduce the topographic effect on Landsat data. *Int. J. Remote Sens.* **2** (1981) pp. 213-230.

CHLOROPHYLL MAP OF THE BALTIC SEA - THE RESULT OF AN AUTOMATIC PROCESSING CHAIN FROM DATA RECEPTION UP TO ARCHIVING AND DELIVERY OF VALUE-ADDED PRODUCTS

B. FICHTELMANN, E. BORG, A. GÜNTHER, H.-G. GUDER, R. REIMER, E. WOLFF, H. DAMEROW

German Aerospace Center (DLR), Kalkhorstweg 53, D-17235 Neustrelitz, Germany
E-mail: Bernd.Fichtelmann@dlr.de

The regional monitoring of chlorophyll in the Baltic Sea since March 1998 based on MOS-B data demonstrates a so-called value added product processed by an automatic processing chain at the DFD in Neustrelitz. The processing chain is a result of close cooperation among the "Remote Sensing Technology Institute" Berlin-Adlershof (DLR-MF), the "Baltic Sea Research Institute Warnemuende" (IOW, and the German Remote Sensing Data Center (DLR-DFD). Some important requirements of process automation in remote sensing will be discussed and the different modules of the actual chain described. For the demonstration of monitoring of the Baltic Sea data have been used since 1996. The chlorophyll map can support studies of bioproductivity, the long-time monitoring of ecosystems at minimal cost, and the detection of special phenomena. Such a product should be of interest for fishery and tourism. With a view to the MERIS sensor on board the ENVISAT satellite, further value-added products for coastal monitoring and for the study of land cover and atmosphere are in preparation at the DFD in Neustrelitz in cooperation with the MAPP team, GKSS, FUB, DLR and Brockmann Consult.

1 Introduction

The value-added product "Chlorophyll Map of the Baltic Sea" is a demonstrator [4] for automatic processing of remote sensing data. One of the main tasks of operational use of remote sensing data requires the provision of such data or their corresponding value-added products within a short time interval after data reception.

Furthermore, their use in monitoring systems or as parameters in environmental models requires the automation of data processing based on standardized algorithms. An "ordinary" user is not able to process data step by step in a large number of data sets, and if using a complex processing software, perhaps atmospheric correction tools, that means that the repetition of a processing step will not supply the same result. Also, small differences resulting from subjective influence conflict with monitoring requirements.

Since March 1998 the operational production of a chlorophyll map of the Baltic Sea has been based on data from the MOS sensor on board the Indian satellite IRS-P3. It is the result of close cooperation among the "Remote Sensing Technology Institute" Berlin-Adlershof (DLR-WS - developer of the "Multispectral Optoelectronic Scanner" MOS), the "Institute of Baltic Sea Research Warnemuende" (IOW), and the German Remote Sensing Data Center (DLR-DFD). This regional monitoring of chlorophyll is an example of a so-called value added product processed by an automatic processing chain at DFD. Process automation is essential for minimizing the manpower requirements of operational monitoring to simple control work packages. Automation implies finding an optimum between precision and operationality.

In this paper, we discuss requirements for automatic processing of remote sensing data and present some results of a regional monitoring of the Baltic Sea. Finally we give a short preview of future products expected from the MERIS Advanced Product Project (MAPP) .

2 Processing chain for the chlorophyll map

2.1 *Data availability*

The MOS sensor has a resolution of 500 m and a swath width of 200 km. The IRS-P3 orbit yields a repetition rate of 24 days with the adjacent path to the east viewed after five days. Four consecutive days have a path over the Baltic Sea, the fifth day does not view it. The next four days cover the adjacent path over the Baltic Sea, followed by a non-viewing day. With the 13th day the whole coverage of the Baltic Sea is completed, apart from the problem of cloudiness. During monsoon periods, the Indian satellite is in an economy mode and no data is available. Furthermore, results are not available in wintertime, because low temperatures suppress bioproductivity.

2.2 *Processing chain for the chlorophyll map*

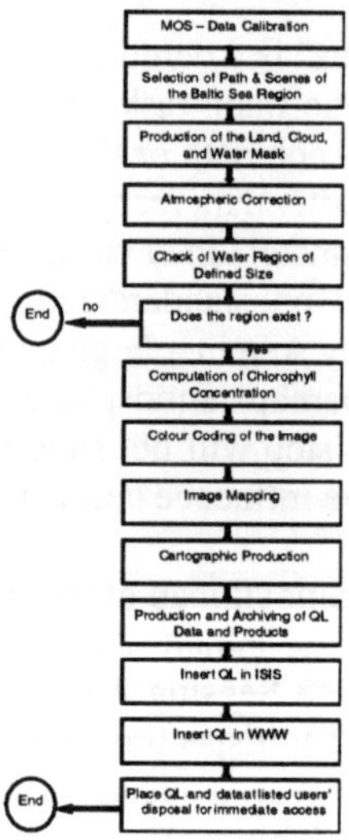

Figure 1. Flow chart of the processing chain

The processing chain [12] shown in the flow chart (Fig. 1) is based on Level-1B MOS-B data; raw data are operationally received and processed to this level at the Neustrelitz ground station. According to the desired path number, indicating the geographical region viewed, Level-1B files are provided for further processing. In the next step of data selection the geographical corner coordinates in the auxiliary data of each scene are checked to determine whether the scene covers the Baltic Sea area. If appropriate scene(s) can be found, higher level processing continues.

The masking and area check of the next step includes the derivation of a water, land and cloud mask. Masking is necessary both for atmospheric correction and for deriving the chlorophyll concentration, because the processing algorithms are only suitable for water. This mask of concatenated scenes is also used to check whether a connected water region of variable shape but defined minimum size exists. This minimum sized area is assumed to be large enough for useful interpretation.

If no suitable area can be found, the processing stops.
The atmospheric correction of the next processing step takes into account the atmospheric model of Gordon [5] and Sturm [11] and Krawczyk *et al.* [6], [7], [8]. The water model (reflectances) is based on the model of Sathyendranath, *et al.* [9]. The needed intrinsic parameter set is based on the region (Baltic Sea) and the solar zenith angle. The output is a scene file of reflectivity values of channels 1 to 6 (408-615 nm). This part of the processing chain was developed by the DLR Remote Sensing Technology Institute.

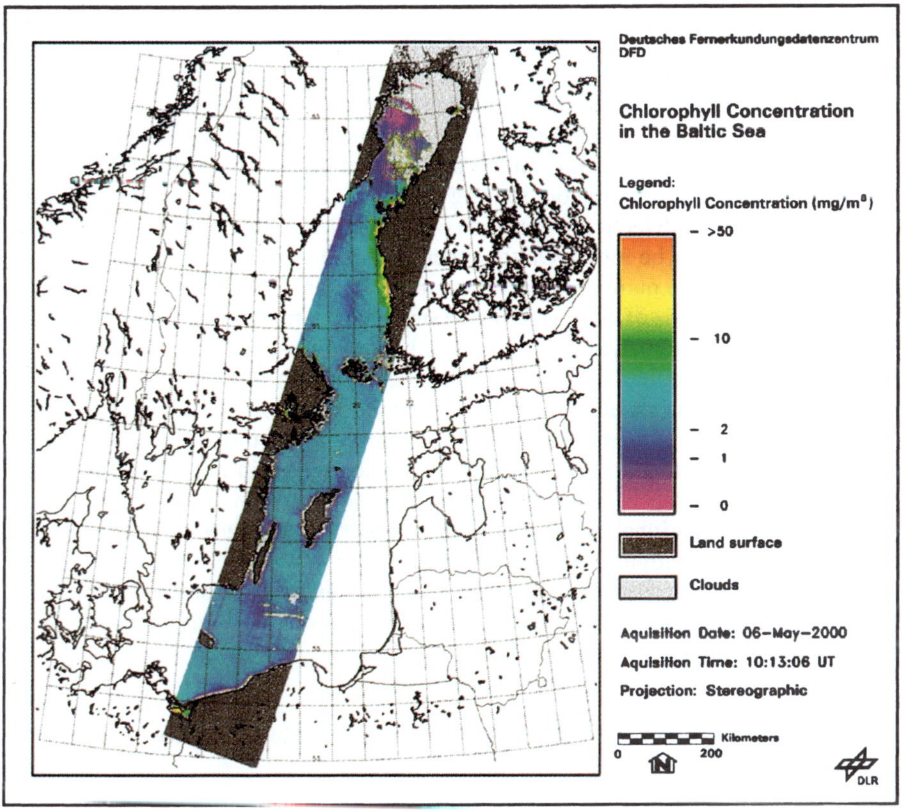

Figure 2. Example of a quicklook showing chlorophyll concentration on 06 May 2000 as available to each unregistered ISIS or WWW user. High concentrations are conspicuous in the coast region of Finland.

The algorithm for deriving the chlorophyll (chlorophyll and phaeopigment) concentration as described in [10] is a contribution of IOW. The output data file is the first product to be archived and available to users.

Because the values resulting from the previous processing steps can not conveniently be shown in a map, a logarithmic stretch was applied. The data of this stretch are mapped by a stereographic projection and framed into the map layout. An example of this second product is given in figure 2. In the last processing steps the products are inserted into the DFD database system and delivered to registered users. A user can view the quicklook in ISIS (isis.dlr.de) and the latest update on the web (dfd.dlr.de).

2.3 Requirements for operational use of remote sensing data

Process automation is not only essential to minimize the time interval from data acquisition to delivery of corresponding products or to avoid the manpower requirements of operational monitoring. It implies finding an optimum between precision and operationality. Perhaps, on the one hand an operator is able to detect special phenomenon within a remote sensing scene and to use corresponding tools for their study. If such a detection algorithm is not included in the processing chain this phenomenon will not be found. On the other hand, a user is not able to process data step by step in a large number of data sets and to guarantee the constant quality of a data product. The basis of automatic processing chains are standardized algorithms.

Furthermore, automation is a step from the qualitative analysis of remote sensing data toward quantitative analysis in connection with a mistake analysis. The latest point is a strong request of the user community. It provides information about the range of validity of a remote sensing product, including the limits of applicability. In this way information based on remote sensing will gain importance if legally binding statements can be included in data interpretation [3]. Such a tendency is suggested by international activities for the protection of nature and the environment in connection with multilateral agreements [2].

In the field of automation in remote sensing a multitude of problems are yet to be solved, but it is the only possibility to supply an increasing user community with global and continuous information.

3 Results of the automatic processing chain for a regional product

3.1 Monitoring of a selected region of the Baltic Sea

Since monitoring the chlorophyll concentration of the Baltic Sea began, the following number of data sets are available for each year: 1998 106(37), 1999 65(36), 2000 87(42) and 2001 73(36) including 24 July 2001. The number in brackets reflects the number of data sets with a cloudless water area of more then 50 percent after rough estimation. Despite normal cloud coverage of more than 40 to 60 percent, and beside the small swath width of the sensor, time intervals (for instance 7 - 19 May 1998, 21 April - 10 May 2000) exists with nearly full coverage of the Baltic Sea. Figure 3 demonstrates the monitoring of a selected area (Gulf of Riga with the

Bay of Parnu) over the last few years. The maps for 1996 and 1997 are also the result of the automatic processing chain's transferring the archived Level-1B MOS-B data to the starting point of the chain. Included in this figure are only the maps with the lowest cloud coverage. In contrast to temporally limited campaigns with in-situ measurements, automation makes it possible to monitor ecosystems over a long time at minimal cost and to detect special phenomena.

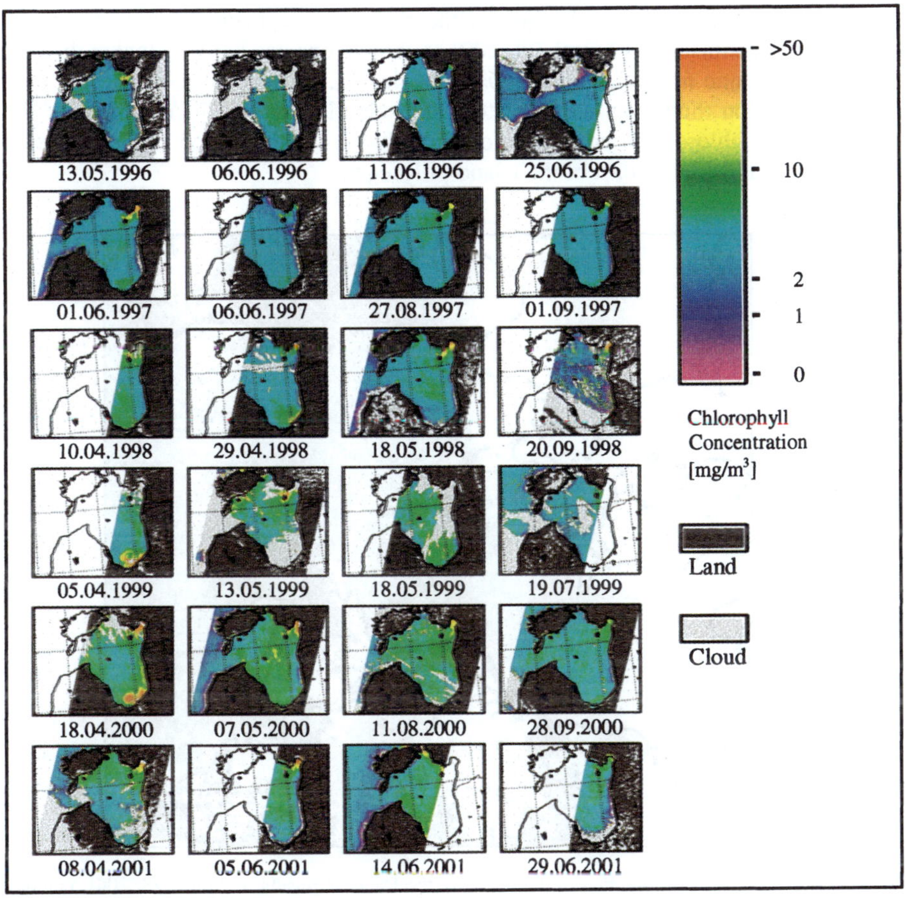

Figure 3. Quicklooks of chlorophyll concentration within the Gulf of Riga from 1996 to 2001 with obviously increased concentrations near the mouth of the Parnu river (to the north) and near Daugawa river (to the south)

3.2 Single events within the monitoring period

In addition to monitoring the mouth regions of large rivers with their high incidence of strong increases in chlorophyll concentration, there are also singular events with such an increase, as shown on available maps of the open sea for 28 July 1999 or of the coastal region of Finland for 06 May 2000 (Fig. 2).

Monitoring could be helpful in assessing the outcome of disaster and monitoring subsequent disturbances. In figure 4 the chlorophyll concentration in the region of the Darss peninsula on March 28, 2001 is shown, one day before a tanker disaster in the Kadetrinne northwest of the Darss peninsula. During the next satellite pass over this region there was unfortunately high cloud coverage. Independent of the disaster, high chlorophyll concentrations are obvious in the region of Greifswalder Bodden and at the east coast of the island of Ruegen.

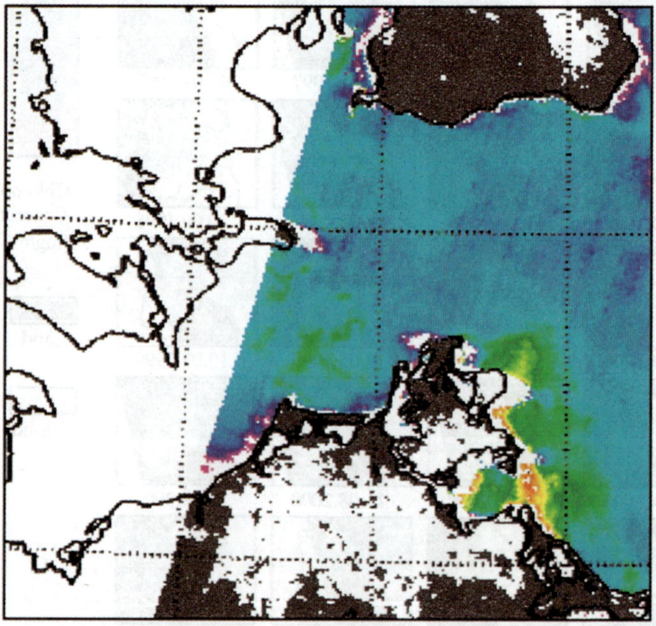

Figure 4. Chlorophyll concentration in the region of the Darss peninsula on March 28, 2001, one day before the tanker disaster. The data have undergone rough geographic correction, in contrast to the original map data.

4 Discussion

The automated chlorophyll map gives a good overview for identifying interesting phenomena including their location, extension and quantity. Their detection is in most cases very difficult without remote sensing. These maps make it possible to optimize campaigns for in-situ measurements with the purpose of determing causal relationships. This information should not only be of special interest to scientists, but also to environmental agencies. Furthermore, these maps are of possible interest to the fishing and tourism industries.

Despite the experimental status of the MOS sensor with its only 200 km swath width, which means that it takes a relatively long time for complete coverage of the Baltic Sea region, and the fact that the algorithm does not differentiate between Case I and Case II water (coast and open sea), and technical disruptions in satellite operations, the processing chain is a good demonstrator for operational and automatic production of a value added remote sensing product.

With the start of MERIS (Medium Resolution Imaging Spectrometer) on board ENVISAT this year, some of the disadvantages of the MOS sensor will be eliminated, perhaps this sensor will provide improved regional coverage with its higher swath width of 1150 km. Data reception is planned in Neustrelitz, using at first the standard ESA processing chain and afterwards the value adding processor of MAPP (MERIS Advanced Product Project) [1] with specified algorithms for open sea and coast and additionally providing products for land cover, atmosphere (cloud, aerosol) and large inland waters for a region of 4000 km x 4000 km.

5 Acknowledgements

The authors wish to thank A. Neumann (DLR, Remote Sensing Technology Institute), his team, and H. Siegel (IOW) for their good cooperation in preparing the automatic processing chain.

References

1. Brockmann C., Doerffer R., Fischer J., Lüdeker W., Neumann A., Schroeder M., van der Piepen H. and Zimmermann G., Untersuchungen zu Sonderprodukten aus Daten des Abbildenden Spektrometers MERIS auf ENVISAT, *Final Report, DARA: 50 EE 9511*, (1996).
2. Blaschke Th., Fernerkundung im Naturschutz. Bedarf an operationellen Fernerkundungsmethoden durch die EU-Gesetzgebung, BfN-Skripten **42** (2001) *Proceedings of the Workshop Naturschutz und Fernerkundung, Island Vilm, May 11-12 2000*
3. Borg E., Thesen zur Nutzung von Fernerkundungsdaten im Bereich des Kulturlandschaftsmonitorings, (1999) *internal concept, unpublished.*
4. Fichtelmann B., Borg, E., Guder H.-G., Klisch A., Operational processing of MOS Data - the demonstrator "Chlorophyll Map of the Baltic Sea", *Proceedings of the 4th International Workshop on MOS-IRS and Ocean Colour, Berlin, May 30-31 2001* (in print).

5. Gordon H., Removal of atmospheric effects from satellite imagery of the oceans, *Applied Optics*, **17**(10) (1978) pp. 1631-1635.
6. Krawczyk H. and Hetscher M., Principal component inversion algorithm for the retrieval of water constituents and ist application, *Proceedings of the 1st International Workshop on MOS-IRS and Ocean Colour, Berlin, April 28-30 1997* (1997) p. 7.
7. Krawczyk H., Neumann A. and Walzel Th., Interpretation Potential of marine environments multispectral interpretation, *Proceedings of the 3rd Thematic Conference in Remote Sensing for Environments, Seattle, Sept. 18-20 1995* (1995) pp. II 51-61.
8. Krawczyk, H., Neumann A., Walzel Th., Hetscher M. and Siegel H., Application of a multispectral interpretation algorithm to remote sensing data over the Baltic Sea, *SPIE* **2963** (1996) pp. 234-239.
9. Sathyendranath S., Prieur L. and Moral A., A three-component model of ocean colour and its application to remote sensing of phytoplankton pigments in coastal waters, *Int. J. of Remote Sensing*, **10**(8) (1989) pp. 1373-1394.
10. Siegel H., Gerth M., Beckert M., Krawczyk H. and Gerasch B., Distribution of water constituents derived from MOS-IRS data in the southern Baltic Sea, *Proceedings of the 1st International Workshop on MOS-IRS and Ocean Colour, Berlin, April 28-30 1997* (1997) p. 6.
11. Sturm B., Ocean colour remote sensing and quantitative retrieval of surface chlorophyll in coastal waters using Nimbus CZCS data, *Marine Science* **13** Oceanography from Space, Proceedings of the COSPAR/SCOR/IUCRM Symposium on Oceanography from Space, Venice, May 26-30 1980 (1981) Plenum Press New York pp. 267-279.
12. Wolff E., Reimer R., Borg E., Günther A., Baruth B., Fichtelmann B.and Preuss H., Bereitstellung einer operationellen Chlorophyllkarte für die Ostsee, *ZPF*, **66**(5-6) (1998), pp. 146-156.

SUSPENDED MATTER IN THE NORTH SEA: A MULTI-TEMPORAL ANALYSIS BASED ON SEAWIFS IMAGERY IN 2000

R. PASTERKAMP, R.J. VOS, AND H.J. VAN DER WOERD

Institute for Environmental Studies,
Amsterdam, The Netherlands
tel: +31 20 4449552 fax: +31 20 4449553
email: reinold.pasterkamp@ivm.vu.nl

H.J. HOOGENBOOM

National Institute for Coastal and Marine Management/RIKZ
The Hague, The Netherlands

A multi-temporal analysis was conducted of total suspended matter (TSM) images retrieved from imagery from NASA's SeaWiFS satellite instrument. Data collection comprised the selection of 129 SeaWiFS images in the year 2000 and subsequent turbid water atmospheric correction and retrieval of total suspended matter concentration for the North Sea near the Dutch coast. Two approaches were tested to visualize and analyze the retrieved concentrations. The first approach was based on the averaging of the images in two-monthly bins, to be able to compare the TSM dynamics in different seasons on a full (1x1 km^2) spatial resolution and scale. The second approach focused on the detailed temporal change by sampling TSM values along a transect reaching 235 kilometer out of the Dutch coast and interpolating these values to give a daily horizontal profile of TSM concentrations.

As presented in this paper, both methods yielded promising results. However, more research is needed in future to provide answers to the research questions discussed in this paper.

1 Introduction and problem definition

Water quality monitoring, management and decision making is generally a responsibility for local, regional or national institutes and authorities. Their standard monitoring program is usually based on measurement of a large number of biological and chemical water quality parameters at a limited number of sampling points measured on a regular basis. The spatial representativeness of these individual sampling points is however restricted to a several square kilometers or even a few square meters. Especially along the coast, near river outflows or estuaries, gradients in water quality parameters can be very steep, both spatially and temporarily. The exact location of the measurement point highly influences the measured parameters, i.e. if the sampling would have been performed at a different location (but still very near the original sampling point), or at a different time (with different atmospheric conditions, tidal stage, etc.) results could have been totally dissimilar. In a sense the individual *in situ* concentrations are thus arbitrary within a certain range. The main virtue of regular sampling strategies lies in the establishment of long term series of measurements, averaging out the short-term fluctuations and enabling the detection and analysis of seasonal or annual trends in water quality parameters. With the launch of NASA's Sea-Viewing Wide Field of View Sensor (SeaWiFS), and the prospected MEdium Resolution Imaging Spectrometer Instrument (MERIS) on board ESA's Envisat, full spatial coverage of the North Sea on a diurnal basis is established and is foreseen to continue in future. Both sensors have the capability to accurately measure essential water quality parameters such as the total suspended matter concentration and the chlorophyll-a concentration in the turbid North Sea. The high spatial resolution of these sensors (SeaWiFS ~1 km^2, MERIS ~0.09 km^2), in combination with the present knowledge of long term trends at the monitoring stations yields a promising opportunity to increase the understanding of patterns and fluxes of sediment in the North Sea, optimize dredging operations, and closely observe eutrophication problem areas and algal blooms. The aim of the work presented in this paper was to investigate the merits of the use of multi-temporal maps of suspended matter in the Dutch and Belgian North Sea, derived from SeaWiFS images 2000.

2 Proposed method

2.1 Processing of SeaWiFS imagery to maps of total suspended matter

The processing of the SeaWiFS images to maps of total suspended matter, follows the concept as developed within the Pre-Operational Water and Environment Regional Service (POWERS) [1; 2], an ESA DUP project. Practically, the processing scheme is divided in the following steps

1. Selection and downloading of the SeaWiFS images, using cloud cover and region of interest (i.e. the North Sea) criteria;
2. Applying the turbid water atmospheric correction [3], developed by the Belgian Management Unit of the Mathematical Model of the North sea (MUMM) as an extension on SeaDAS 4.0 (http://www.mumm.ac.be/OceanColour/; http://seadas.gsfc.nasa.gov/);
3. Output of images to HDF, including subsurface irradiance reflectance, geolocation information, geometry, and atmospheric parameters;
4. Applying of the algorithm to yield total suspended matter concentrations (TSM) based on the subsurface irradiance reflectance in SeaWiFS band 5 (~555 nm), and adding of this data set to the HDF product;

Archiving of the intermediate product of atmosphere corrected images with additional ancillary products, enables the use of updated algorithms for e.g. chlorophyll-a estimation in turbid coastal waters without the need to recalculate the whole atmospheric correction (which is quite a (computational) effort).

2.2 *2.2 Overview of relevant SeaWiFS images for the North Sea in 2000*

The SeaWiFS images covering the North Sea with no or partial cloud-coverage are presented graphically in Figure 1.

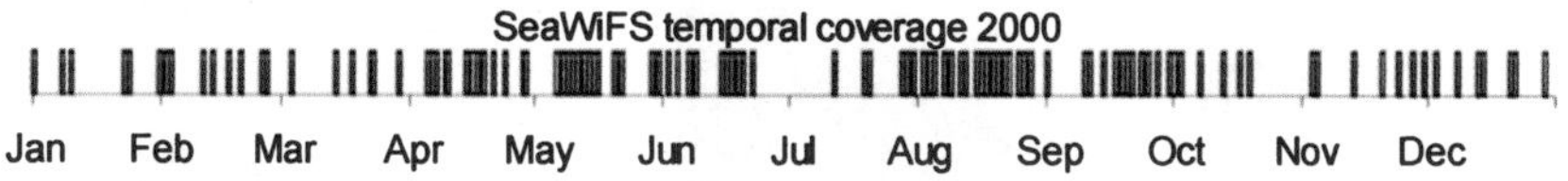

Figure 1. Graphical representation of the temporal coverage of the North Sea with SeaWiFS. Only days with no or small cloud-coverage are shown as a black vertical bar.

The temporal resolution for the North Sea for 2000 is excellent. In total 129 images were selected, and were processed to eventually yield maps of total suspended matter concentration, an average of 11 images per month. Even in winter a number of images was available, which is remarkable because the winter season is often mentioned for its lack of good-quality remote sensing images. Only at the end of June and the beginning of July occurs a gap of 20 days.

2.3 *Multi-temporal analysis*

Initially the TSM images were rectified to a grid of latitudes ranging from 51-55°N (0.01° step size) and longitudes ranging from 0-9°E (0.015° step size) using nearest-neighbor interpolation. Then, to gain insight in the typical seasonal distribution and variation of total suspended matter in the North Sea in 2000, the images were averaged to bins of two month, i.e. Jan-February, March-April, etc.. Mean and variance of TSM and the number of observations per pixel were recorded accordingly. Clouded pixels were masked and were left out of the analysis.
The image analysis presented in the preceding paragraph focuses on the *spatial* distribution for the seasons in the year 2000. However, this way several images have to be compared when analyzing *temporal* TSM dynamics. Therefore another way of analysis was undertaken where the temporal change can be evaluated within one image. Along the transect indicated in figure 4, TSM concentrations were retrieved from all rectified SeaWiFS images. The resulting unevenly spaced grid of TSM values was subsequently interpolated (to daily values of TSM) using one-dimensional piecewise cubic Hermite interpolation along the time-axis.

3 Results

The resulting averaged images are presented in Figure 2-3, in six rows were the first row contains the statistics for January-February, the second row for March-April etc. The left side gives the mean TSM concentration, and the right side gives the standard deviation and the number of observations per pixel. Note that in general the number of observation per pixel is smaller than the total number of

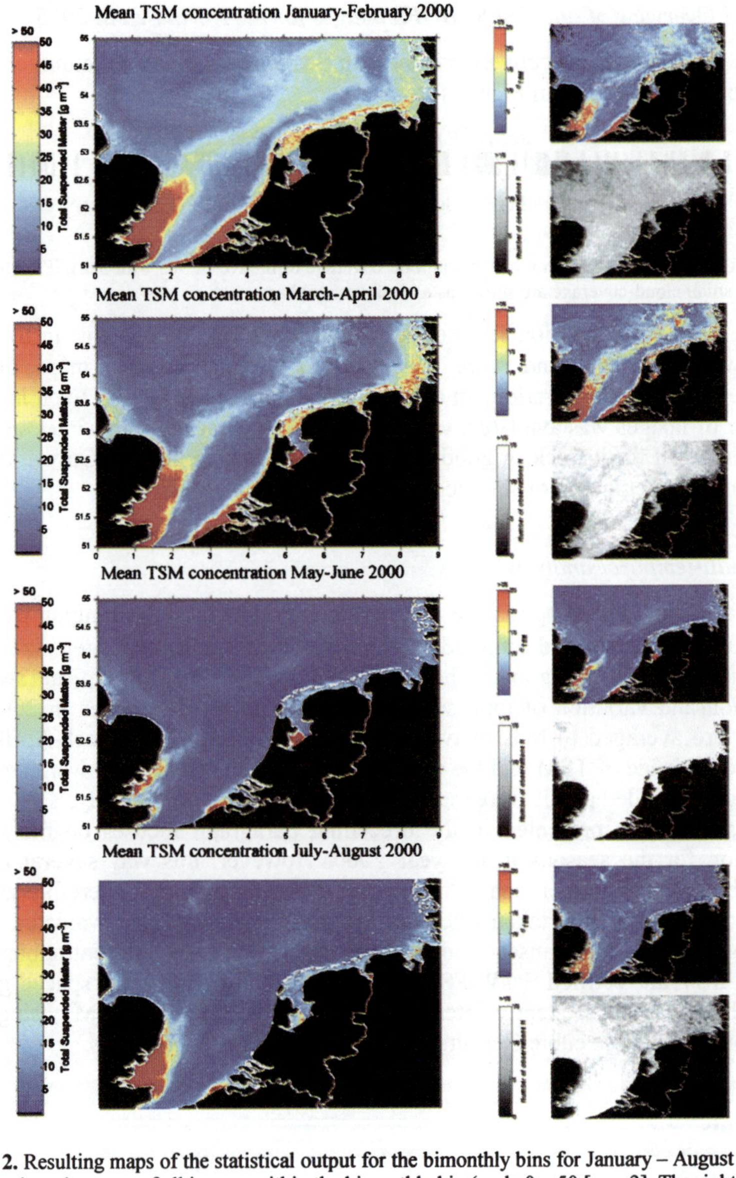

Figure 2. Resulting maps of the statistical output for the bimonthly bins for January – August 2000. The left side gives the mean of all images within the bimonthly bin (scale 0 – 50 [g m-3]. The right side gives the standard deviation (top, scale 0 – 25 [g m-3]) and the number of observations N (bottom, scale of 0-15).

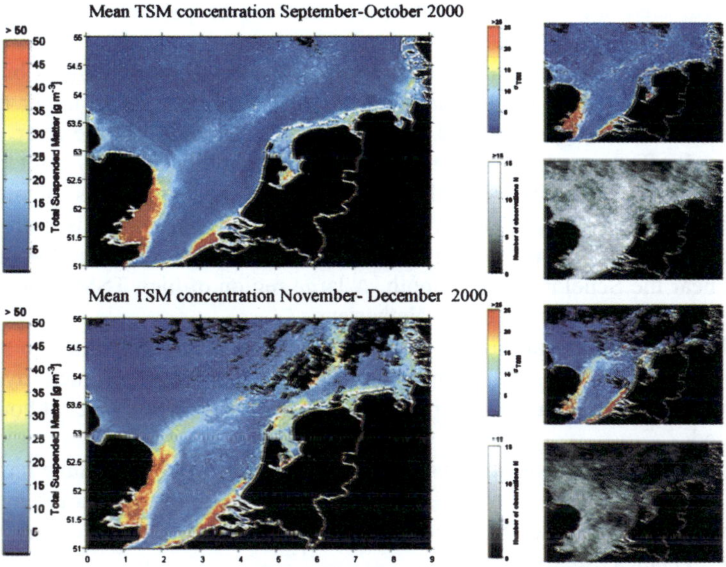

Figure 3. Resulting maps of the statistical output for the bimonthly bins for September – December 2000. The left side gives the mean of all images within the bimonthly bin (scale 0 – 50 [g m-3]. The right side gives the standard deviation (top, scale 0 – 25 [g m-3]) and the number of observations N (bottom, scale of 0-15).

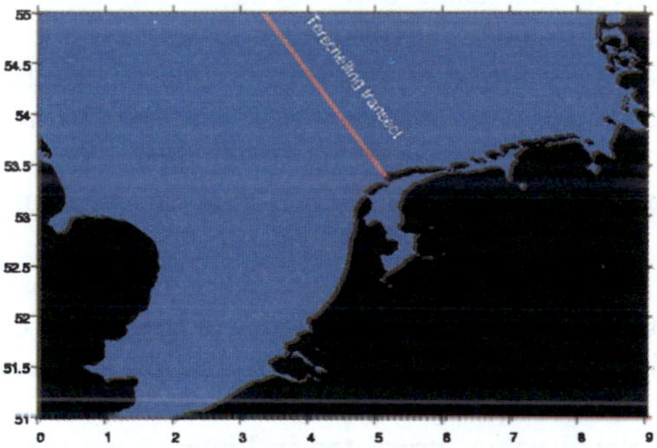

Figure 4. Selected transect used in the analysis. The red line represents the so-called Terschellingen transect, which starts near the 'Wadden' island Terschellingen and runs 235 km off the coast in NW direction

images evaluated for the two months, because of the partially clouded images within the set.
In January-February, the mean TSM concentration is the highest of all six bins, as is expected in winter. Also the standard deviation is very high, even larger than the mean concentration. A pronounced plume is visible beginning in the river Thames estuary and off the coast of Great Britain, running all the way to Denmark. Interesting is a region between the Dutch coastal zone and this plume, were concentrations are notably lower. A zone of elevated concentrations is visible along the Dutch coast, increasing both in concentration and in width to the South, being maximal near the Scheldt estuary mouth. A large region of high TSM concentration and dynamics is visible North of the Netherlands.
In March-April, concentrations are overall lower than in Jan-Feb. But the standard deviation within the bin is still high, indicating high suspended sediment dynamics. The plume from The Thames estuary towards Denmark is still visible, although less pronounced.
In May-June concentrations are much lower than in Jan-Apr. The plume has totally disappeared, and the zone of elevated concentrations along the East coast is reduced to a patch near Zeebrugge. Nonetheless some dynamics is visible near and North-East of the Thames estuary.
The July-August image resembles the May-Jun image, except for the Thames Estuary where significant higher concentrations and a much higher standard deviation are found in the July-August image.
In Sep-Oct, concentrations are increasing. The sediment plume from the British coast is building up, and the Thames Estuary also shows high concentrations. The patch near Zeebrugge is increasing and moving in NE direction along the Dutch coast. This behavior is continued in Nov-Dec.
In Figure 5 the results for the interpolation of TSM values along the transect are visualized. In accordance with the previous analysis, values are found to be particularly low in summer, even very close to the coast. Only from approximately October until March/April elevated TSM concentration are visible close to Terschelling.
Between 30 and 100 kilometers out of the coast several patches are visible in February, March and in October. These patches might be due to the sediment plume. Concurrently, the patch in April might also be attributed to algal blooms, which are known to occur in spring near the Frisian Front.
Note that the interpolated values presented in Figure 5 are less reliable when there are large time-gaps in the retrieved TSM values along the transect.

4 Discussion and conclusions

This paper demonstrates the feasibility of using time-series of SeaWiFS derived TSM maps for the analysis of seasonal change in pattern sand dynamics of suspended sediments in the North Sea. The results presented here already provide a

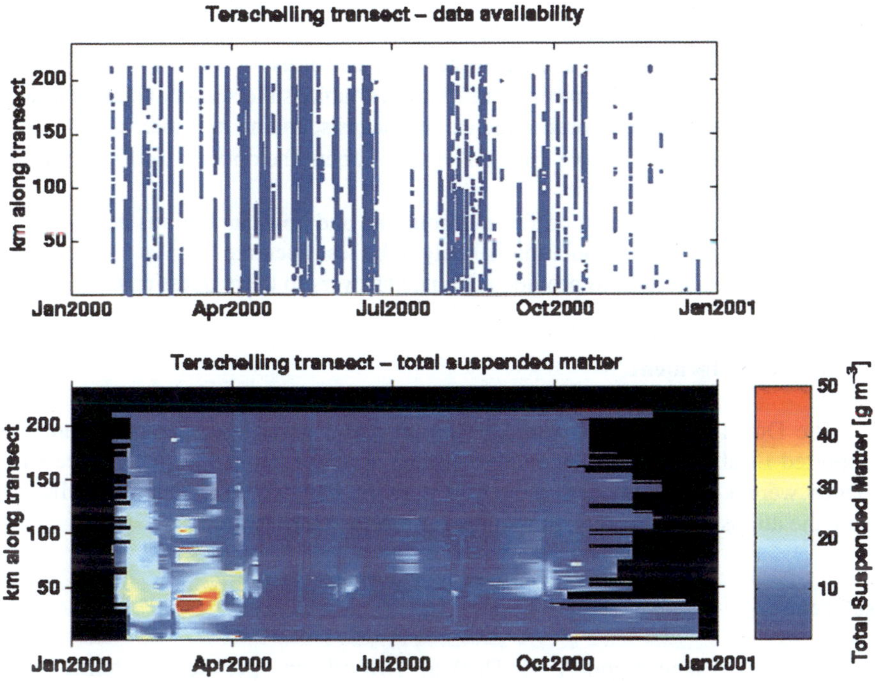

Figure 5. Multitemperal analyis of total suspended matter along the Terschelling transect. The vertical axis gives the distance from the coast in kilometers along the transect (where 1 kilometer is close to the coast, and 235 kilometer is at the end of the transect). The horizontal axis gives the date in the year 2000. In the top figure the data-availability from the SeaWiFS imagery is indicated by blue dots. For each blue dot a TSM value was retrieved from the SeaWiFS images. Large white areas indicate poor data availability due to cloud-cover. In the bottom figure the interpolated total suspended matter concentration is shown on a color scale from blue (low TSM) through green and yellow to red (high TSM). Black values indicate that no data was available. Note that large white area's in the corresponding top figure indicate unreliable interpolation results in the bottom figure.

valuable research technique because of the full spatial coverage at high resolution (1 km^2), and optimal availability of SeaWiFS images, both in winter and in summer. However several questions remain:

- How much is the statistical analysis biased by the availability of SeaWiFS images within a certain time-bin. Or, what is the optimal time-period for the statistical analysis of images to result in unbiased, representative averaged values. In this study a time-period of two month was used, but maybe averaging over month will give better results.
- What is the influence of intense events such as storms?
- What are the results for other years? How variable is the seasonal change when looking at a multi-temporal sequence of images spanning multiple years?
- From this analysis, can we draw conclusions about the optimal locations for *in situ* measurement stations?
- What is the optimal method for interpolation, and how can regions be indentified where interpolation is likely to be less accurate?

Future research is needed to provide adequate answers to these questions.

5 Acknowledgement

The Dutch National Institute for Coastal and Marine Management financially supported Establishment of the archive of time-series of SeaWiFS images. SeaWiFS images were kindly provided by NASA's GSFC/DAAC. We thank MUMM for using the atmospheric correction module for turbid coastal waters.

References

1. H. van der Woerd and A. G. Dekker, Towards an operational satellite-based water-quality information service, *The 27th International Symposium on Remote Sensing of Environmen*t. (1998).
2. H. van der Woerd, H., Hakvoort, J. H. M, Hoogenboom, H. J., Omtzigt N., Pasterkamp, R., Peters, S. W. M., Ruddick, K. G., de valk C., and Vos, R. J., Towards an operational monitoring system for turbid waters," *O-00/16. IVM\VU Amsterdam* (2000).
3. Ruddick,K.G., Ovidio,F., Rijkeboer,M., "Atmospheric Correction of SeaWIFS Imagery for Turbid Coastal and Inland Waters," *Applied Optics* **39** (2000) pp. 897-912.

THE USE OF MULTI-TEMPORAL SAR IMAGES FOR DETECTING MAIN HYDROLOGICAL PARAMETERS

M.BIANCHINI, G. MACELLONI, S.PALOSCIA, P. PAMPALONI, E.SANTI

Istituto di Ricerca sulle Onde Elettromagnetiche – CNR VIA PANCIATICHI 64, 50127 FIRENZE, ITALY

Abstract - In this paper the sensitivity of backscattering to the hydrological parameters, such as vegetation biomass, soil moisture and surface roughness, is discussed. Recent results obtained using multi-frequency SAR data measured by JPL/AIRSAR, SIR-C, ERS-1/2 and JERS-1 sensors are described. The sensitivity of L- and C-bands to soil moisture for individual fields is not very high, since surface roughness and spatial variability affect the radar signal. However, observing data collected at different dates and averaged over a relatively wide area that includes several fields, the correlation to soil moisture becomes significant, since the effects of spatial variations are smoothed. Lastly, a multi-temporal analysis on coherence maps obtained from repeat-pass interferometry was carried out.

1 Introduction

The sensitivity of the backscattering coefficient to the geophysical parameters involved in hydrological processes – namely vegetation biomass, soil moisture and surface roughness – is well known. The estimate of these parameters, needed in modeling the geophysical processes of the hydrological cycle, has been the subject of several investigations carried out in the past years [e.g., 1-3]. The most significant information was obtained by combining different frequencies, polarizations and incidence angles [4,5,3]. However, most satellites have onboard single-frequency, single-polarization sensors. From these data fairly good correlations with the geophysical parameters have been established, though significantly affected from the spatial variability. A method able to minimize this effect and extract the maximum information from such a type of data is based on the use of multi-temporal images [6].

An overview of SAR data analyzed in this paper and collected on the agricultural area of Cerbaia, close to Florence, is shown in Table I. An analysis of sensitivity of radar data to soil moisture content (SMC) is carried out by considering both the effects of spatial averaging and of multitemporality.

Moreover, the potential of ERS SAR repeat pass interferometry has been tested on a semi-arid area, located on the South-Est of Morocco, within the framework of the European Community Project FLAUBERT (FLood in Arid Units By Earth Remote Techniques). Interferometry is an innovative and promising technique which is based on the complex analysis of multitemporal images, taking into account amplitude and phase of the backscattering coefficient. The capabilities of this technique in land cover classification has been pointed out in several works carried out for instance by Askne et al. [7], and Wegmüller and Werner [8]. Recent studies [9-10] demonstrates that for desert areas, where surfaces feature are stable over long time intervals, coherence (which is the parameters obtained from a couple of complex SAR images) remains high enough to allow the interpretation of interferometric pairs collected at intervals of several months or years.

2 Analysis of backscattering coefficient from bare and vegetated surfaces

From several experiments carried out worldwide in past years a reasonable sensitivity of the backscattering coefficient to soil moisture (SMC) has been observed, although the radar signal was found to be considerably affected by both vegetation cover and surface roughness (e.g., Ulaby et al. 1986, Wang et al. 1987, Ferrazzoli et al. 1992, Dubois et al. 1995) [1,11,12,2].

	Frequency Band	Polarization	Observation Angle	Ground resolution (m)	Dates
AIRSAR	P, L, C	Quad	20° 35°- 50°	12.2 x 6.6	22-29/6/1991 14/7/1991
SIR-C/X-SAR	L, C, X	Quad VV	23°-55°	25 x 25	12-17/4/1994 3-14/10/1994
EMISAR	L, C	HH, VV, HV	45°- 50°	5 x 5	23/6/1994 5-6/7/1995
ERS-1	C	VV	23°	30 x 26.3	29/5/92 07/8/92 24/4/94
JERS-1	L	HH	35°	18.3x24.6	24/6/92 14/4/94

Table 1 – Multifrequency SAR data used in this paper

The data-set collected over Montespertoli agricultural area, close to Florence, (Table I) was analyzed in order to better investigate the radar sensitivity to *SMC* of agricultural fields. The backscattering coefficient $\sigma°_{HH}$, measured at *L*–band and at 35° incidence angle (θ), was correlated with the gravimetric *SMC* of the 0-5 cm layer (Fig. 1). In order to reduce the effects of vegetation, we selected an ensemble of bare or scarcely vegetated fields (i.e., with a Leaf Area Index, *LAI*, < 1 and a plant water content, *PWC*, < 0.5 kg/m^2). Although the increasing trend of $\sigma°_{HH}$ with moisture is rather clear, a considerable scattering of data is present, which results in a fairly low correlation ($r^2 = 0.44$).

The influence of vegetation and soil roughness can be reduced if radar observation is performed at small incidence angles. However, SAR images lose spatial resolution with decreasing incidence angle, which can make the processing

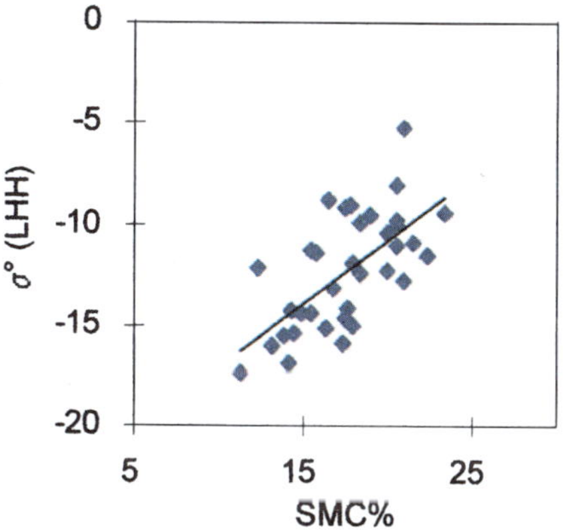

Figure 1 The backscattering coefficient (σ°) at L-band (HH pol., θ=35°) as a function of the SMC%

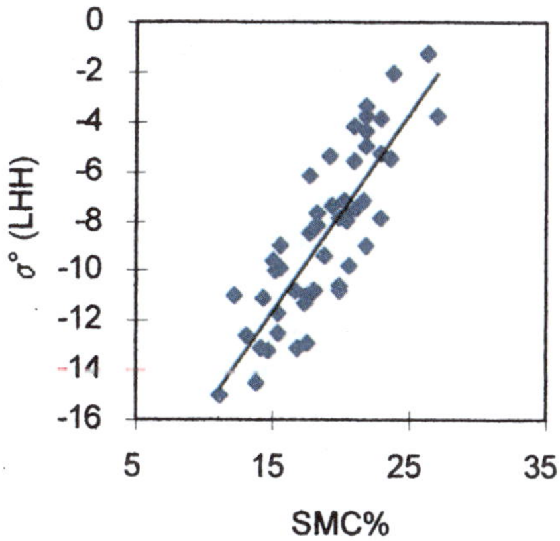

Figure 2 The same as figure 1, but θ=25°

of the image difficult due to the worst identification of field borders. In Fig. 2, where $\sigma°_{HH}$ at θ=25° over a sub-set of the bare or scarcely vegetated fields is shown vs. SMC, we can note that the correlation is indeed increased (r^2 = 0.7) by the

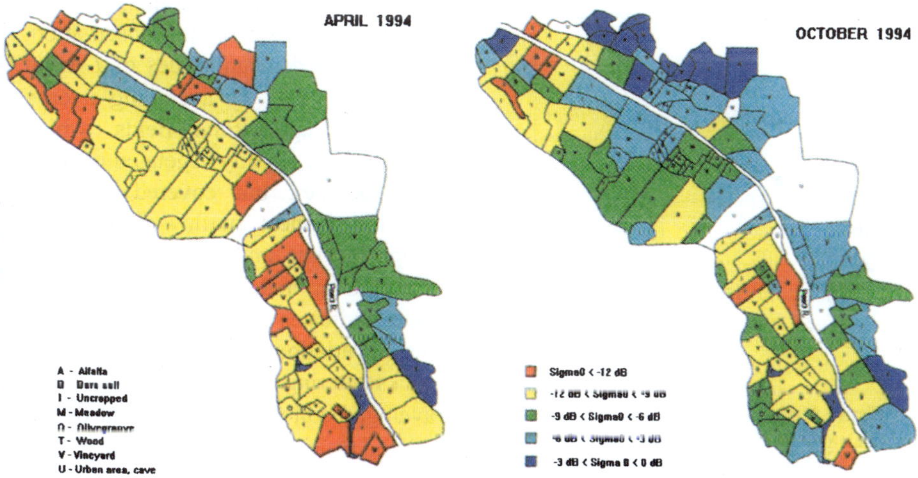

Figure 3 Maps of $\sigma°_{hh}$ at L-band taken in April and October 1994 on the agricultural area along the Pesa River of Cerbaia site

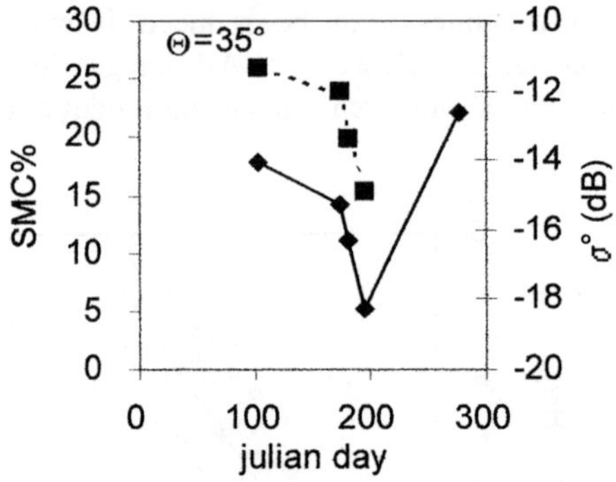

Figure 4a temporal trend of SMC% (continuous) and σ° at L-band (HH pol., θ=35°) (dashed)

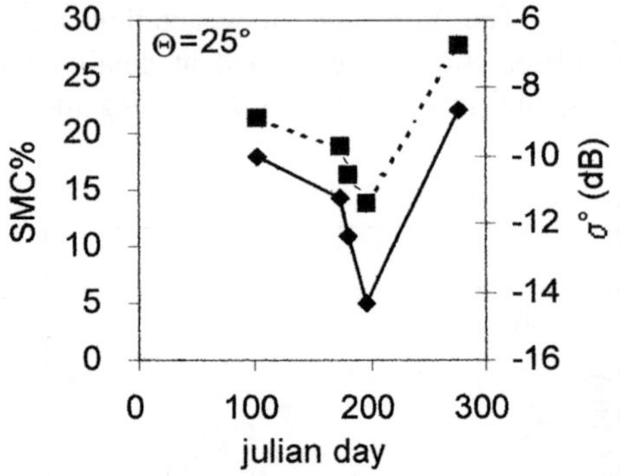

Figure 4b the same of fig 4a, but θ=25°

decreased incidence angle [6]. These results point out that the dependence of backscattering on the peculiar surface features of individual fields sets an upper limit to the sensitivity of radar to surface moisture. However, if we consider the evolution with time of backscattering from the area, a significant correlation of σ° to *SMC* is evident. The images shown in Fig. 3 represent a part of the agricultural area of Montespertoli and refer to data collected at L-band (HH pol., θ=35°) by SIR-C in April and October 1994, when fields were bare or lightly vegetated and the spatially averaged SMC had values of 19% and 24% respectively. Although the change of moisture between these two dates is rather moderate, its effect on the general level of backscattering is clearly appreciable even by a visual inspection.

On the basis of this observation, we computed mean values of *L*–band $\sigma°_{HH}$ and *SMC* by averaging over a group of agricultural fields, which were bare or slightly vegetated in April and October, but covered by well developed wheat, sunflower and alfalfa crops in June and July. Fig.4, which shows $\sigma°_{HH}$ and *SMC* as functions of date, indicates an agreement between the variations of area-averaged backscattering and those of area-averaged soil moisture, both at θ=35° (a) and θ=25° (b) [6].

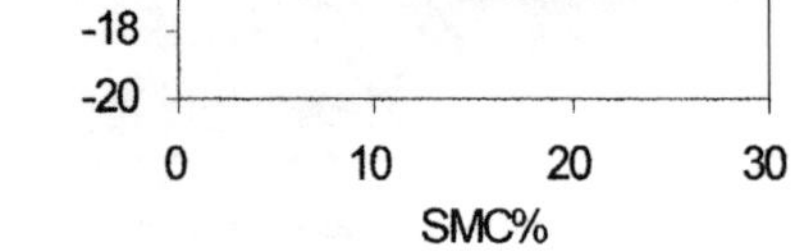

Figure 5 σ° (L-band, HH pol.) at θ=25° (upper line) and θ=35° (lower line) as a function of SMC%. Lines represent the correlation lines.

Moreover, if σ°_{HH} is put into direct relation with *SMC* (Fig. 5), the correlation becomes surprisingly high at both incidence angles: $r^2 = 0.96$ at θ=25° and r^2=0.99 at θ=35°. This result indicates that the disturbing effects of the peculiar backscattering features of individual parcels are smoothed out by averaging over the same set of fields, so that the spatial average of backscattering is essentially controlled by the spatial average of *SMC*, even if well developed crops are present in summer.

Although the best sensitivity to SMC is expected to be found at L-band, we have to consider that most present and future satellites (e.g. ERS, RADARSAT and ENVISAT) have onboard C-band sensors. Therefore, an analysis of the capabilities of C-band data in measuring SMC was also tested, using data collected with ERS-2 satellite on an agricultural area in Southern France, close to Avignon, within the framework of the EC RESEDA (REmote SEnsing Data Assimilation) project. From the analysis of the direct relationship between σ°_{vv} (θ=23°) and SMC of individual fields of wheat and sunflower, C-band does not seem to be able to detect the soil moisture variations. However, if we give a look at the temporal trends of both σ°_{vv} and SMC for a single field, a rather good agreement between the two parameter behavior can be noticed (Fig. 6). Also in this case, we spatially averaged over all the agricultural fields both σ°_{vv} and SMC values. The direct relationship between the spatial-averaged values of σ° and SMC is shown in Fig. 7. We can see that, as expected, the correlation in this case can be considered rather good, showing a correlation coefficient r^2=0.54 [13]. Moreover, the same procedure was used for ERS-C-band data collected over Montespertoli area and the result is shown again in Fig. 7.

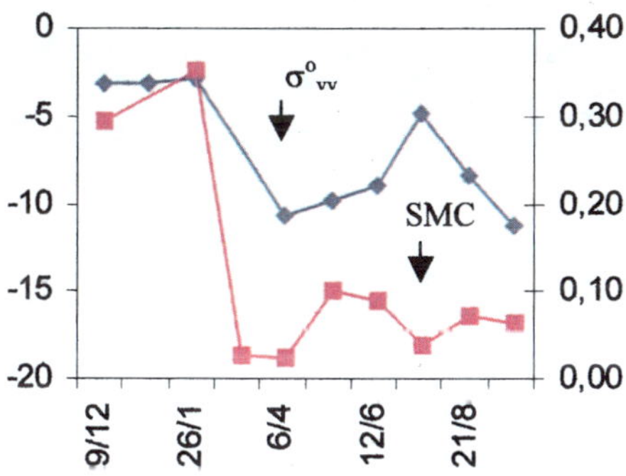

Figure 6 Temporal trends of gravimetric soil moisture (mg) and corresponding value of σ°_{vv}

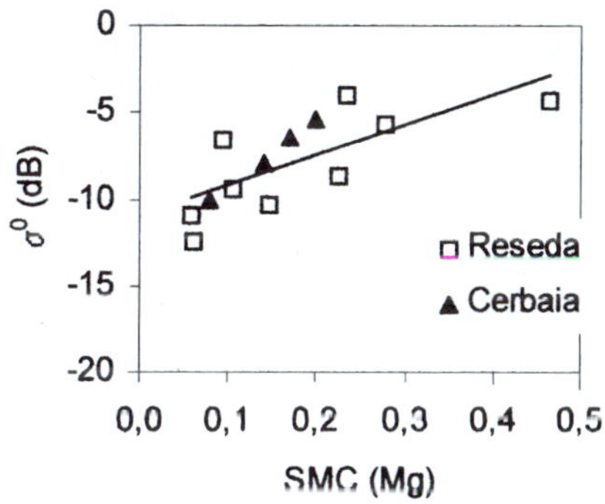

Figure 7 A direct comparison between ERS σ° and SMC spatial-averaged data collected on the agricultural areas of Avignon and Montespertoli (Cerbaia)

3 SAR Interferometry

In order to evaluate the potentiality of the ERS SAR repeat pass interferometry, a flat, arid region of the South-Est of Morocco was chosen. Two couples of ERS images, collected on April and September (couple 1) and July and September (couple 2) respectively, were processed.

After the co-registration, interferogram and coherence maps were realized. From the coherence maps we note that the coherence is in general low for the couple 1 whilst is rather high for the couple 2. The high mean value of coherence for couple 2, is due to the fact that both images were acquired during the summer period where vegetation and soil remained practically the same, due to the absence of rain events. Vice versa, for the couple 1, the images were collected in April, when vegetation and soil were in the most wet period of the year, and in September when soil and vegetation were in very dry conditions [10].

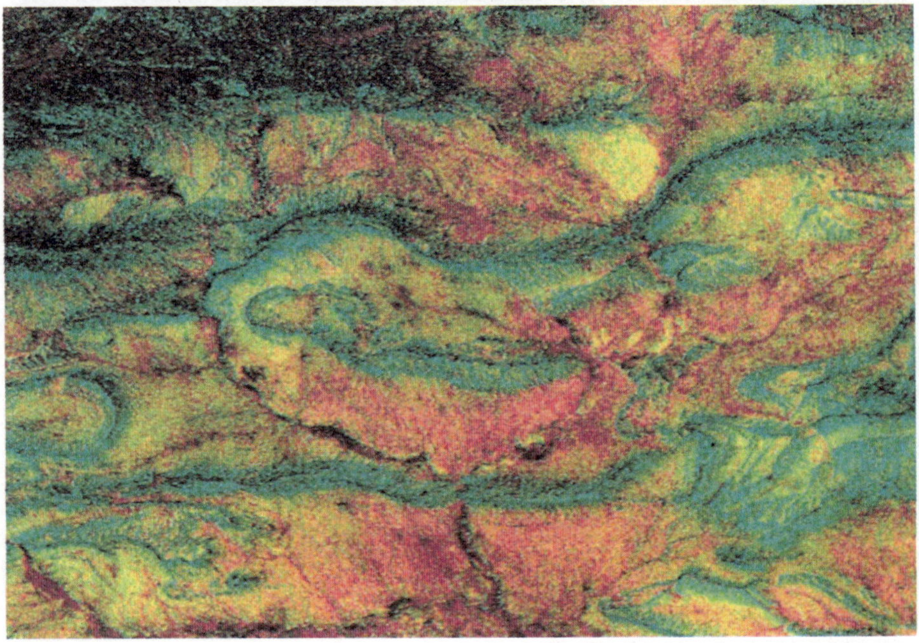

Figure 8 interferometric classifier of the Morocco test area. The colors correspond to the following land type:**Azure**: mountains (rocky) - **Yellow**: Bare soil/deserted areas - **Magenta**: spontaneous semi-arid vegetation (Alfa and Mugworth) - **Brown**: moist cultivated areas along the rivers.

To better investigate these aspects, a color composite image (RGB) was obtained combining the coherence of the first couple of ERS images (R channel), the coherence of the second one (G channel) and their difference (B channel), as in Fig. 8 [9].

We can observe that the yellow indicates an area where the coherence is high in both images, magenta indicates low coherence in couple 1 and high coherence in couple 2, brown corresponds to low coherence in both images, black to very low coherence in both images, and blue to area with high value in couple 2 and high difference between both images. On the basis of ground data and physical consideration, five classes of different surface types were identified, corresponding to rocky mountain areas, semi-arid vegetation, cultivated crops, desertic zones, ridges or rivers. To validate this classifier a comparison with an optical classifier was carried out. The comparison shows that the classifiers are in good agreement.

4 Conclusions

Multitemporal analyses allow a better estimate of biophysical parameters when only single-frequency, single-polarization data are available, since in this way the effect of spatial variability is smoothed. SAR Interferometry, which is a new technique just based on the image multitemporality, gave new tools for land classification and the extraction of certain parameters involved in the hydrological cycle.

5 Acknowledgments

This work was partially supported by the Italian Space Agency (ASI) and the EC Projects ReSeDA and FLAUBERT under the contracts ENV4-CT96-0326 and IC18-CT97-0154, respectively.

References

1. Ulaby F.T., R.K. Moore, and A.K. Fung, 1986, Microwave Remote Sensing: Active and Passive. Vol.III - Surface Scattering and Emission Theory, Artech House, Dedham, MA.
2. Dubois P.C., J.J. van Zyl and T. Engman, 1995, "Measuring Soil Moisture with Imaging Radars", I.E.E.E. Transactions on Geoscience and Remote Sensing, **33**, 896 - 904.
3. Dobson, M.C., F.T.Ulaby, L.E. Pierce, T.L. Sharik, K.M. Bergen, J.Kellindorfer, J.R. Kendra, E. Li, Y.C.Lin, A. Nashshibi, K. Sarabandi, and P. Siqueira, 1995, "Estimation of Forest Biophysical Characteristics in Northern Michigan with SIR-C/X-SAR", I.E.E.E. Transactions on Geoscience and Remote Sensing, 33, 877 – 895

4. Baronti, S., Del Frate, F., Ferrazzoli, P., Paloscia, S., Pampaloni, P., Schiavon, G., 1995, "SAR polarimetric features of agricultural areas", International Journal of Remote Sensing, **16,** 2639-2356
5. Ferrazzoli, P., S.Paloscia, P. Pampaloni, G.Schiavon, S.Sigismondi and D. Solimini, 1997, "The potential of multifrequency polarimetric SAR in assessing agricultural and arboreous biomass", I.E.E.E. Transactions on Geoscience and Remote Sensing, **35,** 5-17
6. Macelloni G., S. Paloscia, P. Pampaloni, S. Sigismondi, P. de Matthæis, P.Ferrazzoli, G.Schiavon, D.Solimini, 1999, "The SIR-C/X-SAR experiment on Montespertoli: sensitivity to hydrological parameters", International Journal of Remote Sensing, vol.20, n.13, pp. 2597-2612
7. Askne J., P.B. Damnert, L.M. Ulander and G. Smith, ``C-band repeat-pass interferometric SAR observations of the forest," IEEE Trans. Geoscience and Remote Sensing, vol. 35, pp.25-35, January, 1997.
8. Wegmüller U. and C.L. Werner, ``SAR interferometric signatures of forest," IEEE Trans. Geoscience Remote Sensing, vol. 5, pp. 1153-1161, 1995.
9. Wegmüller U., T. Strozzi, T. Farr, and C. Werner, "Arid Land Surface Characterization with Repeat Pass SAR Interferometry," *IEEE Trans Geosci. And Remote Sensing,* vol 38, pp. 776-781, 2000.
10. S. Paloscia, M. Bianchini, G. Macelloni , P. Pampaloni, E. Santi, N. Aqid, D. Ouazar, M. Zribi," Estimating vegetation cover and soil roughness in semi-arid regions from multi-temporal dual-frequency SAR data", *International Geoscience and Remote Sensing Symposium (IGARSS 2001)*, Sidney , July 9-13 2001,in press.
11. Wang, J.R., T.E. Engman, T.J. Schmugge, and J.C. Shiue, 1987, "The effects of soil moisture, surface roughness, and vegetation on L-band emission and backscatter". I.E.E.E. Transactions on Geoscience and Remote Sensing, **25**, 825-833.
12. Ferrazzoli, P., S. Paloscia, P. Pampaloni, G. Schiavon, D.Solimini, and P.Coppo, 1992, "Sensitivity of Microwave Measurements to Vegetation Biomass and Soil Moisture Content: A Case Study", I.E.E.E. Transactions on Geoscience and Remote Sensing, **30**, 750-756.
13. G.Macelloni, S.Paloscia, P.Pampaloni, R.Ruisi, M.Dechambre, R.Valentin, A.Chanzy, L.Prévot and N.Bruguire, "Modelling Radar Backscatter From Crops During The Growth Cycle" Physics and Chemistry of the Earth, ISSN 1464-1917,(accepted November 2000)

Land-cover Dynamics

Multi-temporal Remote Sensing Analytical Approaches for Characterizing Landscape Change

ROSS S. LUNETTA

U.S. Environmental Protection Agency, National Exposure Research Laboratory, Research Triangle Park, North Carolina, 27711, USA (lunetta.ross@epa.gov)

Changes in landscape composition and function result from both acute land-cover conversions and chronic landscape changes. Land-cover conversions are typically mediated by human land-use activities (e.g., conversion from forest to agriculture), while more subtle chronic landscape changes can result from either natural processes (i.e., insect infestations, successional processes, and climatic changes) or human land-use activities (e.g., forest thinning). Landscape conversions are relatively rare events at regional-to-national scales of analysis. Over large geographic regions within the United States it has been estimated that land-cover conversions have been estimated to occur at a rate of approximately 0.5% per annum. Numerous efforts have recently been undertaken in an attempt to document land-cover conversions using moderate-resolution remote sensor data for both the United States and Mexico. However, current spectral-based change detection techniques using either pre- or post-classification approaches have tended to be performance limited in biologically complex ecosystems, owing (in large part) to vegetation phenology induced errors. Phenology errors are largely associated with the temporal displacement inherent with sequential satellite data collections. This temporal displacement can result in an unacceptable level of both omission (type 1) and commission (type 2) errors when the data are processed using traditional spectral-based approaches. Ongoing research is focusing on the development of automated land-cover change detection methods based on "vegetation dynamics" to identify change locations and minimize errors.

INTRODUCTION

The characterization of land-cover (LC) type, extent, and distribution over time, is an important landscape characterization element required to monitor ecosystem condition. LC is an important variable required for the study of numerous ecosystem process including habitat suitability, wetland functions, the allocation of non-point nutrient sources, and the study of erosion and sedimentation potentials. Of particular importance, is the application of LC data for the generation of landscape based assessments metrics to evaluate relative ecosystem condition over a range of analytical scales (watershed - national) to assess the impacts attributable to human land-use activities (Jones et al., 1997). Past, present, and emerging remote sensor technology provides a cost-effective high quality source of data to characterize changing LC conditions and to monitor changes, at multiple scales, adequate to support a wide-range of assessment applications.

To date, large area LC characterization efforts have primarily focused on the development of baseline data (Loveland et al., 1991; Vogelmann et al., 1998, Lunetta et al., 2001). With numerous baseline inventories now complete, many researchers are now focusing on the development of regional scale change detection methods to both update the existing LC data sets, and to determine the type, extent, and distribution of LC changes. Two impediments that must be overcome to develop practical methods for large area change detection monitoring using remote sensing include: (1) the elimination or significant reduction of vegetation phenology mediated omission (type 1) and commission (type 2) errors; and (2) robust methods to assess the accuracy of LC change results.

As part of the natural processes associated with vegetation dynamics, plants undergo intra-annual variations or cycles (phenology). Physical drivers including temperature, and solar radiation mediate these cycles in large part. For example, in the northern latitudes, plant cycles are principally driven by temperature and photoperiod. While, in the arid southern latitudes, water availability and photoperiod are the main drivers. During different stages of vegetation growth, plant structures and associated pigment assemblages can vary significantly. Because our ability to identify vegetation classes using remote sensor systems is a result of wavelength specific foliar reflectance (0.76-0.90 Φm), pigment absorptions (0.45-0.69 Φm), and foliar moisture content (1.55-1.75 Φm), the same vegetation type can appear significantly different at various stages during its intra-annual growth cycle.

BACKGROUND

Remote sensing change detection techniques can be broadly classified as either post-classification change methods or pre-classification spectral change detection. The post-classification approach involves the analysis of difference between two independent categorization products using GIS-based analytical tools. Originally, this approach was considered to be the most reliable approach and was used to evaluate emerging methods (Weismiller et al., 1977). Factors that limit the application of post-classification change detection techniques can include: (a) cost; (b) consistency; and (c) error propagation. Thus, this approach can produce large numbers of erroneous change indications since an error on either data gives a false indication of change (Singh, 1989). Categorization errors associated with post-classification techniques can be characterized using a multiplicative (Stow et al., 1980) or error propagation model (Lunetta et al., 1991).

Numerous pre-classification change detection approaches have been developed and refined to provide optimal performance over the greatest possible range of ecosystem conditions. Among the first, semi-automated digital data processing approaches, was image-based composite analysis. This technique is based on a single analysis of a multi-date data set using standard pattern recognition and spectral classification techniques to identify LC change areas (Weismiller et al.,

1977). Although this technique requires only a single classification, data must be collected under similar conditions (i.e., vegetation phenology, solar zenith angles, and atmospheric conditions), and the interpretation of LC versus change classes can be very complex. A variation to the image composite analysis technique that minimizes manual interpretation requirements involves the use of principal components analysis (PCA). PCA is a powerful data transformation technique useful for information extraction with multi-spectral or multi-temporal data (Lillesand and Keifer, 1972). Experiments have demonstrated that when PCA is applied in a multi-temporal data analysis, the PC3 and later principal components tended to contain the changed LC information (Byrne et al., 1980; Richards, 1984).

Most recently, research has focused on the development of several image differencing techniques. These fall under the following two general categories: (1) simple image subtraction and thresholding; and (2) data transformation, subtraction, and thresholding. The most commonly applied data transformations include: (a) band ratioing; (b) normalized difference vegetation index (NDVI); and (c) the tasseled-cap transformation. Simple image subtraction and thresholding requires input imagery first be normalized (or otherwise corrected) for solar illumination, atmospheric scattering and absorption, and detector performance. Some transformations (e.g., NDVI) incorporate data normalization directly in their calculation. While others, such as the tasseled-cap transformation, require data normalization prior to the transformation calculation. Also, the tasseled-cap transformation has been determined to be influenced by sensor calibration, thus is sensor dependent (Crist, 1985).

An emerging analysis technique that can be applied to interpret data transformation results is change vector analysis (CVA). More powerful than the simple thresholding technique, the change vector can be used to compare the differences in the time-trajectory of a biophysical indicator such as the NDVI or tasseled-cap transformation. The change vector represents the multidimensional vector difference between successive time-trajectories, with length indicating the magnitude of change and direction indicating the nature of change (Lambin and Strahler, 1994). When the time trajectory of an indicator departs from that expected, a LC change can be detected (Lambin and Strahler, 1994). This provides a particularly powerful technique to analyze many successive dates of data.

PAST RESULTS

Past research evaluating the comparative performance of competing methods for LC change detection over large geographic areas has been limited. Cohen and Fiorella (1998) found that composite analysis out performed both image differencing and CVA in a two-date experiment for detecting conifer forest change using Landsat Thematic Mapper (TM) imagery in the Oregon Cascades. Lyon et al. (1998) reported that the Normalized Difference Vegetation Index (NDVI) was the best performing vegetation index (VI) for detecting general LC changes in the biologically complex vegetation communities of Chiapas, Mexico. Subsequently,

Lunetta et al. (2002) determined that NDVI differencing performed equal to CVA (normalized) in a biologically complex vegetation community using a two-date experimental design in North Carolina (NC). However, both NDVI and CVA resulted in unacceptable levels of commission and omission errors. Analysis of LC data for the country of Mexico using three dates of Landsat Multi-spectral Scanner (MSS) imagery over a two-decade (1972-1992) study period demonstrated that ecosystem variability combined with classification errors precluded scene-wise or pixel-wise change detection (Lunetta et al., 2001). Finally, Elvidge et al. (1998) clearly demonstrated the value of using high temporal resolution MSS imagery to monitor changes in wetland vegetation.

One of the greatest difficulties in evaluating the performance of change detection methods is the inability to adequately characterize outcome accuracies (Khorram et al., 1999). Because change (conversion) is a relatively rare occurrence on a unit area basis, standard accuracy assessment methods tend to over estimate the actual performance. Rather than relying on the traditional percent accurate statistic, a better representation of performance is the Kappa coefficient that directly incorporates the percentage of omission and commission errors (Table 1). The second impediment to the development of high quality change detection products has been the unavailability of high temporal resolution imagery at a spatial and spectral resolution adequate to capture vegetation patch dynamics. Hopefully, this impediment will be overcome (for many applications) with the coming of NASA's operational Moderate Resolution Imaging Spectroradiometer (MODIS) in late 2001.

FUTURE DIRECTIONS

A simplistic approach currently being studied uses cumulative NDVI values based on annual growth cycles. Preliminary analysis using bi-weekly AVHRR data indicate that cumulative NDVI values tended to be quite uniform from year-to-year (Figure 1) for all non-intensively managed cover types located in the Neuse River Basin, NC. For the deciduous forest example shown below, if the forest stand were to be clear-cut, the cumulative NDVI slope would rapidly go to zero. This change in slope could easily be detected and coupled to an automated computer alarm system. Future availability of MODIS derived 250-meter calibrated NDVI 8-day composite data will provide a greatly improved data source to support the development of new phenology based change detection methods.

Change Method	Percent Accurate	Kappa Coefficient	Change Omission %	Change Commission %
NDVI	90.5	0.39	63.8	44.4
NDVI (Stratified)	91.3	0.42	60.5	42.5
CVA	85.8	0.23	78.5	53.5
CVA (Stratified)	84.5	0.23	62.1	50.2
CVA (ASCR)	91.2	0.38	62.1	50.2
CVA (Stratified + ASCR*)	91.7	0.40	59.8	50.4

Table 1. Overall accuracies and Kappa coefficients based on the probability that any point is correct and percent of change omission and commission errors for several change detection methods in the Neuse River Basin of North Carolina, USA. (*Automatic Scattergram-Controlled Regression).

SUMMARY

Current spectral-based change detection techniques have tended to be performance limited over large geographic regions. Future methods must be sufficiently robust to reduce omission and commission errors to an acceptable level. Also, because LC conversions are a rare event over large regions, methods are needed to identify only change areas (alarm) to facilitate the updating of baseline maps on an annual basis. Existing methods that require complete data processing (e.g., image differencing) tend to not be cost-effective, timely, or sufficiently accurate for large area applications. With past methods the change signals of interest tended to be masked by the inherent radiometric variability (noise) attributable to vegetation dynamics (phenology). It is anticipated that new methods incorporating vegetation dynamics will facilitate the detection or modeling of LC conversions in near real-time.

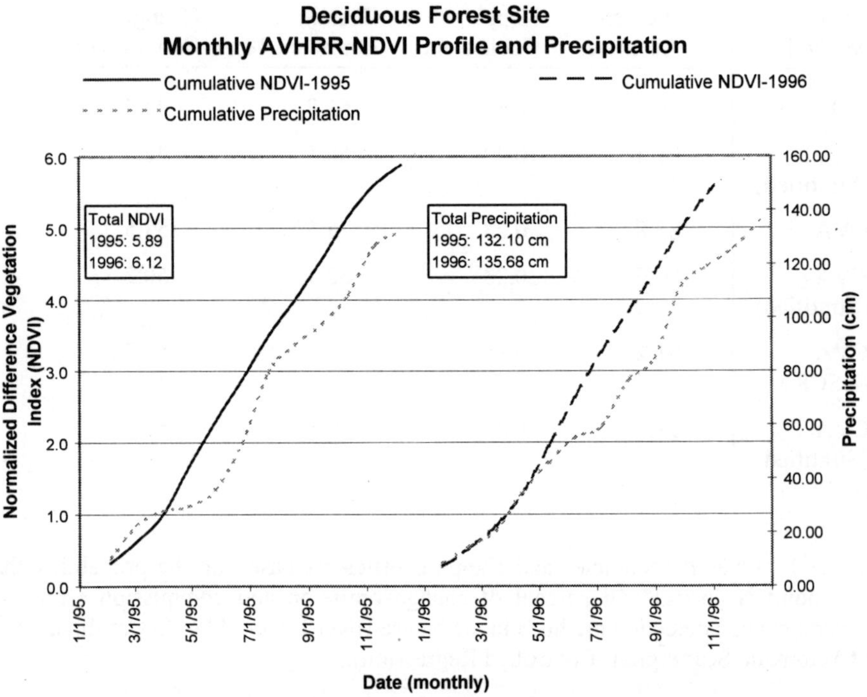

Figure 1. Cumulative Normalized Difference Vegetation Index (NDVI) values and precipitation over a two-year period (1995-1996) for a deciduous forest stand in North Carolina, USA.

REFERENCES

Byrne, G.F., P.F. Crapper, and K.K. Mayo, 1980, Monitoring land-cover change by principal component analysis of multitemporal Landsat data, *Remote Sens. Environ.*, 10, 175-184.

Crist, E.P., 1985, A TM tasseled cap equivalent transformation for reflectance factor data, *Remote Sens. Environ.*, 17, 301-306.

Cohen, W.B., and M. Fiorella, 1998, Comparison of methods for detecting conifer forest change with Thematic Mapper imagery. In: *Remote Sensing Change Detection: Environmental Monitoring Methods and Applications* (R.S. Lunetta and

C.D. Elvidge, Editors), ISBN 1-57504-037-9, Ann Arbor Press, Chelsea, MI, 318 p. + Illustrations, and co-published in Europe by Taylor and Francis, UK.

Elvidge, C. D., T. Miura,W.T. Jansen, D.P. Groeneveld,and J. Ray, 1998, Monitoring trends in wetland vegetation using a Landsat MSS time series. In: *Remote Sensing Change Detection: Environmental Monitoring Methods and Applications* (R.S. Lunetta and C.D. Elvidge, Editors), ISBN 1-57504-037-9, Ann Arbor Press, Chelsea, MI, 318 p. + Illustrations, and co-published in Europe by Taylor and Francis, UK.

Jones, B.A., K.H. Ritters, J.D. Wickham, R.D. Tankersley Jr.,R.V. O'Neill, D.J. Chaloud, E.R. Smith, and A.C. Neale, 1997, *An Ecological Assessment of the United States Mid-Atlantic Region: A Landscape Atlas*, U.S. Environmental protection Agency, Report No. EPA/600/R-97/130, U.S. Printing Office, Washington, DC, 104 p.

Khorram, S.K., G.S. Biging, N.R. Chrisman, D.R. Colby, R.G. Congalton, J.E. Dobson, R.F. Ferguson, M.F. Goodchild, J.R. Jensen, and T.H. Mace, 1999, *Accuracy Assessment of Remote Sensing-Derived Change Detection*, American Society for Photogrammetry and Remote Sensing, Monograph Series, ISBN 1-57083-058-4, 64 pp.

Lambin, E.F., and A.H. Strahler, 1994, Change-vector analysis in multispectral space: A tool to detect and categorize land-cover change processes using high temporal-resolution satellite data, *Remote Sens. Environ.*, 48, 231-244.

Lillesand, T.M., and R.W. Keifer, 1972, *Remote Sensing and Image Interpretation.* John Wiley and Sons, Second Edition, 721 pp.

Loveland, T.R., J.W. Merchant, D.O. Ohlen, and J.F. Brown, 1991, Development of a land-cover characteristics data base for the conterminous U.S., *Photogrammetric Engineering and Remote Sensing*, 57(11), 1453-1463.

Lunetta, R.S., R. Alvarez, C.M. Edmonds, J.G. Lyon, C.D. Elvidge, R. Bonifaz, C.García, G. Gómez, R. Castro, A. Bernal, and A.L. Cabrera, 2001, An assessment of NALC/Mexico land cover mapping results: implications for assessing landscape change, *Int. J. Remote Sensing*, in press.

Lunetta, R.S., R.G. Congalton, L.K. Fenstermaker, J.R. Jensen, K.C. McGwire, and L.R. Tinney, 1991, Remote sensing and geographic information system data integration: error sources and research issues, *Photogrammetric Engineering and Remote Sensing*, 57(6), 677-687.

Lunetta, R.S., J. Ediriwickrema, D. Johnson, J.G. Lyon, and A. McKerrow. 2002, Impacts of vegetation dynamics on the identification of land-cover change in a biologically complex community in North Carolina, USA *Remote Sens. Environ., in* review.

Lyon, J.G., D. Yuan, R.S. Lunetta, and C.D. Elvidge, 1998, A change detection experiment using vegetation indicies, *Photogrammetric Engineering and Remote Sensing*, 64(2), 143-150.

Richards, J.A., 1984. Thematic mapping from multitemporal image data using the principal components transformation, *Remote Sens. Environ.*, 16, 25-46.

Singh, A., 1989. Digital change detection techniques using remotely-sensed data, *Int. J. Remote Sensing*, 10(6), 989-1003.

Stow, D.A., L.R. Tinney, and J.E. Estes, 1980. Deriving land use/land cover change statistics from Landsat: a study of prime agricultural land. *Proceedings of the 14th International Symposium on Remote Sensing of the Environment*, Ann Arbor, Michigan, pp. 1227-1237.

Vogelmann, J.E., T. Sohl, and S.M. Howard, 1998, Regional characterization of land cover using multiple sources of data. *Photogrammetric Engineering and Remote Sensing*, 64(1), 45-57.

Weismiller, R.A., S.J. Kristof, D.K. Scholz, P.E. Anuta, and S.A. Momin, 1977. Change detection in coastal zone environments, *Photogrammetric Engineering and Remote Sensing*, 43(12), 1533-1539.

Notice: This paper has been reviewed in accordance with the U.S. Environmental Protection Agency's peer and administrative review policies and approved for presentation and publication. Mention of any trade names or commercial products does not constitute endorsement or recommendation for use.

SIMULATION OF FOREST ECOSYSTEM PROCESSES BY THE ANALYSIS OF MULTITEMPORAL NOAA-AVHRR NDVI IMAGES

MARTA CHIESI, LAURA BONORA, FABIO MASELLI

IATA-CNR, P.le delle Cascine 18, 50144Florence, Italy. E-mail:foremms@iata.fi.cnr.it

MARCO BINDI

DISAT – University of Florence, P. ledelle Cascine, 12, 50144 Florence, Italy

LUCA FIBBI

LaMMA- FMA, Via Einstein, 35, CampiBisenzio, Florence, Italy

The current work deals with the utilization of multi temporal NOAA-AVHRR NDVI images and ancillary data to derive Leaf Area Index (LAI) profiles of two forest stands in the Radicondoli test area (Central Italy). This profiles are then inserted into a bio-geochemical model, Forest-BGC, in order to obtain multitemporal estimates of forest evapo-transpiration and photosynthesis. The statistical analysis of these estimates versus the original NDVI and LAI values showed that the latter are the main determinant of vegetation processes in this kind of environment. These findings imply that multitemporal remotely sensed estimates of LAI can, after suitable calibration, be used as a surrogate of evapo-transpiration and photosynthesis estimates in most operational applications.

1 Introduction

The use of remotely sensed data for providing inputs to ecosystem simulation models has recently seen increasing spread. Bio-geo-chemical models such as the well known Forest-BGC of Running and co-workers [9] are frequently used to derive estimates of ecosystem processes over wide spatial and temporal scales. A major problem for this objective is however given by the availability of suitable inputs about vegetation parameters, such as Leaf Area Index (LAI), which are extremely variable both in space and time. The derivation of these parameters from remotely sensed data is therefore a suitable alternative which is becoming very popular [12].

The current work was aimed at using Forest-BGC to estimate long-term vegetation processes (evapo-transpiration and photosynthesis) in a Mediterranean area where the model calibration had already been performed [2]. Multitemporal NOOA-AVHRR NDVI data were previously transformed into LAI profiles to provide these essential inputs to the model. The estimated monthly evapo-transpiration and photosynthesis were finally regressed against the original NDVI and LAI values to evaluate their possible direct use for flux estimation.

In particular, the work was carried out as follows:

1. derivation of meteorological data collected at the meteorological station in Larderello (400 m a.s.l.) and their derivation for the area to which the model is applied
2. derivation of the characteristics of the area, like altitude, latitude and others by the Forest Plan of Management [4]
3. collection of all the ecophysiological parameters useful to describe the behaviour of the species characterising the forest stands
4. collection of the values of Leaf Area Index that are used to determine and drive daily the model Forest-BGC.

Only limited information will be currently given about the collection of the ecophysiological parameters and the meteorological data as well as the validation of the model, that were already described in a parallel publication [2]. Much emphasis is instead given to the methodology used to derive the daily values of NDVI by the elaboration of the images and their integration within the model to derive daily values of transpiration.

2 The Model Forest-BGC

The model Forest-BGC [9] can be used to determine and quantify the most important cycles that describe the behaviour and the response of the ecosystem to environmental changes.

Its use requires different information about the area at which it has to be applied and in particular:

1. climate data
2. environmental data and information about the stand
3. ecophysilogical data about the species that characterise the ecosystem
4. daily values of LAI driving the model.

All these data can be collected by different sources and integrated within a GIS in order to extend the application of the model to a large area. The use of information derived from remote sensing data is particularly effective, thanks to their sinopticity and high acquisition frequency. This is the subject of the current work, which particularly emphasizes the inclusion of monthly LAI estimates within the model. In fact the original version of the model has been modified to make it more adaptable to forests characterised by the presence of deciduous species [2].

3 Study Area and Data

The study area, Radicondoli, is close to Siena (43.18-43.29° N, 10.95-10.57° E , south of Tuscany, a region of Central Italy). It is characterised by an hilly landscape dominated by the alternation of agricultural fields, pastures and forest areas. The weather is Mediterranean sub-humid [8] and it is characterised by the alternation of dry and warm summers and cold winter. The mean annual rainfall is around 800 mm while the average annual temperature is around 15 °C. The vegetation is dominated by the presence of mixed deciduous trees; in particular the forests are characterised by oaks (mainly *Quercus pubescens* Willd. and *Quercus cerris* L.) and other deciduous trees of less importance. Together with this natural formation, there are some artificial forests dominated by the presence of conifers (*Pinus nigra* Arnold, also some cedars and *Cupressus sempervirens* L.) which were planted to prevent soil erosion.

Ancillary data were collected as a topographic map (1:25.000) produced by the Italian Military Geographic Institute (I.G.M.I.) and a digital elevation model (DEM) in the form of raster images with a pixel size of 20x20 m, derived from contour lines every 25 m. Land cover information was derived from the Land Cover classification produced during the EU Project CORINE at a nominal scale of 1: 100.000 [1].

In addition, more detailed information about the forested area were derived digitising a 1: 10000 scale map produced by DREAM [3] for the Forest Plan of Management of the area. Together with it a database describing each forest stand is also available.

Meteorological data were collected from the Larderello meteorological station that is owned by the Italian National Hydrological Service; a long time series is available since 1951. Additional data have been retrieved by a portable station situated in Collitalli (43°15'3'' N lat. - 11°03'29'' E long; 485 m a.s.l). This series starts in 1996 and it is still under acquisition.

LAI values were collected within 14 forest stands characterised by different densities; the values were collected along linear transect of about 50 m and considered as an average of at least 6 different measures. This collection started in 1996 and it is still carried on during each growing season.

NOAA-AVHRR images were finally derived from the archives of Telespazio (Rome) and of the University of Berlin within the framework of the EU projects RESMEDES and RESYSMED. The first archive contained images from 1986 to 1993, while the second images from 1993 to 1998. These images were already in the form of monthly NDVI maximum value composite, produced following the procedure described by Maselli et al., [6] and Koslowsky [4] respectively.

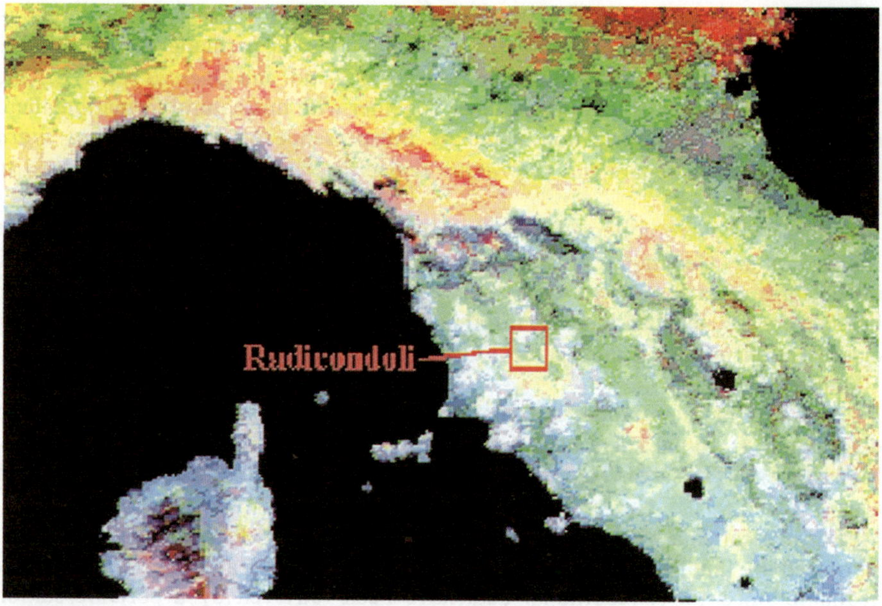

Figure 1: False color composite of three monthly NDVI images (February, May and August 1998), showing the position of the study area (Radicondoli) in Central Italy.

4 Data Processing

4.1 Extraction of NDVI profiles for the two stands

The extraction of the monthly NDVI data for the forest areas considered was carried out by relying on the CORINE Land Cover Map of Tuscany and using a method recently proposed by Maselli [7]. This method is based on locally calibrated multivariate regression analyses which can determine spatially variable NDVI values of all land cover types as derived from a higher resolution map. By the application of the method, per-pixel NDVI estimates can be obtained for each pure class the abundance of which is known from a higher resolution source. In the current case, abundance estimates of five cover classes were derived from the CORINE map, and monthly NDVI "end-member" images were obtained for the forest class over all the period considered (1986-98). The monthly NDVI values corresponding to the two selected stands were then extracted for the determination of LAI.

4.2 Computation of LAI profiles

The determination of the LAI values was obtained applying the equation proposed by Lacaze et al. [5] modified to account for the different characters of each stand considered. The equation is the following:

$$LAI = \frac{1}{K} * \ln \frac{(NDVIback - NDVI)}{(NDVIback - NDVImax)} \quad (1)$$

where NDVIback and NDVImax are the minimum and maximum NDVI values for the forest, which can be extracted directly from the images, and K is linked to the slope of the LAI/NDVI relationship, which was the coefficient to modify. This modification was based on the assumption that NDVI is closely related to LAI only if the forest canopy closure is greater than 90%. In different cases, when the forest is not dense (which is very frequent in Mediterranean areas), the relationship between the two variables changes due to the contribution of the understorey, so that NDVI overestimates Leaf Area Index. The correction of the K coefficients for each stand was carried out using the LAI values measured in the growing seasons from 1996-1998. The NDVI values of the two study stands derived by the previous procedure were log-transformed using the right hand side of equation 1, and were then regressed against the relevant LAI measurements to find out appropriate K values for each stand. Equation 1 was then applied to the NDVI profiles of the two stands to compute LAI multi-year profiles.

4.3 Application of Forest-BGC with variable LAI

The LAI values obtained previously were used within the modified version of the model, which had been already calibrated for the two study stands as described in Chiesi et al., [2]. The meteorological data necessary to the model were simulated for the two stands using MT-Clim [version 4.3; 11], while some physiological parameters describing the two stands were found out during the calibration phase [2]. The application of the model for 14 study years (1985-1998) produced daily estimates of transpiration and photosynthesis for both study stands.

5 Results

The relationships found for the two stands considered between NDVI and LAI are shown in Figure 2A-B. As can be seen, different K values were found for the two stands, which was due to the higher density of the stand dominated by *Quercus cerris* L. with respect to the one characterised by *Quercus pubescens* Willd.. The transformation from NDVI to LAI was changed according these figures, thus using a K coefficient of 0.417 for *Quercus pubescens* Willd. and of 0.283 for *Quercus cerris* L..

The relationships between monthly LAI and estimated transpiration and photosynthesis for both stands considered are shown in Figures 3 e 4 A-B.

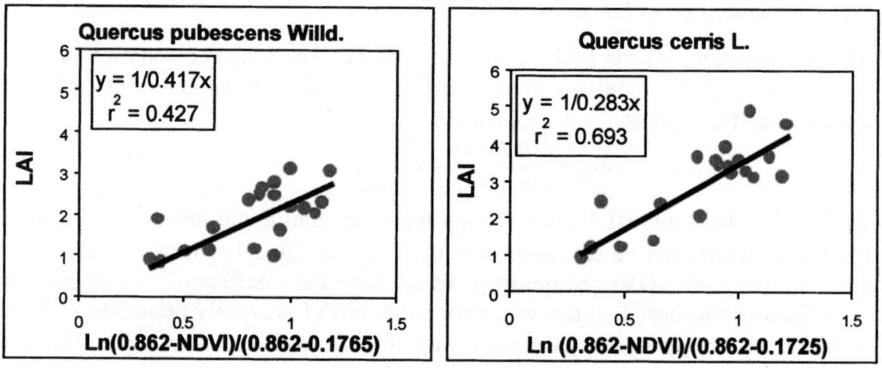

Figure 2A-B: relationship between LAI and NDVI for the two different stands.

As can be seen, there is a quite good agreement between LAI and the two dependent variables, and particularly for photosynthesis. Actually, r^2 higher than 0.8 were found in all cases, indicating a very high predictive value of LAI.

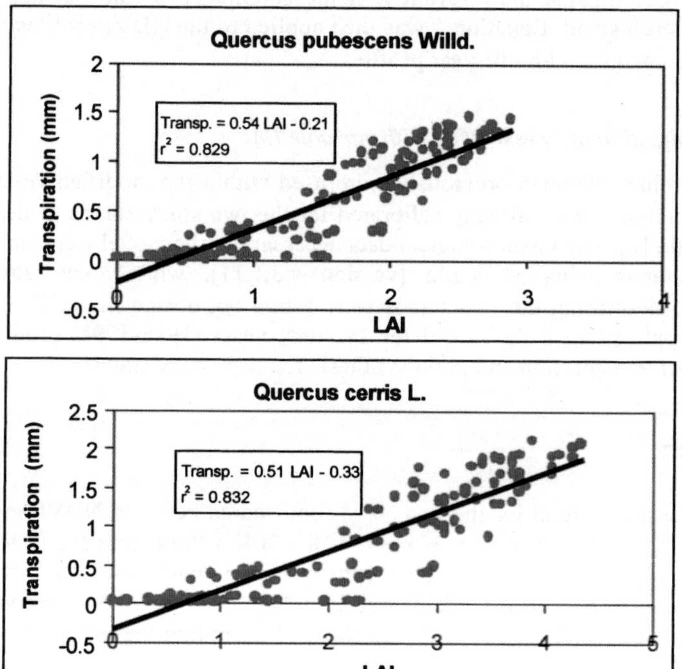

Figure 3 A-B: relation between monthly LAI and estimated values of transpiration for both the species.

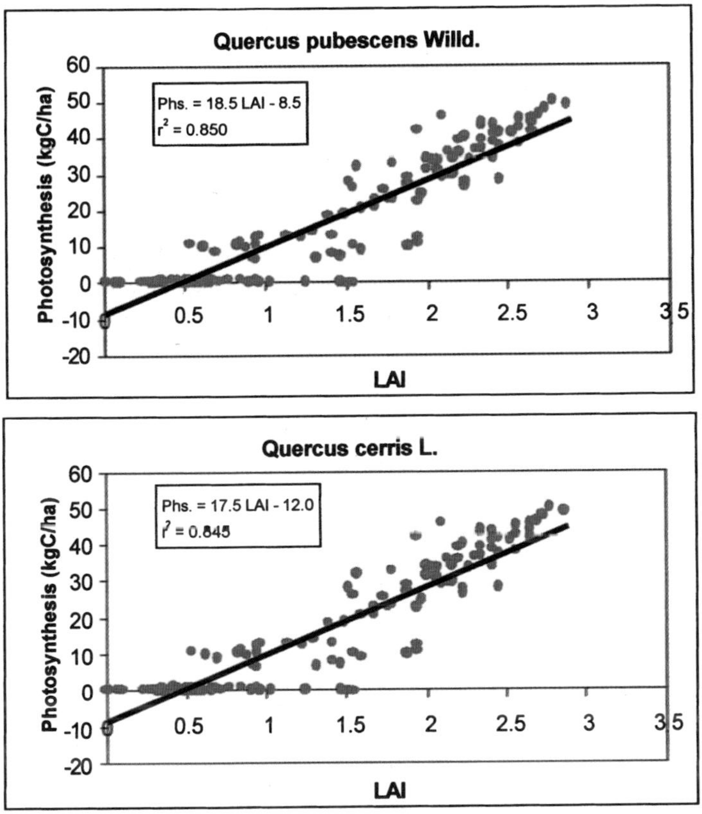

Figure 4 A-B: relation between monthly LAI and photosynthesis shown for both the species.

6 Conclusions

The current work was based on previous studies, which indicated that conventional and remote sensing data can be integrated within the Forest-BGC model to accurately estimate evapo-transpiration and photosynthesis of Mediterranean forests. The results obtained show that remotely sensed estimates of LAI have a high predictive value with respect to simulated water and carbon fluxes. This can be interpreted considering that LAI is a major determinant of Forest-BGC functions, in accordance with the concepts suggested by the model's authors [12]. It can be concluded that the use of multi-temporal images is of utmost importance to detect long-term changes of the main forest ecosystems processes. Hence, the integration of remotely sensed data and modelling of ecological processes can be a valid support to all decision makers dealing with the management of forest ecosystems.

Acknowledgements
The work was partially supported by the Agenzia Spaziale Italiana under contract ASI ARS00.

References

1. Annoni, A. and Perdigao, V., 1997. Technical and methodological guide for updating CORINE Land Cover Data Base. European Commission, EUR 17288EN, Space Application Institute of Joint Research Centre, Ispra, Italy.
2. Chiesi M., Maselli F., Bindi M., Fibbi L., Bonora L., Raschi A., Tognetti R., Cermak J., Nadezhdina N. (2001). "Calibration and application of Forest-BGC in a Mediterranean area by the use of conventional and remote sensing data". Submitted to *Ecological Modelling.*
3. D.R.E.A.M. (1994), "Complesso forestale Le Carline – La Selva. Piano di assestamento forestale 1994 – 2004", Regione Toscana (a cura di), Italy.
4. Koslowsky D. (1998), "Daily extended 1-km AVHRR data sets of the Mediterranean", Proceedings 9th Conf. Sat. Meteor. And Oceanogr. UNESCO, Paris25-29 May, AMS, Boston, MA, 38-41.
5. Lacaze B., Caselles V., Coll C., Hill H., Hoff C., de Jong S., Mehl W., Negendank J. F.,Riesebos H., Rubio E., Sommer S., Teixeira Filho J., Valor E. (1996), "De Mon – Integrated approaches to desertification mapping and monitoring in the Mediterranean basin", finale report of De-Mon I Project, Joint Research Centre of European Commission, Ispra (VA), Italy.
6. Maselli F., Gilabert M. A., Conese C. (1998), "Integration of high and low resolution NDVI data for monitoring vegetation in the Mediterranean environments", *Remote Sens. Environ.*, 63: 208-218.
7. Maselli F. (2001), "Definition of spatially variable spectral endmembers by locally calibrated multivariate regression analyses", *Remote Sens. Environ.*, 75: 29-38.
8. Rapetti F., Vittorini S. (1995), "Carta climatica della Toscana centro-settentrionale, centro-meridionale ed insulare", C.N.R. – Centro di Studio per la Geologia Strutturale e Dinamica dell'Appennino, Pisa, Italy.
9. Running S. W., Coughlan J.C. (1988), "A general model of forest ecosystem processes for regional applications", *Ecological Modelling*, 42: 1425-154.
10. Running S. W., Nemani R.R., Peterson D. L., Band L.E., Potts D. F., Pierce L. L., Spanner A. M. (1989), "Mapping regional forest evapotranspiration and photosynthesis by coupling satellite data with ecosystem simulation" *Ecology*, 70:1090 – 1101.
11. Thornton, P.E., Running S.W. (1999), "An improved algorithm for estimating incident daily solar radiation from measurements of temperature, humidity, and precipitation." *Agricultural and Forest Meteorology*, 93:211-228.
12. Waring R.H., Running S. W. (1998), "Forest ecosystem. Analysis at multiple scales", Academic Press.

COMPARISON OF SINGLE YEAR AND MULTI YEAR NDVI TIME SERIES PRINCIPAL COMPONENTS OVER A GRASSLAND-FOREST ECOTONE

MRYKA HALL-BEYER

Department of Geography, University of Calgary, Calgary, Alberta T2N 1N4, Canada
E-mail: mhallbey@ucalgary.ca

Standardized Principal Components Analysis is performed on NDVI time series a 10° by 10° area of western Canada including grassland, parkland and forest ecosystems. When used with input from a single year growing season (April-October 1992, 1993 and 1995), early components correspond closely to ecoregional mapping based on long-term vegetation composition. Later components isolate areas showing particular agricultural practice or climatic stress particular to the year. When all three years are included in a single analysis, similar patterns occur in the first component. However, new components occur early in the series that isolate areas having different seasonality patterns associated with different phenology. Areas dominated by deciduous vegetation are prominent in a different component from those primarily grasslands. Multi year SPCA includes an early component apparently related to latitudinal variation in day length, which does not appear in single year components. There is also a major component unrelated to any known geographical or climatological pattern or event, which may relate sensor degradation. Both single year and multi year SPCA isolate NDVI variability that is limited in space and/or in time in single later components. This allows interpretation of transient or localized events using detailed local data, separately from regional trends that occur in earlier components. These results demonstrate the possibility of refining ecoregion mapping to take into account regional patterns based on selected early components. This would incorporate actual interannual variability for selected time periods as well as the long term stable elements of biogeoclimatic regions. Longer time series could potentially quantify observed unidirectional trends resulting from climate change.

1 Introduction

Global vegetation dynamics have been followed using NDVI time series. The primary method of doing so has been to use metrics such as date of greenup, date and magnitude of maximum vegetation density, integrated NDVI over time, and rates of greenness change [6, 7, 8, 10]. Metrics allow progressive greenup and dieback to be traced on a continental scale and comparisons to be made between one year and another. The desired outcome is to track, on a regional to global scale, annual effects of climate variability recorded by vegetation phenology [7, 10]. However, the algorithm used to extract many of these metrics, and its interpretation, is biome-specific. In particular, the onset of NDVI greenup in deciduous forests corresponds to leafout and is relatively independent of remnant snow cover under the trees, whereas boreal forests' "greenup" may be snowmelt dates [7]. In grasslands, NDVI greenup occurs when green grass surpasses the height of standing or flat-lying dead grass. It may also be sensitive to the amount of snow clover trapped by dead grasses. In addition, greenup algorithms based on changes in the slope of the NDVI vs time plot may fail to respond to the very gradual rate of increase in greenness that often occurs in grasslands. The greenup date derived thus may depend on grassland grazing practice in different areas rather than on regional climatological patterns.

Principal Components Analysis provides an alternative method of tracking vegetation response to climate variability and to changes in agricultural practice such as irrigation. Since eigenvectors are extracted from the series itself and have no necessarily universal meaning, each series must be interpreted through a combination of geographical knowledge, climatological data and loadings (correlations between individual components and the original input date NDVI). Nevertheless, common patterns occur. Individual components early in the series correspond to pervasive and durable geographic differences and temporal variability, while those occurring later in the series highlight variability more limited in space, in time, or in both. Noise components such as those due to long-duration cloud events tend to be isolated in later series members.

Unlike the technique for deriving components of multispectral datasets, extraction of eigenvectors for time series commonly is based on the correlation matrix ("standardized" or SPCA), rather than the covariance matrix. SPCA eliminates heavy weighting of summer images caused by their inherently much larger dynamic range [3]. SPCA allows inter-seasonal variability and short-term fluctuations to be readily visible in the resulting images.

This paper examines the date loadings into the first 3 components, and also the geographic pattern of component values, when computing SPCA for single-year growing seasons. These were compared with the first 6 components resulting from a continuous 3-year growing season analysis. The years 1992, 1993 and 1995 were used, thus including the unusual 1993 pattern caused by climatic variation associated with both a strong El Niño event and with volcanic upper-atmosphere dust. The data for the Province of Alberta covers a wide latitudinal and elevation range and includes ecotones between grasslands, deciduous forest, coniferous forest and alpine regions.

2 Methods

2.1 Data

Time series of 1 km resolution NDVI AVHRR datasets were acquired over the western Canadian prairies (lat 49 to 60N, long 110 to 120W) for 1992, 1993 and 1995. The images were processed by the EROS Data Center and are freely available to researchers [5]. The images were composited at 10-day intervals; compositing selects individual pixels for the largest NDVI over each period. This eliminates most clouds and biases in favor of pixels closer to sensor nadir. Further processing provided partial correction of bidirectional effects and atmospheric scattering [4, 9]. Standardized principal components analysis was performed using images from early

April (still in the dormant season) through late September. The same dates were used for all years with the exception of Sept. 11-20 and 21-30 1993, which had quality problems. Winter images at these latitudes suffer from very low light, but in these climates are not necessary to interpret vegetation dynamics, as plants are dormant or snow covered during the October-March period.

Eigenvectors were calculated based on the correlation matrix for all 18 (single year) or 52 (multi-year) potential image components. Output images were prepared for the first 6 (single year) and 12 (multi-year) components. Band (date) loadings were calculated for each output component.

3 Results and discussion

3.1 Single-year SPCA

The years 1992, 1993 and 1995 are appropriate for comparing single-year time series. 1992 was a strong El Niño year, and weather records show a very early spring warming. However, unusually cold weather occurred later in the growing season, as the result of atmospheric ash from the eruption of Mt. Pinatubo in summer 1991. 1993 did not deviate from long-term normals. 1995 also followed an El Niño event, less intense than in 1991-92, however Alberta in 1995 was slightly cooler and dryer than 1961-1980 normals [1]. The presence or absence of El Niño has been shown to affect the growing season and precipitation on the Canadian prairies [2].

SPCA was performed separately for each of the three years. Each component was interpreted using a combination of loadings for each date and geographic pattern. Figure 1 shows that the loadings for 1993 and 1995 perform similarly, with high corre-lation between component 1 and all dates.

High component 1 pixel values correspond to areas with the maximum integrated yearly vegetation density, grading down to low values in areas of permanent non-vegetated cover. Figure 2 illustrates this geographical pattern,

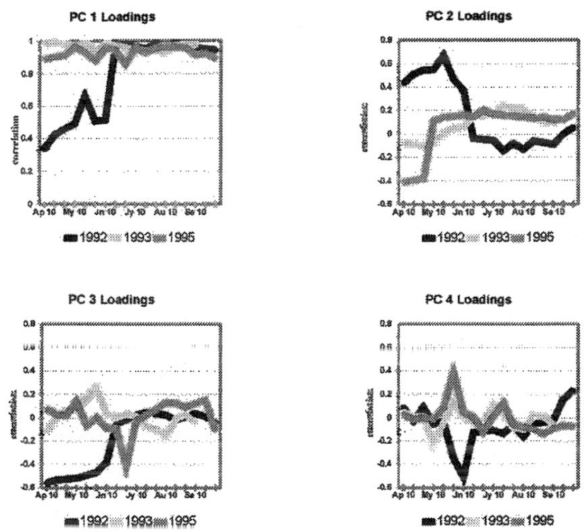

Figure1. Loadings (correlations) between components and dates

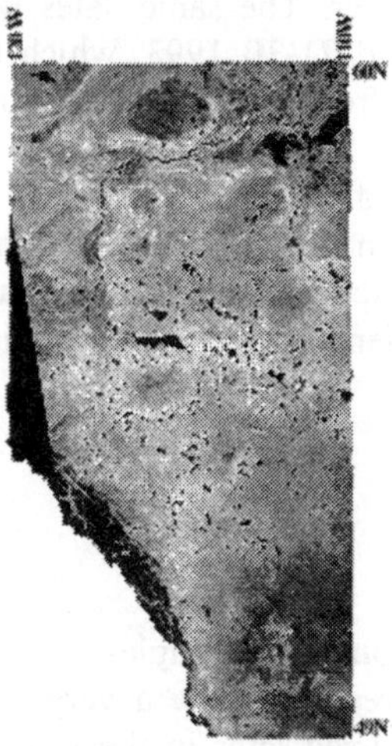

Figure 2. PC1, 1995

which closely resembles ecoregional mapping for the area (see Figure 8 at end of paper). The brightest component 1 values are in central and northern forested areas, where little land has been converted to agriculture. Slightly lower values occur over less dense forest, in northern areas (especially in the Porcupine Hills, the ovoid structure at center north) and at higher elevation in the southwest. There is little distinction between primarily deciduous and primarily coniferous forest. PC1 values become lower still where forest is sporadic and where much land has been converted to annual crops, in the south central area. Lowest values occur in the high mountains along the southwest border, and in dry grasslands of the south and east. Some extensively irrigated districts in this area have a contrastingly higher value.

PC1 patterns are those expected from ecoregion mapping, which is based on long-term stable climate and topography. In 1992 however, the first component derives this pattern only from summer and fall vegetation density, with spring values presenting a different axis of variance.

Component 2 (Figure 3 illustrates 1995) shows high pixel values in mainly deciduous forest or in areas where bare ground in spring and fall contrasts to uniform annual crop in summer. Coniferous forests are dark, and grasslands intermediate. The main element loaded into component 2 is the contrast between spring and summer NDVI values. The date of loadings transition from positive to negative corresponds to general deciduous leafout, which varies surprisingly little over the latitudinal range of the area.

Figure 3. PC2, 1995

PC2 for 1992 is different. In 1992, PC3 loadings resemble the PC2 loadings in the other two years, while PC2 in 1992 picks up the anomalous spring behavior that was missing from PC1. Bright PC2 pixels are those with particularly high NDVI values in spring. The spatial pattern (Figure 4) shows a marked greening of the northern areas, particularly in the Athabasca River valley cutting diagonally across the northwest. There is some early greening in the southern grassland, but little effect in the dark band of the central mixedwood area.

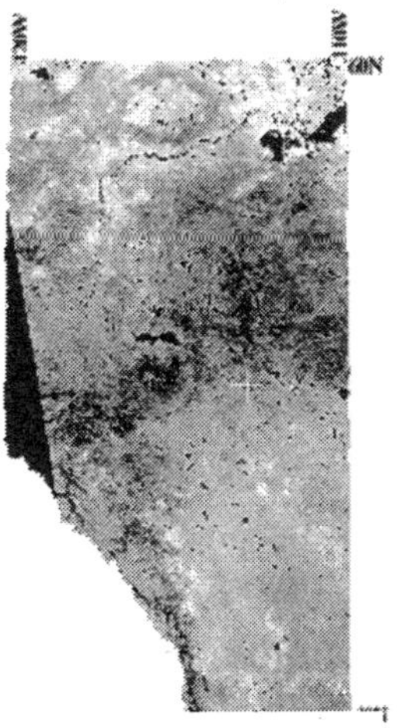

Figure 4. PC2, 1992

Components 3 and later for "normal" years, and PC4 and later for 1992, are more date specific (Figure 1; images not illustrated). They highlight the contrast between certain dates and those NDVI values that are more stable over the long term. The loaded dates include unseasonable snow cover or cloud cover of unusual duration, or short-term drought. As such they are of particular interest for local applications, particularly in forestry and agriculture.

To sum up single-year SPCA,

- early components allow the definition of areas whose vegetation development behaves in a similar fashion. This corresponds to biome or to some extent to land use.
- later components single out areas that deviate from the regional value either in magnitude of greenness or in timing of events.
- A widespread climatic anomaly is effectively isolated into a particular component.

3.2 Multiple year SPCA

Widespread anomalous events occurring in a single year will contribute significantly to multi-year variance, so are expected to be isolated in individual components.

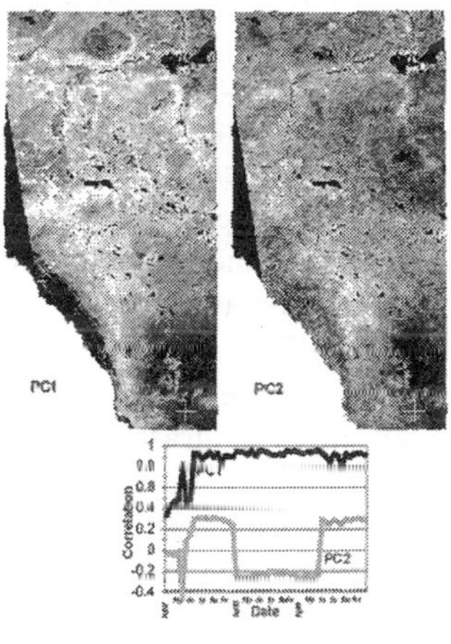

Figure 5. Multiple-year PC1 and PC2 images and loadings

PC1 (Figure 5), as in single-year analysis, correlates strongly with all dates across all years, with the exception of spring 1992. The image very closely duplicates the single-year PC1 images, as expected, since it represents the inter-regional biome differences. PC1 accounts for 81% of total dataset variance. Multi year PC2, unlike single year PC2, does not represent seasonality. Instead its bright areas show where there is a great deal of contrast between 1993 and spring 1995 on the one hand, and the rest of the dates on the other. This occurs over much of the Province, with the exception of dry grasslands of the southeast. No natural phenomenon has been found that would account for this type of variance, accounting for 6.6% of the total dataset variance. Instrumental variation is being investigated as an explanation.

Figure 6 shows data for PC3 (3.6% of dataset variance) and PC4 (1.4%). PC3 represents the degree of seasonality, which was expected to show up in PC2 [3]. Loadings alternate between positive for spring and fall and negative for summer. The crossover between positive and negative varies somewhat with the year, and is associated with particular climatic conditions. Note especially the rise to positive in early August 1995, corresponding to a notoriously cool and wet August in that year. This caused early plant maturation [1]. Brightest pixels, meaning areas of maximum contrast between summer and spring/fall, occur in the central mixedwood and in the foothills regions, which are covered with similar vegetation. These represent areas of the Province with the greatest proportion of deciduous for-ests. It is some-what surprising to see a zone within the southeastern grasslands that is moderately bright in PC3, even though this area is devoted more to pasture than to the annual crops that might be expected to show greater seasonal contrast. Irrigated areas show low values in PC3. PC3 can thus be interpreted as isolating the scene variance due primarily to deciduous leafout and senescence, rather than to seasonality in general.

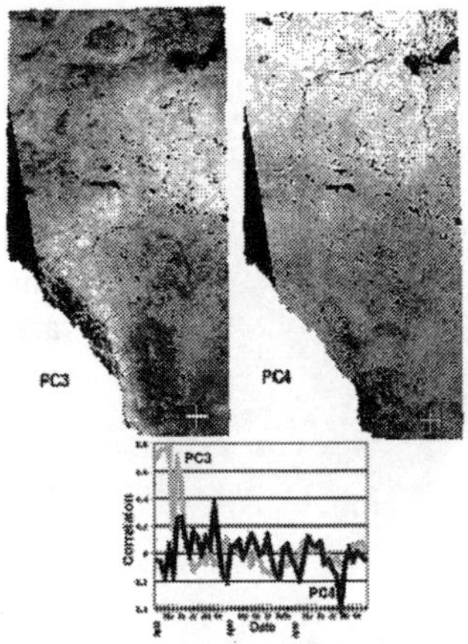

Figure 6 Multiple year PC3 and PC4 images and loadings

PC4 loadings are highly erratic and the geographic pattern shows strict latitudinal variation with no evidence of topographic or biome influence. It is interpreted as showing variance due to sun angle differences across the image.

PC5 (0.9% of variance) and PC6 (0.7%) (Figure 7) loadings both exhibit seasonal cycling, with a slightly different periodicity than PC3. Within anomalous 1992, PC5 records a strong contrast between the warm April and the cold later spring. Geographically, bright areas occur in the central mixedwood zone as for PC3, but also in the central mountains and the high elevation western grasslands which showed little contrast in PC3. These are areas exhibiting the greatest seasonal advance in spring 1992 followed by greatest effects of the cold snap. PC6 is bright primarily in high elevation areas. Loadings show seasonal periodicity, but with crossovers from positive to negative loading in mid August and late June. Thus

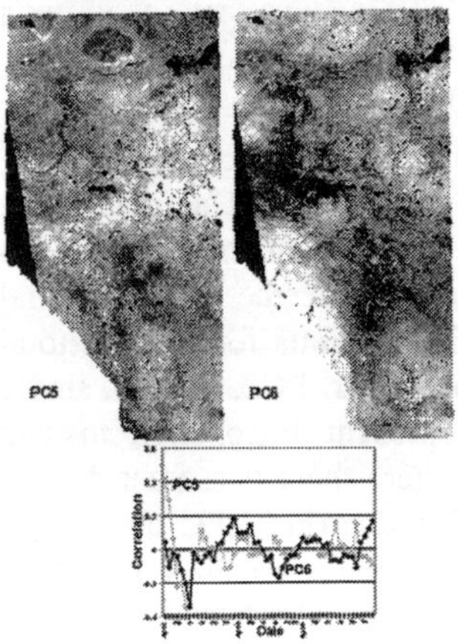

Figure 7. Multiple year PC5 and PC6 images and loadings

PC6 highlights areas with a short growing season. The grasslands have moderately bright values in PC6, corroborating the late appearance of green through the dead grass litter and grass' early senescence in response to the end of the rainy season in July.

Components beyond PC6 were examined, and primarily show events localized in time and/or space. Detailed analysis would rely on local data, not feasible in a regional synthesis. Later components may isolate small regions meriting such close scrutiny.

To sum up multiple year SPCA,

- Component 1, as in single-year SPCA, allows the definition of areas whose vegetation defines a single biome or ecoregion.
- Later components single out areas that have in common a particular pattern of seasonality. Areas with common timing of greenup and senescence appear together in a single component., while those dominated by different timing appear in a different component. Causes of this timing difference may be degree of dominance by deciduous trees or by grasses and elevation differences. This information is not available from single year SPCA.
- Components appear early in the sequence that relate to inter-year contrast or to sun angle variation. These are not apparent in single year analysis.
- Later components, as in single year analysis, relate to vegetation events limited in time and/or space that must be interpreted with detailed local knowledge.

4 Conclusions

Principal Components Analysis of NDVI time series provides a way of mapping geographical variation of vegetation behavior. Primary variation is related to elements of long-term inter-regional differences, in turn related variables associated with ecoregional mapping. In an area including the grassland-forest transition, later components highlight seasonality and differences between years that are caused by large-area influences such as climatic anomalies. Interpretation of broad trends is not hampered by the problems of curve smoothing inherent in NDVI metrics, since Principal Components isolate short-term anomalies into separate components. Inclusion of several years in the time series allows tracing of inter-year contrasts. The judicious selection of components may provide a means of mapping ecoregional boundaries based on dynamic as well as long-term factors.

References

1. Alberta Agriculture, Food and Rural Development. 1995. Alberta Agricultural Weather Summary, January-December, 1995. Volume 9, Issue 35 ISSN 0835-7307.
2. Budikova, D. 2001. El Nino and the Southern Oscillation - Surface Air Temperature Implications for Western Canada. Unpublished PhD Thesis, University of Calgary.
3. Eastman, J.R.; Fulk, M. 1993. Long sequence time series evaluation using standardized principal components. *Photogrammetric Engineering and Remote Sensing* 59:8 1307-1312.
4. Eidenshink, J.C.; Faundeen, J.L. 1994. The 1-KM AVHRR global land data set: first stages in implementation. http://edcdaac.usgs.gov/1KM/paper.html accessed 20010329
5. EROS 1996. http://edcwww.cr.usgs.gov/landdaac/1KM/comp10d.html accessed 20010730
6. Markon, C.J.; Fleming, M.D.; Binnian, E.F. 1995. Characteristics of vegetation phenology over the Alaskan landscape using AVHRR time-series data. *Polar Record* 31:177, 179-190.
7. Moulin, S.; Kergoat, L.; Dedieu, G. 1997. Global-scale assessment of vegetation phenology using NOAA/AVHRR satellite measurements. *Journal of Climate* 10:6, 1154-1164.
8. Reed, B.C.; Brown, J.F.; Vanderzee, D.; Loveland, T.R.; Merchant, J.W.; Ohlen, D.O. 1994. Measuring phenological variability from satellite imagery. *Journal of Vegetation Science* 5:5, 703-714.
9. Zhu, Z.-L.; Yang, L. 1994. Characteristics of the North America 1-km AVHRR Data Set. http://edcdaac.cr.usgs.gov-/1KM/zhu.html accessed 20010329.
10. Yang, L.; Wylie, B.K.; Tieszen, L.L.; Reed, B.C. 1998. An analysis of relationships among climate forcing and time-integrated NDVI of grasslands over the US northern and central Great Plains. *Remote Sensing of Environment* 65:1, 25-37.

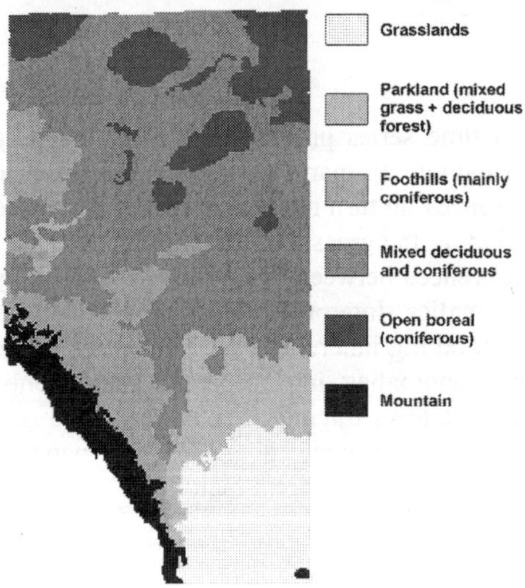

Figure 8. Simplified vegetation of Alberta

LONG TERM CHANGE DETECTION USING CONTINUOUS FIELDS OF TREE COVER FROM 8KM AVHRR DATA FOR THE YEARS 1982-2000

MATTHEW C. HANSEN[1]
Email: mhansen@geog.umd.edu

JONATHAN C-W CHAN[1]
JULIE PAGIS[2]
RUTH DEFRIES[1,3]
DI LUO[1]

[1] *Laboratory for Global Remote Sensing Studies, University of Maryland, College Park, MD 20742*

[2] *Carnegie Institution of Washington, Department of Plant Biology, Stanford, CA 94305*

[3] *Earth Systems Science Interdisciplinary Center, University of Maryland, College Park, MD 20742*

This paper describes the use of the long term 8km Pathfinder Advanced Very High Resolution Radiometer (AVHRR) dataset in detecting changes in global tree cover. The procedure involves the use of a regression tree in predicting percent tree cover for the years 1982-2000. Training data are created from high-resolution imagery and used with phenological metrics derived from the annual AVHRR time series. Using the 18 years of predicted tree cover, change can be measured using a simple regression. Results are qualitatively promising for tropical regions. Countries exhibiting extensive reductions in tree cover include Brazil, Paraguay, Bolivia and Indonesia. Venezuela exhibits the most extensive gains in tree cover due to plantations in the eastern part of the country. Confidence maps have also been generated and allow for filtering the change pixels for more probable areas of change.

1 Introduction

The archive of the Advanced Very High Resolution Radiometer (AVHRR) represents a unique data set within the satellite record. This is the only source of satellite data that exists as a continuous global record for the past two decades. As such, using it to monitor change could provide information on global trends of land cover change. However, many difficulties exist in using the data to such ends. Problems associated with its coarse resolution, resampling, compositing, among others can make it difficult to extract meaningful information. The work here builds on a new approach to analyzing these data, that of estimating land cover proportions using continuous field characterizations [4,5].

Continuous fields of vegetation cover represent an improvement over discrete land cover classifications. Continuous fields depict sub-pixel land cover proportions and are more appropriate representations of thematic information in areas of spatial heterogeneity [1]. Due to the fact that many areas of human-induced land cover change are marked by increased spatial heterogeneity and fragmentation, continuous fields offer a tool for detecting changes which would not be possible using comparisons of consecutive discrete classifications. This study reports the generation of a series of continuous fields of global tree cover from 8km Pathfinder AVHRR data [7] for the years 1982-2000 excepting 1994 due to incomplete data. These products have been measured for trends, and change within the continuous fields is indicated. The technique is based on the methodology to be used in generating the MODIS 500 meter continuous field products and offers a future tool for change detection with data of improved spectral and spatial characteristics.

2 Algorithm

A regression tree algorithm was used to generate the annual tree cover layers. Regression trees have been used in remote sensing by others [2,8,9]. They are a non-linear tool which recursively splits a continuous training variable into subsets, called nodes, which minimize the overall residual sum of squares. The regression tree algorithm takes the form of the following equation:

$$D = \sum_{cases(j)} (y_j - u_{[j]})^2$$

where D is the deviance as measured by the corrected sum of squares for a split. This is calculated from all j cases of y and the mean value of those cases, u. The split which minimizes the residual sum of squares is chosen as the best split.

In this case, the dependent variable is percent tree cover. The 8km training data for percent tree cover were derived from high-resolution images which were characterized into 4 tree cover strata (0-10%, 11-40%, 41-60%, and 60-100%). The approach uses the same database described in DeFries et al. [3] and Hansen et al. [6]. The mean value for the tree strata was then assigned to each high-resolution pixel and aggregated to 8km resolution. This results in a continuous tree cover training dataset. The independent variables used in the splitting process are annual AVHRR metrics created to represent the salient points of the phenological curve. These metrics are similar to those from Hansen et al. [6]. Annual phenological metrics of multi-spectral and temperature information are insensitive to time of year, allowing for their use in global analyses of land cover.

The regression tree was used with the metrics to predict tree cover based on a training data set of 61 222 pixels. Regression trees are very robust and can fit to noise. In order to better generalize the relationships between dependent and independent variables, a two step procedure is undertaken. First, a regression tree is grown using a 50 percent sample of the training data. The remaining half of training pixels are then used to prune, or better generalize the initial tree. For this procedure, the initial tree for each year of data was pruned based on where the set aside of training data's residual sum of squares was minimized within the initial tree's structure. This final tree was then run on the global data and the mean values of the training data present in each node were assigned to the output 8km pixels.

Once the 18 year sequence was produced, a per pixel regression model was run to detect trends in tree canopy cover during the time sequence. Based on the regression outputs, areas of deforestation and afforestation can be discerned. Additionally, a confidence layer was made using the node statistics from each pixel and year.

3 Results

Figure 1 shows sample years for the tree cover outputs. Fairly stable results are generated year-to-year and reflect the findings of DeFries et al. [4] that for core areas, the percent tree cover predicted is fairly consistent. DeFries found an exception to this to be high-latitude forests and preliminary findings here reflect that finding. Due to the shorter growing season and the possibility of snow cover contamination, high-latitude tree cover depictions may be inconsistent. Areas with low tree cover such as savannas are difficult to consistently depict as well. However, the results show that it is possible to detect change with continuous fields for other areas. Figure 2 shows the regression trends (slopes) over the eighteen year record. Figure 3 is an image of the magnitude of change from 1982 to 2000 derived from the regression outputs for a subset South America centered on the Amazon basin. Centers of deforestation indicated here include Para, Toncantins, Mato Grosso and Rondonia states in Brazil. Bolivia, Paraguay and Ecuador also have significant contiguous areas of tree cover loss. Landsat Enhanced Thematic Mapper Plus and Thematic Mapper data are shown for areas of deforestation in Bolivia and Paraguay along with the 8km trend analysis in Figure 4. Other large areas of deforestation present in the data include Madagascar and Java and Borneo in Indonesia. Some areas of marked tree cover increase are also present, such as large-scale plantations, an example of which can be found in eastern Venezuela.

From these data, it is obvious how noisy the results are as well. Sources of such errors are myriad, including artifacts within the data, deleterious compositing effects, resampling problems, cloud contamination and others. However, some large area change patterns do correlate with high-resolution data as in Figure 4. This lends some credence to the approach. Additional work using data from change analyses derived from higher resolution sources will be studied to validate these initial findings.

One approach to filtering errors in the data is by generating a confidence map based on node statistics. Nodes which are highly variable can be flagged in the output map and used to filter the data for only high-confidence change sites. Figure 5 shows a preliminary confidence map for this analysis. Present work is focused on using tools such as this to reduce errors of omission and commission.

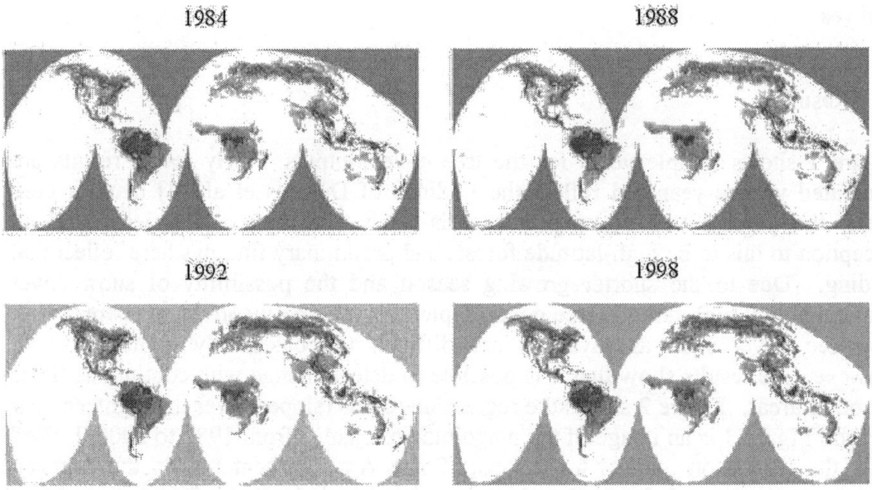

Figure 1. Selected years of continuous field percent tree cover maps. High and low values are represented by dark and light shades respectively.

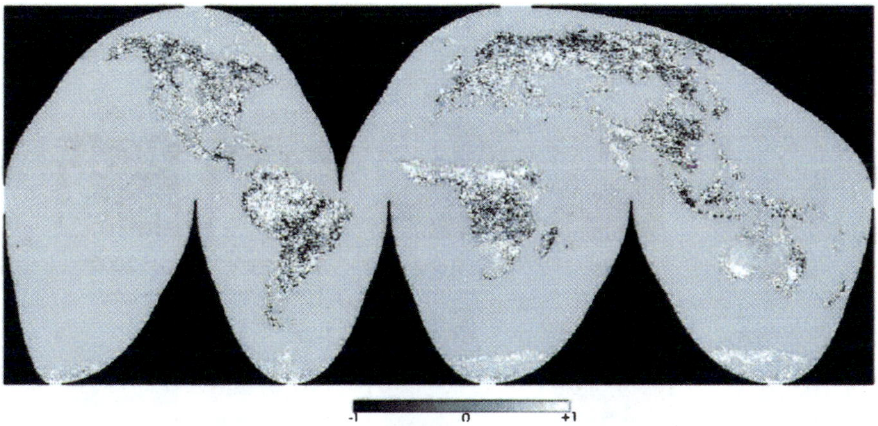

Figure 2. Normalized regression slopes for 1982-2000 tree cover depictions, low values indicate loss of tree cover, high values gains.

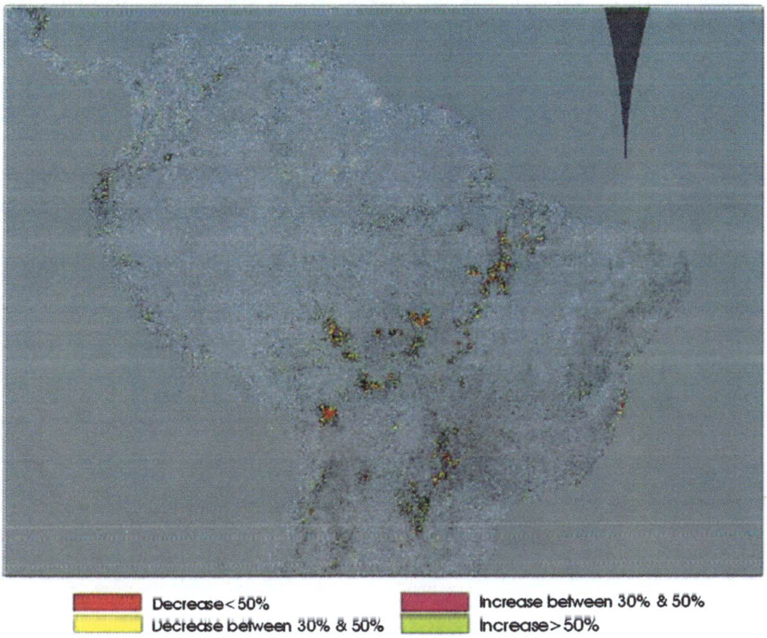

Figure 3. Magnitude of tree cover loss and gain from regression predictions for a subset of South America.

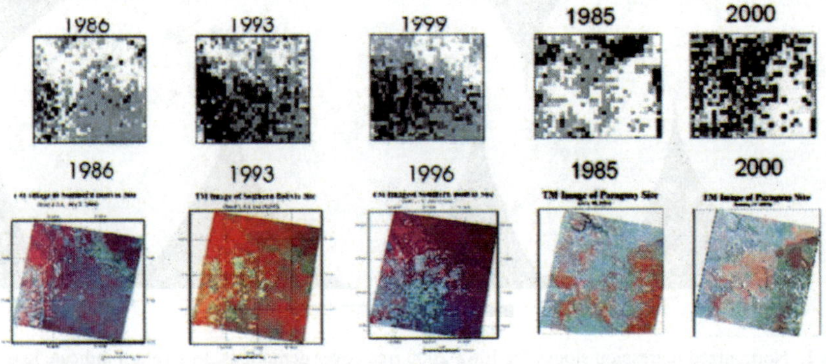

Figure 4. Qualitative comparison between annual continuous field tree cover maps (white is high tree cover, dark is low tree cover) and high resolution Landsat data in 4-3-2 displays (left three are for Bolivian subset near Santa Cruz and the right two are for a Paraguay subset near Iguazu).

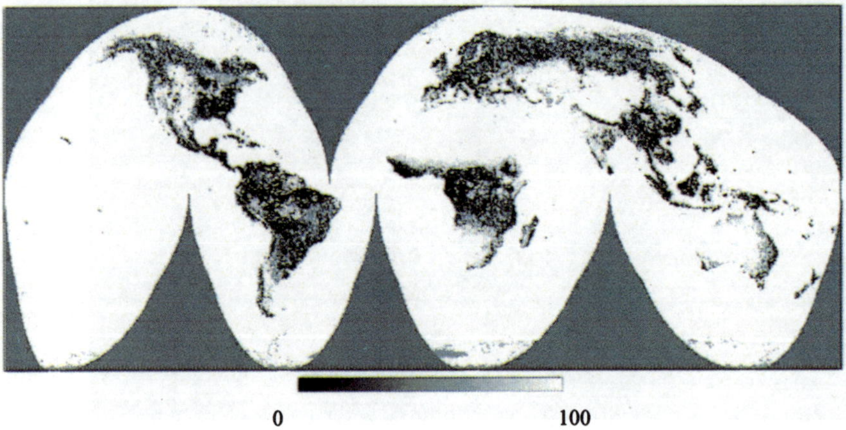

Figure 5. Normalized confidence level from low to high (0 to 100) as measured by the averaged residual sum of squares in the terminal nodes.

4 Discussion

This initial foray into long-term record change detection using coarse satellite data has resulted in promising findings of tree cover change. The procedure is based on a supervised algorithm where the variable of interest, tree canopy cover, is trained on directly. In so doing, the algorithm explicitly predicts the percent cover present in each pixel and allows for a change comparison. Each of the 18 years of percent tree cover was generated automatically, with no interpreter input. As such, it represents an operational approach to global forest monitoring.

Preliminary indications are that for some areas the change result found here is real. For the tropical zone, Brazil, Bolivia, Paraguay and Indonesia account for the majority of deforestation. Many of these areas correspond with known development of croplands and pastures following forest removal. Qualitative comparisons of the AVHRR maps with high-resolution data reveal concurrency between the patterns of change. The next step is to create quantitative comparisons between both spatial depictions of forest cover as well as country level statistical databases. Understanding the errors through such validation efforts and filtering out the most reliable cases using confidence indices are the next areas of study. Difficulties exist when using coarse resolution data, as mentioned before, and it may not be possible to characterize change in a way which eliminates most of the false change.

However, if the results prove accurate, then these data and techniques could be used in modeling carbon flux due to large-scale clearing for crops and pasture as well as ecosystem and hot spots of change monitoring. Future work using better calibrated sensors such as MODIS should result in an improved capability for using this approach. With the improved spatial resolution, geolocation and spectral characteristics of MODIS, annual updates of tree cover will be made and should allow for superior map products.

5 Acknowledgements

This research was supported by the National Aeronautics and Space Administration under contract NAS596060 and grant NAG59339.

References

1. DeFries, R. S., Field, C. B., Fung, I., et al., Mapping the land surface for global atmosphere-biosphere models: towards continuous fields of vegetation's functional properties. *Journal of Geophysical Research* **100** (1995) pp. 867-920.
2. DeFries, R. S., Hansen, M. C., Steininger, M., Dubayah, R., Sohlberg, R. S., and Townshend, J. R. G., Subpixel forest cover in Central Africa from multisensor, multitemporal data. *Remote Sensing of Environment* **60** (1997) pp. 228-246.
3. DeFries, R. S., Hansen, M. C., Townshend, J. R. G. and Sohlberg, R. S., Global land cover classifications at 8km spatial resolution: the use of training data derived from Landsat imagery in decision tree classifiers. *International Journal of Remote Sensing* **19** (1998) pp. 3141-3168.
4. DeFries, R. S., Hansen, M. C., and Townshend, J. R. G., Global continuous fields of vegetation chararcteristics: a linear mixture model applied to multi-year 8km AVHRR data, *International Journal of Remote Sensing* **21** (2000) pp. 1389-1414.
5. DeFries, R. S., Hansen, M. C., Townshend, J. R. G., and Janetos, A. C., A new global 1-km dataset for percentage tree cover derived from remote sensing. *Global Change Biology* **6** (2000) pp. 247-254.
6. Hansen, M. C., DeFries, R.S., Townshend, J.R.G., and Sohlberg, R., Global land cover classification at 1km spatial resolution using a classification tree approach *International Journal of Remote Sensing* **21** (2000) pp. 1331-1364.
7. James, M.E., and Kalluri, S.N.V, The Pathfinder AVHRR land data set: an improved coarse resolution data set for terrestrial monitoring, *International Journal of Remote Sensing* **15** (1994) pp. 3347-3364.
8. Michaelson, J., Schimel, D. S., Friedl, M. A., Davis, F. W. and Dubayah, R. O., Regression tree analysis of satellite and terrain data to guide vegetation sampling and surveys. *Journal of Vegetation Science* **5** (1994) pp. 673-696.
9. Prince, S. D., and Steininger, M. K., 1999, Biophysical stratification of the Amazon basin. *Global Change Biology* **5** (1999) pp. 1-22.

AN APPLICATION OF KALMAN FILTERING FOR MONITORING FOREST GROWTH AIDED BY SATELLITE IMAGE TIME SERIES

STEVE JOYCE

Department of Forest Resource Management and Geomatics, Swedish University of Agricultural Sciences, SE-901 83, Umeå, SWEDEN
E-Mail: Steve.Joyce@resgeom.slu.se

This paper presents a framework for updating estimates of forest parameters based on historical inventory data and growth models using a time sequence of satellite remote sensing measurements. If a forest growth model can be constructed in linear state equation form, with a vector of forest parameters representing the current state estimate, then it is shown how available remote sensing measurements can be used to improve the current state estimate and reduce its variance using Kalman filtering. Differences in the conditions of image acquisition and image quality are handled implicitly by constructing acquisition-specific models relating spectral measurements to forest parameters. The technique is applied and evaluated on a selection of permanent sample plots from the Swedish National Forest Inventory and a time sequence of 7 Landsat TM scenes over an 11-year period. When the initial forest state is known with some certainty, then the remote sensing measurements give only modest improvement over growth modeling alone, except perhaps when unexpected disturbances occur. However, it is straightforward to apply the same technique when the initial state is totally unknown or initial measurements are poor. In this case, forest parameter estimation and monitoring with remote sensing time series using Kalman filtering may be a useful complement to field inventory activities.

1 Introduction

Forest managers and forest authorities require detailed information about the current state of forest holdings in order to effectively manage the total wood supply and plan the timing and nature of various silvicultural treatments. Increasing emphasis is also being placed on collecting inventory information for habitat assessment and environmental protection. This inventory represents a 'snapshot' of the forest state at one point in time, and since the forest is a dynamic system affected by growth and disturbance, the current state must be estimated through use of growth models. Due to random error in sampling and growth modeling, this estimate becomes more and more uncertain over time, often necessitating a complete re-inventory at regular intervals,.

Analysis of remotely sensed data has long been proposed as an alternative or complement to field-based inventory for assessment of forest resources. The advantage of using remote sensing is broad spatial coverage for relatively low cost and the possibility of assessing growth and change using multi-temporal data. In practice, the predictive capability of single date forest parameter models based on optical satellite data is rather poor [1,2], and multi-temporal data acquired under unique conditions of atmospheric clarity and sun-sensor geometry may require extensive calibration and atmospheric correction for temporal monitoring applications. In addition, while the detection of large unexpected changes such as clearfelling between two dates is fairly well understood in remote sensing, methods for long term monitoring of more gradual changes associated with forest growth using longer time series are not well developed [3].

Kalman filtering is a time series analysis method used for estimating the state of a system or process based on a time series of noisy measurements from a related process [4]. It is commonly used in real-time control system applications where the process of interest can't be measured directly but a model for its dynamic behavior is known, and noisy measurements are available from some sensor. One advantage over other time series analysis methods is that estimates of the process state are updated as the data are observed, rather than when the whole time series is available.

This paper presents and evaluates a framework for making forest state estimates aided by field inventory data, growth models, and multi-temporal remote sensing. The application context is within the Swedish National Forest Inventory (NFI), where a large number of temporary and permanent sample plots are measured in the field and image data from Landsat TM are available approximately once per year during the vegetation season. Two specific applications are outlined and evaluated: (1) improved estimates of forest state based on historical inventory data and growth models, and (2) estimation of current state when the initial state is unknown.

2 Material and Methods

2.1 NFI sample plot and Landsat image data

The Swedish NFI is a sample plot based inventory with a mix of permanent (10m radius) and temporary (7m radius) plots arranged in clusters on a regular grid over the entire country. A large number of tree and site variables are observed, measured, and derived for each sample plot. The permanent plots have a nominal revisit interval of 5 years, temporary plots are visited only once. Most plots visited after 1997 are position with GPS, other plots have digitized map coordinates or nominal grid coordinates. For this study, plots from coastal Västerbotten in northern Sweden that fell within the overlap area of the available satellite imagery over a period from 1985 to 1997 were used.

There were 7 different Landsat scenes over 11 years used for this investigation (table 1). All images are from the Thematic Mapper sensor aboard Landsat-5 and were precision corrected to the Swedish National Grid and resampled to 25m by the SSC Satellitbild (now Satellus AB). The scenes were selected to be relatively cloud-free in the area of interest and as close as possible to the peak of the vegetation season. No suitable scenes were available in the missing years.

Table 1: Landsat TM scenes used

Year	Date	Path/Row
1985	June 26	193/15
1989	June 21	193/15
1990	June 8	193/15
1992	June 6	194/15
1994	July 5	193/15
1995	June 13	194/15
1996	August 11	193/15

2.2 *Process modeling and Kalman filtering*

Suppose we can describe the forest state on a sample plot at a particular time t by a vector of n state variables:

$$\mathbf{x}_t' = [x_1, x_2, \ldots, x_n]_t$$

where the n state variables are commonly measured or derived forest parameters such as height, basal area, density, species composition etc, normally stored in inventory databases for the purposes of forest management planning. Now suppose the behavior of this forest plot over time can be described by a 1st order linear difference equation with a one-year time increment of the form:

$$\mathbf{x}_{t+1} = \mathbf{A}\mathbf{x}_t + \mathbf{w}_t \ , \qquad \mathbf{w}_t \sim N(0, \mathbf{Q})$$

where $\mathbf{A}$ is an nxn matrix describing the state transition from $\mathbf{x}_t$ to $\mathbf{x}_{t+1}$, and $\mathbf{w}_t$ is a vector of error terms having a normal distribution with zero mean and a known variance-covariance matrix $\mathbf{Q}$.

In the absence of any additional measurements (from field surveys or remote sensing), the uncertainty about the current forest state effectively grows with each application of the state transition model. Specifically, if the variance of a state estimate at time t is known to be $\mathbf{P}_t$, then the variance of the forecast at the next time point is given by:

$$Var(\hat{\mathbf{x}}_{t+1} - \mathbf{x}_{t+1}) = \mathbf{P}_{t+1} = \mathbf{A}\mathbf{P}_t\mathbf{A}' + \mathbf{Q}$$

If we have available a time sequence of measurements from remote sensing comprised of m variables (spectral bands) over a number of years, there is a possibility to generate better estimates of the current state, as well as reduce its variance. Call the remote sensing variable vector $\mathbf{z}_t$, and suppose it is linearly related to the state variables described above by the following equation:

$$\mathbf{z}_t = \mathbf{H}_t \mathbf{x}_t + \mathbf{v}_t , \qquad \mathbf{v}_t \sim N(0, \mathbf{R}_t)$$

where $\mathbf{H}_t$ is a *mxn* matrix of coefficients and $\mathbf{v}_t$ is an *m*x1 vector of error terms that are normally distributed with zero mean and variance-covariance matrix $\mathbf{R}_t$.

As described above, we can use our growth model to forecast the state of the forest at time t, given our previous estimate at time t-1. Given the forecast state from the growth model, we should expect to observe a certain remote sensing measurement associated with it (i.e. $\mathbf{H}_t\mathbf{x}_t$). We can then compare this with the observed measurement ($\mathbf{z}_t$) to adjust the state estimate. This is essence of the Kalman filter: to adjust the state estimate that is predicted from the state equation, by an amount proportional to the difference between the *predicted* measurement and the *actual* measurement. The algorithm has two steps: a prediction (the *a priori* estimate) and a correction (the *a posteriori* estimate). Both the state, and its error covariance are estimated at both steps.

Specifically, as described above, the *a priori* estimates are given by

$$\mathbf{x}_t^- = \mathbf{A}\mathbf{x}_{t-1} , \qquad \mathbf{P}_t^- = \mathbf{A}\mathbf{P}_{t-1}\mathbf{A}' + \mathbf{Q}$$

where the superscript minus denotes the prior (to the measurement) estimate.

The measurement update equations (after observing $\mathbf{z}_t$) are given by:

$$\mathbf{x}_t = \mathbf{x}_t^- + \mathbf{K}_t(\mathbf{z}_t - \mathbf{H}_t\mathbf{x}_t^-) , \quad \mathbf{P}_t = (\mathbf{I} - \mathbf{K}_t\mathbf{H}_t)\mathbf{P}_t^-$$

The *nxm* matrix $\mathbf{K}_t$ is the Kalman Gain Factor, and it can be shown that the gain that minimizes the a posteriori estimation error is given by:

$$\mathbf{K}_t = \mathbf{P}_t^-\mathbf{H}_t' \ (\mathbf{H}_t\mathbf{P}_t^-\mathbf{H}_t' + \mathbf{R}_t)^{-1}$$

The Kalman filter is applied at each time step where remote sensing measurements are available by forecasting the previous state estimate, forecasting the error covariance, then correcting the state estimate with the measurement, and finally correcting the error estimate. Essentially a separate filter is run for each sample plot but the filter parameters are common for all plots. For dates when no remote sensing measurement is available, the state estimate and error covariance are simply forecast ahead one step without posterior correction.

3 Results

3.1 Forest growth model

For a simple pine growth model, a state transition matrix (**A**) as well as the error covariance (**Q**) were estimated using data from permanent NFI sample plots measured between 1985 and 1996. Only pine-dominated plots were selected (as indicated by a pine fraction greater than 70%) that were not divided. Since the growth model was required with a one-year time increment, but available permanent plots were re-visited after 5 or 6 years, all variables were interpolated to a common one-year increment before model fitting. Variables were selected according to their perceived importance in growth prediction, and transformed so the relationships were linear and to make the residual variance more uniform (table 2). Coefficients for the state transition matrix were estimated using multiple linear regression, retaining only parameters with coefficients that were significant at the 5% level (table 3). The error covariance matrix was estimated directly from the residuals (table 4). True site index and stem density are assumed to be constant over time but measured with error. The stand age model is simply incremented by one year per year but also has error associated with measurement. Including a '1' in the state vector allows for models with an intercept.

Table 2: Forest variables selected for modeling

Variable	symbol	unit	trans	description
Constant	k	-	-	Constant allows for model intercept
Volume	v	1/10 m^3/ha	^1/3	Pine stem volume
Basal area	g	1/100 m^2/ m^2	^1/2	Pine basal area
Height	h	1/10 m	-	Average tree height on plot
Site Index	SI	m	-	Site productivity est. from site factors
Stand age	a	years	-	Age since establishment
Stem density	st	Count/ha	-	Total # stems >1.3m height

Table 3: Fit coefficients for the state transition matrix (**A**). All non-zero parameters are significant at the 5% level.

	k	v	g	h	SI	a	st
k	1.0000	0.0000	0.0000	0.0000	0.0000	0.0000	0.0000
v	0.2020	0.9320	0.0114	0.0000	0.0075	0.0000	0.0000
g	0.0000	0.3926	0.9243	-0.0113	0.0536	0.0000	0.0000
h	0.0000	1.6543	-0.2348	0.9327	0.1281	0.0189	0.0000
SI	0.0000	0.0000	0.0000	0.0000	1.0000	0.0000	0.0000
a	1.0000	0.0000	0.0000	0.0000	0.0000	1.0000	0.0000
st	0.0000	0.0000	0.0000	0.0000	0.0000	0.0000	1.0000

Table 4: Estimated residual covariance matrix (**Q**) for state transition matrix in table 3.

	k	v	g	h	SI	a	st
k	0.000	0.000	0.000	0.000	0.000	0.000	0.000
v	0.000	0.015	0.062	0.014	0.005	0.004	7.526
g	0.000	0.062	0.355	-0.020	0.023	-0.027	23.794
h	0.000	0.014	-0.020	4.120	0.098	0.519	-12.554
SI	0.000	0.005	0.023	0.098	0.247	-0.186	-0.335
a	0.000	0.004	-0.027	0.519	-0.186	4.339	35.438
st	0.000	7.526	23.794	-12.554	-0.335	35.438	140670.200

3.2 *Remote sensing measurement models*

Pixel values from the satellite images were extracted at the sample plot locations using bilinear interpolation for each corresponding date where field data were available. Permanent and temporary sample plots were pooled together to build the remote sensing measurement models ($\mathbf{H}_t$) to increase the number of available data points (results for 1985 only are shown in table 5). In each case, an attempt was made to find the best linear regression model relating raw spectral values to forest parameters for each date separately (in terms of residual standard error, significance of parameters and pattern of residuals). The error covariance matrices ($\mathbf{R}_t$) were estimated directly from the residuals (table 6).

Table 5: Example coefficients for the remote sensing measurement equations ($\mathbf{H}_t$) with model fir statistics from 1985. All non-zero coefficients are significant at the 5% level. (s.e= residual standard error, R2=coefficient of determination, df=degrees of freedom). Data for other years not shown.

1985

	k	v	g	h	SI	a	st	s.e	R2	df
tm3	27.7	-0.6935	0.0000	0.0000	0.0000	0.0000	0.0000	3.25	0.44	94
tm4	43.0	0.0000	-0.2393	0.0000	0.5657	0.0000	0.0012	6.17	0.32	92
tm5	68.2	-2.8918	0.0000	0.0000	0.0000	0.0000	0.0000	11.54	0.52	94

Table 6: Example estimated residual covariance matrix ($\mathbf{R}_t$) for the remote sensing measurement equations from 1985. Data for other years not shown

1985	tm3r	tm4r	tm5r
tm3r	10.4797	-0.6246	27.6562
tm4r	-0.6246	43.6331	31.1565
tm5r	27.6562	31.1565	131.7644

3.3 *Kalman filtering*

A total of 31 permanent sample plots from 1997 were used to evaluate the performance of growth modeling aided by Kalman filtering since these data were not used directly for model fitting. The measured 1986 values were forecast through all years to 1997 and the remote sensing measurements in the years 1989 to 1996 used to filter the estimates as described above. The resulting estimates for the 1997 compared to the field-measured values are shown in figure 1, with some performance results in table 7 below.

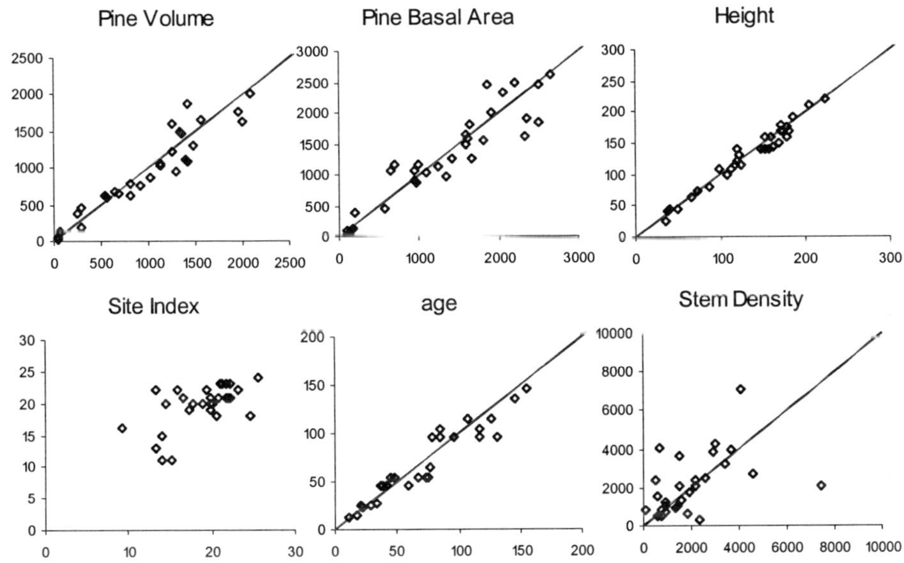

Figure 1: Scatterplots of estimated (x-axis) vs. measured (y-axis) forest parameters on 1997 permanent sample plots based on growth models from 1986 data and Kalman filtering with remote sensing measurements (volume and basal area are back-transformed to original units).

To test the performance when the initial state is unknown (table 7), a set of both permanent and temporary plots from 1997 and 1998 were used (total 89 plots). In this case the state from 1985 was initialized for all plots using mean values and the initial error covariance was set to the sample covariance from all 1985 plots.

Table 7: Performance summary for Kalman filter estimation results showing correlation between estimated and measured values (corr), residual bias, and root mean square error (RMSE).

	1997 permanent plots, known state 1986						1997, 1998 plots, unknown state 1985					
	v	*g*	*h*	*SI*	*a*	*st*	*v*	*g*	*h*	*SI*	*a*	*st*
corr	0.93	0.92	0.98	0.62	0.95	0.34	0.70	0.67	0.81	0.08	0.63	0.40
bias	-2.71	7.43	2.43	-0.83	3.78	-454.05	-70.20	314.31	20.71	-0.32	21.21	-1125.6
RMSE	254.69	313.75	9.53	3.21	12.85	2183.58	519.81	717.70	43.60	4.11	41.35	3744.29

4 Discussion and conclusions

The performance results when the initial state is known (table 7) are only marginally better than using the growth model alone with no remote sensing data (data not shown). Unexpected changes for individual plots can however be detected with a statistical test on the measurement innovation $\mathbf{z}_t-\mathbf{H}_t\mathbf{x}_t$. Modeling results are generally poor overall for site index, which is difficult to assess in the field, and stem density, which includes all stems including any understory. It is expected that results would be improved significantly by making estimates for stand means rather than plot totals, due to inherent variability in plot-level data and problems with precisely locating plots in the imagery. While these results are modest, there are several useful concepts presented here that could be developed further. First, recognizing the forest state as a dynamic system that is constantly changing due to growth and compositional change as well as management disturbance. Second, providing a strategy for integrating multi-temporal remote sensing with field sampling and growth modeling that explicitly takes the estimation error of each data source into consideration. Finally, compensation for differences in atmospheric clarity, viewing geometry, and data quality in the multi-temporal remote sensing data is handled indirectly through the use of time-specific statistical models rather than extensive pre-processing and calibration.

5 Acknowledgement

This work is part of the Swedish Remote Sensing for the Environment programme (RESE) under project 2: Monitoring of Forest Ecosystems (Håkan Olsson, Principal Investigator). RESE is funded by MISTRA, the Swedish Foundation for Strategic Environmental Research.

References

1. J. Hyyppä, H. Hyyppä, M. Inkinen, M. Engdahl, S. Linko, and Y-H Zhu, "Accuracy comparison of various remote sensing data sources in the retrieval of forest stand attributes", *Forest Ecology and Management*, vol.128, pp, 109-120, 2000.
2. R. Salvador and X. Pons, "On the reliability of Landsat TM for estimating forest variables by regression techniques: a methodological analysis", *IEEE Trans. on Geosci. Remote Sensing*, vol 36, no. 6, pp.1888-1897, Nov. 1998.
3. P. R. Coppin, P. R., and M. E. Bauer, "Digital change detection in forest ecosystems with remote sensing imagery", *Remote Sensing Reviews*, vol. 13, pp. 207-234, 1996.
4. R. E. Kalman, "A new approach to linear filtering and prediction problems", *Trans. ASME, Series D, J. Basic Eng.*, V. 82, pp. 35 – 45, Mar.1960.

THE CONTRIBUTION OF C_3 AND C_4 GRASSES TO INTERANNUAL VARIABILITY IN TIME-INTEGRATED NDVI OVER THE U.S. GREAT PLAINS

CARLO RICOTTA

Dept. of Plant Biology, University of Rome "La Sapienza", Piazzale Aldo Moro 5, 00185 Rome, Italy
E-mail: carlo.ricotta@uniroma1.it

BRADLEY C. REED

Raytheon ITSS, USGS EROS Data Center, Sioux Falls, South Dakota 57198 USA
E-mail: reed@edcmail.cr.usgs.gov

LARRY T. TIESZEN

U.S. Geological Survey, USGS EROS Data Center, Sioux Falls, South Dakota 57198 USA
E-mail: tieszen@edcmail.cr.usgs.gov

Time integrated normalized difference vegetation index (ΣNDVI) derived from NOAA AVHRR multitemporal imagery over a 10-yr period (1989-1998) was used as a surrogate for primary production to investigate the impact of interannual climate variability on grassland performance for central and northern U.S. Great Plains. First, the contribution of C_3 and C_4 species abundances to the major grassland ecosystems of the U.S. Great Plains is described. Next, the relation between mean ΣNDVI and the ΣNDVI coefficient of variation (CV ΣNDVI) used as a proxi for interannual climate variability is analyzed. Results suggest that the differences in the long-term climatic control over ecosystem performance approximately coincide with changes between C_3- and C_4-dominant grassland classes. Variation in remotely sensed net primary production over time is higher for the southern and western plains grasslands (primarily C4 grasslands), whereas the C_3-dominated classes in the northern and eastern portion of the U.S. Great Plains generally show lower CV ΣNDVI values.

1 Introduction

In the last decade, NOAA satellite imagery has provided coarse-scale NDVI data to monitor phenological characteristics of terrestrial vegetation with a daily overpass schedule. In this way, the NDVI (i.e., the normalized difference of the bidirectional reflectances in the red and near-infrared bands) can be used for accurate description of vegetation cover due to its strong relationships to various plant canopy functional and biophysical attributes, such as leaf area index (LAI), photosynthesis and chlorophyll concentration, and the fraction of absorbed photosynthetically active radiation (fAPAR). In addition, since the amount of photosynthetically active radiation absorbed by the vegetation canopy during the growing season is the major driving factor on net primary productivity, a direct relationship between the seasonal net primary production and time integrated values of NDVI (ΣNDVI) has been documented [8].

Studies of grasslands ecosystems in the U.S. Great Plains suggest that the performance and productivity of grasslands are clearly under climate control. Lauenroth [4] and Sala *et al.* [9] identified precipitation as the single most important factor controlling broad-scale grassland production throughout the central United States. More recently, Tieszen *et al.* [12] documented substantial reductions in ΣNDVI in grasslands of U.S. Great Plains during dry years and significant effects associated with changes in the beginning of the growing season.

The dominant grass species in these grasslands take up and assimilate CO_2 by two different photosynthetic pathways, C_3 and C_4, with different environmental requirements. C_4 photosynthesis can fix CO_2 at higher rates than C_3 photosynthesis if light energy is abundant. In addition, C_4 grasses have substantially higher water use efficiencies because they maintain a large diffusion gradient for CO_2 and can operate at lower leaf conductances. Thus, we can expect them to be adaptive under conditions of limiting water and higher temperature. For instance, the floristic analysis of Teeri and Stowe [11] documented that 68-82% of the species in south Texas were C_4, and this decreased to ca. 35% in South Dakota. In addition, due to their different environmental requirements, these two physiological types respond differently to interannual climate variability. In this study we first describe the contribution of C_3 and C_4 species abundances to the major grassland ecosystems of the U.S. Great Plains. Second, we analyze the relationships between satellite-derived ΣNDVI and interannual climate variability (expressed as the ΣNDVI coefficient of variation) over a 10-year period.

2 Study area

The area of study in the U.S. Great Plains is defined as the portion shown in Figure 1 between Canada and Mexico and generally east of the foothills of the Rocky Mountains to the state boundaries at Minnesota, Iowa, Missouri Arkansas, and Louisiana. To avoid confusion with intensive agricultural management, we restricted our analysis to a subset of ten land cover classes, seven grassland classes and three grassland/shrubland classes, from the 1990 conterminous United States land cover characteristic database with spatial resolution of 1 km [6]. The database portrays regions identified from multitemporal AVHRR NDVI data collected in 1990 and ancillary information including soil, climate and ecoregions. Tieszen *et al.* [12] suggest that these land cover classes exhibit coherent ecosystem properties rather than simply equivalent species composition. The analyzed land cover classes were recoded from 1-10. The dominant species and the percentage of C_4 composition to primary production of each class during normal years are reported in Table 1.

In the eastern part of our study area, Class 7 defines two clusters, the first in the Flint Hills of Kansas and extending into Oklahoma, and the second west of the corn and grain belt in the eastern part of South Dakota. Class 7 is generally dominated by C_4 species (Tab. 1) and represents the remnants of the tallgrass region after conversion of most grasslands to agricultural uses. Characteristic species include big bluestem (*Andropogon gerardi*), switch grass (*Panicum virgatum*), and indian grass (*Sorghastrum nutans*), whilst the small area in central South Dakota most likely includes elements of the northern mixed-grass prairie [12].

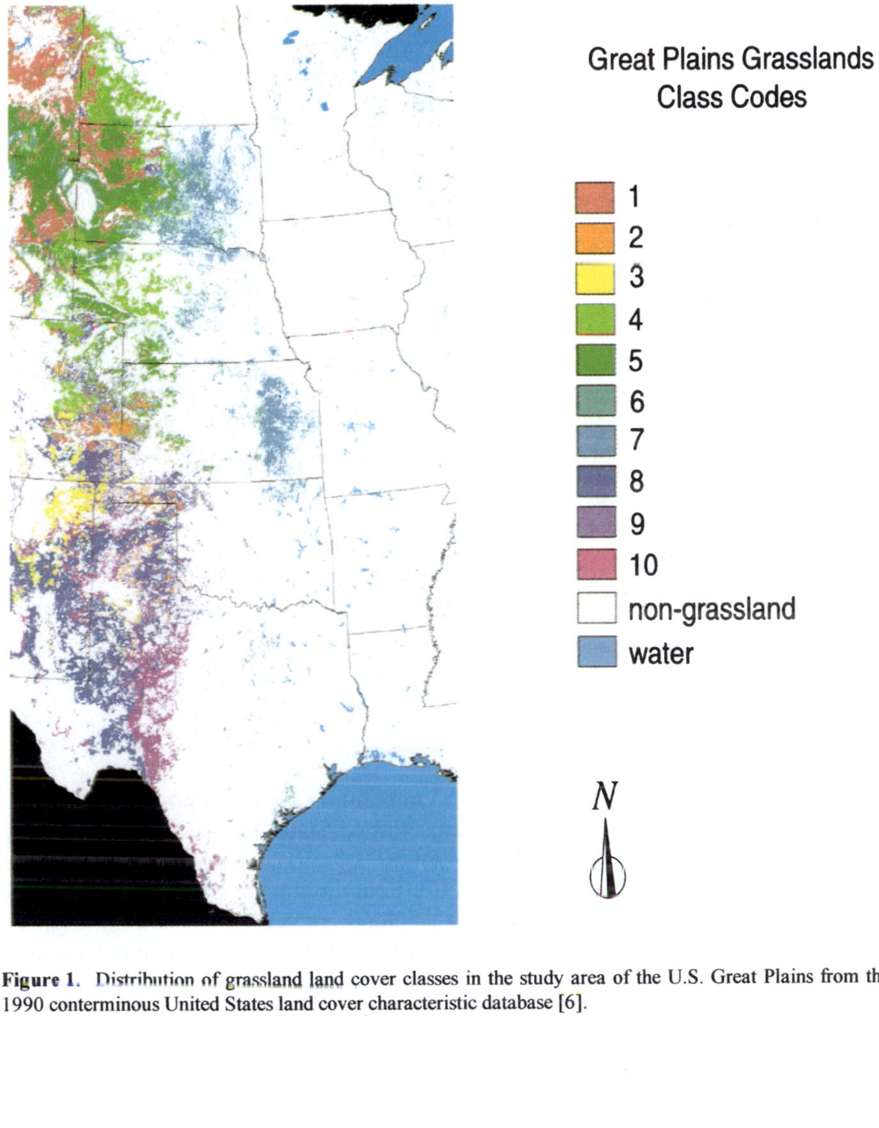

Figure 1. Distribution of grassland land cover classes in the study area of the U.S. Great Plains from the 1990 conterminous United States land cover characteristic database [6].

The northwest portion of the study area is dominated by classes 1, 4, 5 and 6 containing primarily C_3 grasslands with varying elements of wheatgrass (*Agropyron* spp.), needlegrass (*Stipa* spp.) and grama grasses (*Bouteloua* spp.) with class 5 encompassing shrub elements of sage (*Artemisia* spp.) and greasewood (*Covillez* spp., *Atriplex* spp.). Class 4 also includes blue grama (*Bouteloua gracilis*) as a characteristics species as indicated by the higher contribution of C_4 species (Tab. 1). Only two grasslands classes, both dominated by C_4 species occur in the southern portion of the study area: class 2 with little bluestem (*Schizachyrium scoparium*) and blue grama, and class 3 with blue grama, wheatgrass and buffalograss (*Buchloe dactyloides*). In the southwestern part of the Great Plains three grassland/shrubland classes, 8, 9 and 10, all possess blue grama and are differentiated by a variety of shrub components, such as big sage (*Artemisia tridentata*), sand sage (*Artemisia filifolia*), and saltbush (*Atriplex polycarpa*).

Table 1. Seasonal land cover classes from the U.S. Great Plains, dominant species, and C4 composition expressed as the percentage of potential contribution to primary production during normal years. Modified from [12].

Class	Dominant Species	C_4 %
1	*Agropyron smithii, Stipa comata, Stipa* spp.	29
2	*Schizachyrium scoparium, Bouteloua gracilis*	79
3	*Bouteloua gracilis, Buchloe dactyloides, Agropyron* spp.	84
4	*Bouteloua gracilis, Stipa comata, Agropyron* spp.	51
5	*Bouteloua gracilis, Stipa comata, Artemisia tridentata, Agropyron* spp.	39
6	*Stipa comata, Stipa* spp., *Agropyron* spp.	45
7	*Andropogon gerardi, Panicum virgatum, Sorghastrum nutans*	79
8	*Artemisia filifolia, Bouteloua gracilis, Buchloe dactyloides, Agropyron* spp.	97
9	*Bouteloua gracilis, Buchloe dactyloides, Artemisia tridentata, Atriplex polycarpa*	80
10	*Artemisia filifolia, Bouteloua gracilis, Buchloe dactyloides*	94

3 Methodology

A sequence of 10 years of biweekly NDVI maximum value composite (MVC) data (1989 through 1998) for the Conterminous United States [2] is used in this study. The MVC method considers the maximum NDVI value assumed by a pixel in a time period as a good estimate of the entire period reducing problems common to single date remote sensing studies, such as cloud contamination and atmospheric attenuation [3]. Nevertheless, while maximum value compositing increases data quality, the biweekly NDVI composites can still be biased by atmospheric haze and clouds that generally reduce the NDVI values [5]. To address these effects and to facilitate subsequent time-series analyses, further processing, in effect a smoothing, of the temporal NDVI MVC signal must be performed. The smoothing algorithm must serve as a rough interpolation between observations and, in order to remove the effects of the remaining NDVI-reducing phenomena, upwardly bias the results [7].

To correct for signal contamination, an upwardly-biased, weighted least squares linear regression [10] was implemented. The approach uses a moving window operating on the temporal NDVI to calculate a best-fit regression line. The window is moved one temporal period (two-weeks for the conterminous US AVHRR data) at a time, resulting in a family of regression lines associated with each point. This family of lines is then averaged at each point and interpolated between points to provide a continuous temporal NDVI signal. In addition, since the factors that cause contamination usually result in reduced NDVI values, a weighting factor that favors peak points over sloping or valley points was applied.

Once the database is smoothed of temporal discontinuities, the onset and the end of the growing season were extracted from the NDVI MVC time series. The methodology involves applying a moving average filter to the smoothed NDVI MVC time series, which essentially creates a new moving average time-series (MATS) with a time lag that can serve as a predicted NDVI based on the previous *n* (user-defined) observations. When the actual (smoothed) NDVI MVC values are greater than the value predicted by the MATS, then a departure of the observed NDVI data from a trend established by the moving average (i.e., the start of the growing season) is occurring. The end of the growing season was determined in a similar manner only with the moving average applied to the NDVI MVC time series in reverse chronological order [13]. The ΣNDVI was then computed for each pixel as the summation of the NDVI MVC values of each biweekly period from the point of onset to that of end of the growing season. Finally, the mean yearly ΣNDVI along with the ΣNDVI coefficient of variation (CV) was computed for the analyzed grassland classes over the 1989-1998 time period.

4 Results and discussion

Figure 2 shows the mean ΣNDVI of the analyzed grassland land cover classes over the 1989-1998 time period plotted versus the corresponding CV ΣNDVI. If mean ΣNDVI is assumed as a surrogate measure of mean long-term primary production, this component of ecosystem performance varies four-fold from low values in class 8 in the southwest (Fig. 1) to the highest values in the tallgrass prairie (class 7). This easternmost class is followed by classes 6, 4, 5, and 1 that are generally distributed in the northwestern portion of the study area. Conversely, the grassland and mixed grassland/shrubland land cover classes with their main distributions concentrated in the southwest have the lowest mean ΣNDVI values. This wide range of mean ΣNDVI among classes characterized by a common growth form (largely graminoids) suggest a high degree of adaptability within this growth form to environmental conditions that may be sufficiently extreme to impose substantial controls on ecosystem performance [12]. Variation in remotely sensed net primary production over time, mapped as CV ΣNDVI, is generally low for the classes in the northern and eastern portion of the study area, that are expected to be relatively stable in comparison to the other grasslands since precipitation is not as limiting, and shows higher values for the classes in the southwest, the region with the last dependable precipitation regime.

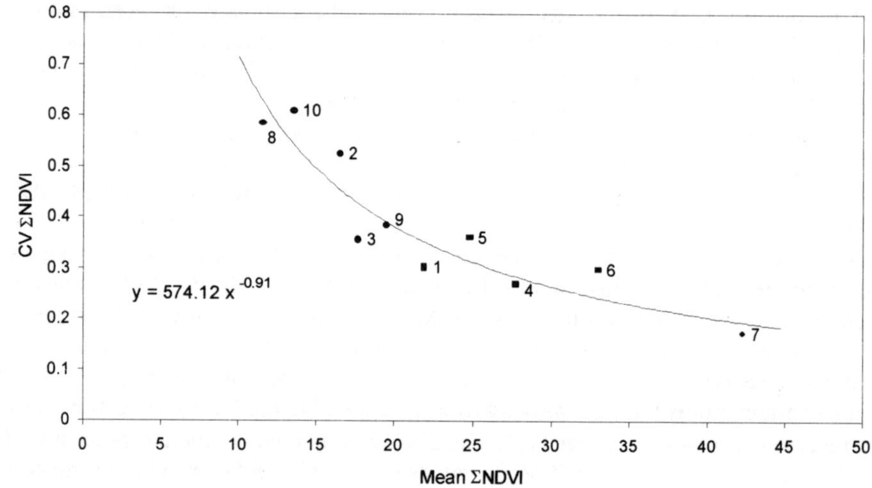

Figure 2. Mean ΣNDVI of the analyzed grassland land cover classes over the 1989-1998 time period plotted versus the corresponding CV ΣNDVI. Labels are grassland land cover classes. Circles are C_4-dominated grassland classes (class 8, 10, 2, 3, 9, and 7); squares are C_3-dominated (class 1 and 5), and C_3/C_4 mixed grassland classes (class 4 and 6).

As shown in Figure 2, the decrease in CV ΣNDVI as a function of increasing mean ΣNDVI can be best described by an empirical power law. Interestingly, the differences in the long-term climatic control over ecosystem performance approximately coincide with changes between C_3- and C_4-dominant grassland classes (Fig. 2, Tab. 1). Within this context, the exception of class 7 that possess the highest ΣNDVI and the lowest CV ΣNDVI being generally dominated by C_4 species was already reported by Briggs and Knapp [1] and Yang *et al.* [13] analyzing the primary production-climate relationships in the tallgrass prairies of the U.S. Great Plains. For instance, the high ΣNDVI of this class is largely due to the higher precipitation in the eastern portion of the Great Plains, and its lower CV is likely a consequence of its current distribution only on relict areas not suitable for agriculture. Thus the present remnants of this once widespread class are likely now restricted in its variability both in terms of performance, ΣNDVI, and possibly in floristic or vegetation diversity.

The results of this study suggest that NDVI MVC time series provide a powerful tool for monitoring grassland ecosystem performance at regional scales. For example, in this study, the main differences in the long-term climatic control over ecosystem performance appeared to be related to different physiological characteristics of C_3 and C_4 species. Nevertheless, the ecosystem responses to climate forcing in the U.S. Great Plains are far from being trivial. It appears therefore that, besides ecosystem functional attributes, a quantitative description of these responses should not overlook the dynamical and competitive relationships that ultimately determine ecosystem composition and persistence.

5 References

1. Briggs J. M. and Knapp A. K., Interannual variability in primary production in tallgrass prairie: climate, soil moisture, topographic position, and fire as determinants of aboveground biomass. *Am. J. Bot.* **82** (1995) pp. 1024-1030.
2. Eidenshink J., The 1990 conterminous U.S. AVHRR data set. *Photogramm. Eng. Remote Sens.* **58** (1992) pp. 809-813.
3. Holben B. N., Characteristics of maximum-value composite images from temporal AVHRR data. *Int. J. Remote Sens.* **7** (1986) pp. 1417-1434.
4. Lauenroth W. K., Grassland primary production: North American grasslands in perspective. In *Perspectives in Grassland Ecology,* ed. by N. R. French (Ecological Studies, Vol. 32, Springer Verlag, New York, 1979) pp. 3-24.
5. Los S. O., Justice C. O. and Tucker C. J., A global 1° by 1° NDVI data set for climate studies derived from the GIMMS continental NDVI data. *Int. J. Remote Sens.* **15** (1994) pp. 3493-3518.
6. Loveland T. R., Merchant J. W., Ohlen D. O. and Brown J. F., Development of a land-cover characteristics database for the conterminous U.S. *Photogramm. Eng. Remote Sens.* **57** (1991) pp. 1453-1643.
7. Reed B. C., Brown J. F., VanderZee D., Loveland T. R., Merchant J. W. and Ohlen D. O., Measuring phenological variability from satellite imagery. *J. Veg. Sci.* **5** (1994) pp. 703-714.
8. Ricotta C. and Avena G. C., The remote sensing approach in broad-scale phenological studies. *Appl. Veg. Sci.* **3** (2000) pp. 117-122.
9. Sala O. E., Primary production of the central Grassland Region of the United States. *Ecology* **69** (1988) pp. 40-45.
10. Swets D. L., Reed B. C., Rowland J. R. and Marko S. E., A weighted least-squares approach to temporal smoothing of NDVI. Proceedings of the ASPRS Annual Conference *"From Image to Information"*, Portland, Oregon, May 17-

21, 1999 (American Society for Photogrammetry and Remote Sensing Bethesda, Maryland, 1999) CD-ROM, 1 disc.

11. Teeri J. A. and Stowe L. G., Climatic patterns and the distribution of C_4 grasses in North America. *Oecologia* **23** (1976) pp. 1-12.
12. Tieszen L. L., Reed B. C., Bliss N. B., Wylie B. K. and de Jong D. D., NDVI, C_3 and C_4 production and distribution in Great Plains grassland land cover classes. *Ecol. Appl.* **7** (1997) pp. 59-78.
13. Yang L., Wylie B. K., Tieszen L. L. and Reed B. C., An analysis of relationship among climate forcing and time-integrated NDVI of grasslands over the U.S. northern and central Great Plains. *Remote Sens. Environ.* **65** (1998) pp. 25-37.

FOURIER DECOMPOSITION OF AN AVHRR NDVI TIME SERIES FOR SEASONAL AND INTERANNUAL LAND COVER CHANGE DETECTION

MARK E. JAKUBAUSKAS

Kansas Applied Remote Sensing Program, University of Kansas, Lawrence, Kansas 66045
Tel: 785-864-7316 Fax: 785-864-0392 email: mjakub@ku.edu

DANA L. PETERSON

Kansas Applied Remote Sensing Program, University of Kansas, Lawrence, Kansas 66045

DAVID R. LEGATES

Department of Geography, University of Delaware, Newark, Delaware 19716

This paper explores the use of Fourier analysis of time-series AVHRR NDVI data for detection of seasonal and interannual change in natural and managed land cover types in the southern Great Plains, USA. Seasonal changes in vegetation greenness, as quantified by the NOAA AVHRR biweekly composite NDVI values typically assume a near-periodic pattern with maximum NDVI observed during midsummer for most North American vegetation types. Such cyclical patterns of NDVI and perturbations to this pattern are highly amenable to Fourier, or harmonic, analysis. Fourier analysis permits a complex curve (seasonal NDVI) to be expressed as the sum of an additive term and a series of cosine waves (terms), with each term defined by an amplitude value and a phase value. Row crops and grassland exhibit strong unimodal seasonal profiles, with high first-term amplitude and phase values indicating a midsummer greenness peak. Changes in vegetation condition or land use/cover over a period of years are manifested as changes in the amplitude, phase, or variance values for a given Fourier term.

1 Introduction

Multitemporal, multispectral remotely sensed imagery are providing new opportunities to identify, map, and monitor regional- to global-scale land cover. Land use/land cover mapping projects using remote sensing are increasingly exploiting seasonal changes in vegetation spectral response to improve the accuracy of vegetation classes. While excellent results have been obtained using a small number of images (typically less than ten), multitemporal analysis is moving toward capitalizing on the information contained in temporally-dense data sets (daily, weekly, or biweekly) that may contain upwards of several dozen to several hundred successive images (hypertemporal analysis). Researchers have evaluated various techniques for analyzing such temporally-dense remotely sensed data sets, from simple classification [1][2], to standardized principal component analysis [3] and textural analysis [4], to the development of phenological metrics that describe seasonal changes in the normalized difference vegetation index, or NDVI [5][6][7]. Few studies have been devoted to developing metrics that exploit the rich information content of temporally-dense image data sets for detecting interannual variability and developing indicators of environmental change.

Vegetation communities are temporally dynamic entities, exhibiting change on multiple time scales. Seasonal changes in vegetation greenness, as quantified by the NOAA Advanced Very High-Resolution Radiometer (AVHRR) NDVI biweekly composites, occur as a function of climate and innate plant characteristics and typically assume a near-periodic pattern with maximum NDVI observed during midsummer for most North American vegetation types. Changes occurring over longer time scales that modify or disrupt this quasi-periodic pattern may occur as a result of natural or anthropogenic change, including crop rotation, land conversion (such as plowing of native prairie for cropland), degradation associated with overgrazing or desertification, and post-disturbance regeneration.

Such a cyclical pattern of greenness, and perturbations to this cyclical pattern, is highly amenable to analysis by Fourier, or harmonic, analysis. In the analysis of dense time series of remotely sensed data, the objective often is to isolate and identify the low frequency patterns associated with vegetation phenology while suppressing high- frequency variation. Because Fourier analysis can act as a filter of time-series data, the desired low frequency patterns useful for land use/cover identification can be separated from the high frequency noise induced by episodic precipitation events and cloud contamination. Lower frequency variation associated with vegetation phenology and change will be contained in the lower Fourier terms while noise is isolated in successively higher terms. Fourier analysis has been used in the earth sciences for analysis of time-dependent climatological and geological data, in digital image processing for analysis of a single image [8][9], and more recently is being used to analyze sets of successive multidate samples of satellite remotely sensed imagery [10].

This paper explores the use of Fourier analysis of time-series AVHRR NDVI data for characterizing seasonal patterns of vegetation greenness, and detecting interannual change in natural and managed land cover types in the southern Great Plains, USA. In this paper, we provide an overview of Fourier analysis as applied to time series image data, describe how seasonal differences in greenness patterns for different land cover types are manifested in each term, and describe methods for using Fourier analysis for detecting interannual and directional change.

2 Methods

Briefly defined, Fourier analysis permits a complex curve to be expressed as the sum of a series of cosine waves (terms) and an additive term. Each wave is defined by a unique amplitude and a phase angle, where the amplitude value is half the height of a wave, and the phase angle (or simply, phase) defines the offset between the origin and the peak of the wave over the range 0 to 2B. Each term designates the number of complete cycles completed by a wave over the defined interval (e.g., the second term completes two cycles). Successive terms are added to produce a complex curve, and each component or term, accounts for a percentage of the total variance in the original time-series data set.

Nine years of NOAA-AVHRR NDVI biweekly composites for 1989 - 1997 for Finney County, Kansas, USA were acquired from the US Geological Survey EROS Data Center in Sioux Falls, South Dakota. Each biweekly NDVI composite image consists of the maximum NDVI value within a defined two- week period for each pixel [11]. Vegetation index data are rescaled by EROS during processing from a range of -1.0 to +1.0, to 0 to 200. Values less than 100 typically represent snow, water, and other non-vegetated earth surfaces. Images for missing biweekly composite periods were created by averaging data for periods bracketing the missing period (e.g., the previous and succeeding periods). Each year of the nine-year data set contained 26 biweekly images.

Per-pixel temporal profiles of seasonal NDVI were decomposed using Fourier analysis into an additive term and component cosine curves with term-specific phase and amplitude coefficients, for each of the nine years. Images of the phase and amplitude values of each pixel for each term and the additive term were produced. Percent variance in each image for each year was calculated by first computing the total variance of all terms j in the harmonic analysis using the amplitude values, where j is each term and n is the total number of terms [12]:

$$\sum_{1\to n}\frac{(amplitude_j)^2}{2}$$

Interannual variation in Fourier terms was assessed by computing year-to-year changes in amplitude and phase values for each term. A derivative measure of inter-annual variability was derived from the phase values produced by the Fourier analysis. Because phase angles are directional, circular variance was calculated as a measure of variation [12]. Circular variance in each image is calculated by first calculating the length of the vector resultant:

$$R=\sqrt{X_r^2+Y_r^2}$$

where X_r and Y_r are equal to the sum of the cosines and sines, respectively, of the individual vectors. To calculate dispersion in the data, the resultant length, R is subtracted from the number of observations, n and then divided by the number of observations, $(s^2_0 = (n - R)/n)$ [12]. Low circular variance indicates that phase values are closely clustered together (e.g., occur within 1-2 periods, indicating a consistent timing of phase during a year), while high circular variance indicates that phase values occur in multiple dispersed periods.

3 Results

3.1 Seasonal change in land cover:

High amplitude values for a given term indicate a high level of seasonal change, and the term in which that variation occurs indicates the periodicity of the event. Land cover types with a simple phenology, such as row crops or grassland, exhibit strong unimodal seasonal profiles (Figure 1), with high first-term amplitude and phase values indicating a midsummer greenness peak. In contrast, winter wheat exhibits a strong bimodal seasonal pattern with high second-term amplitude values (Figure 2), consistent with fall planting, spring greenness, and summer maturity of this crop.

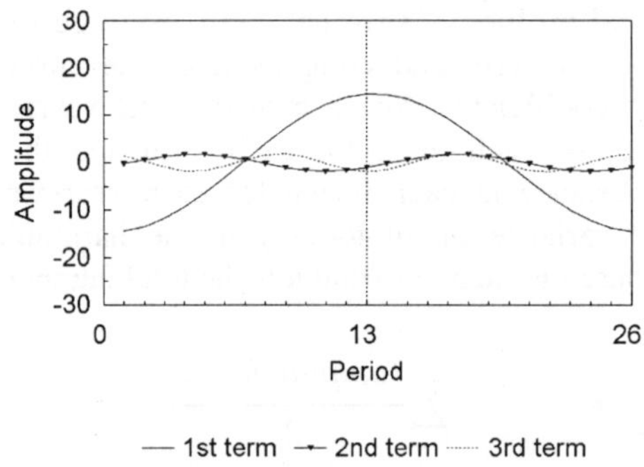

Figure 1. Single-year greenness curves produced for Fourier terms 1-3 for mid-latitude corn.

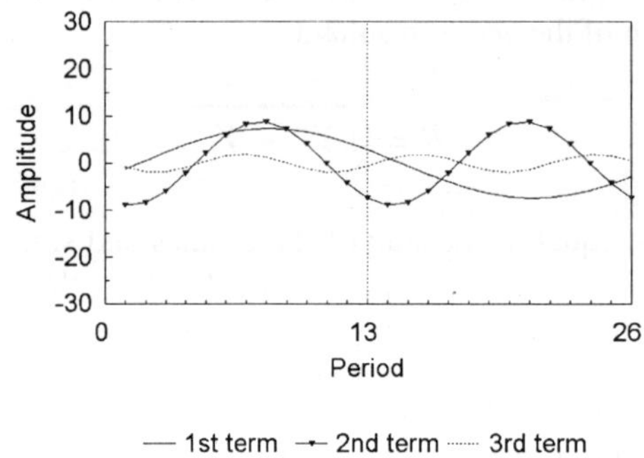

Figure 2. Single-year greenness curves produced for Fourier terms 1-3 for mid-latitude winter wheat.

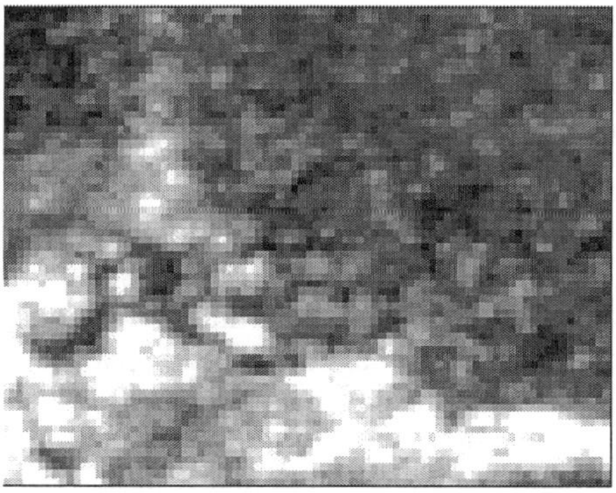

Figure 3. Additive term image for 1992 seasonal NDVI decomposed using Fourier analysis. Values of the additive term (mean seasonal NDVI) range from low in grasslands (dark tones) to high for irrigated crops (light tones).

Figure 4. First-term amplitude image for 1992 AVHRR NDVI data. Single-year amplitude values are high for irrigated crops (light tones) and low for winter wheat and fallow lands (dark tones).

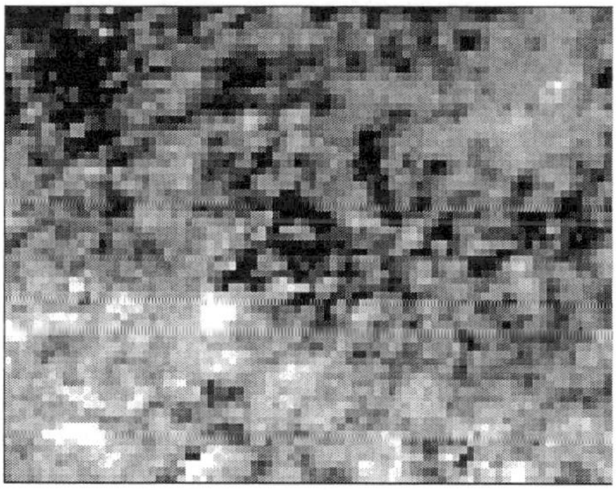

Figure 5. First-term phase image for 1992 AVHRR NDVI data. Light tones on the image indicate mid-summer phase values (grasslands and row crops) compared to dark tones (spring phase) for winter wheat

3.2 Interannual change in land cover

3.2.1 Changes in phase

Year-to-year changes in the amplitude and phase values of a harmonic term are indicative of change in land use or land cover. The annual greenness patterns of row crops and grassland are driven by principally by seasonal climatic factors and tend to be strongly unimodal NDVI curves. Interannual variations in the phase values of a term, with the amplitude remaining unchanged, may indicate climatologically-driven variations in the time of onset of greenness or maximum greenness. Generally, inter-annual variation in the timing, peak, and duration of greenness for row crops is minimal, producing low circular variance for the first-term phase. Small grains (principally winter wheat in this part of the Southern Great Plains) tend to have a high circular variance in the first term phase and a low second-term circular variance. Given the bimodal greenness pattern of winter wheat, more of the overall variance in the data is explained by the second term than the first-term.

3.2.2 Changes in amplitude

Changes in the amplitude of a given term, with the phase value for that term remaining constant, may indicate changes in land use/land cover type or degradation in vegetation condition resulting from drought, flooding, or overgrazing. In Figures 6 and 7 below, the seasonal NDVI of a grassland area in the southern Great Plains is plotted over a period of nine years. The amplitude of the seasonal NDVI pattern becomes progressively more pronounced with time (Figure 6), and this increase is manifested as progressively higher first-term amplitude values (Figure 7).

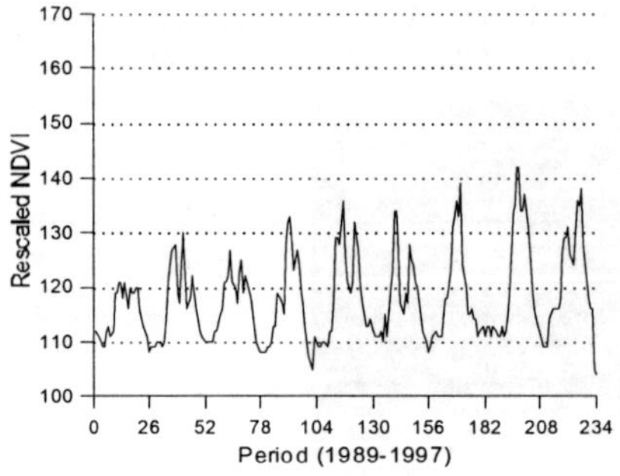

Figure 6. Progressive increase in seasonal NDVI values for a grassland in the southern Great Plains, USA.

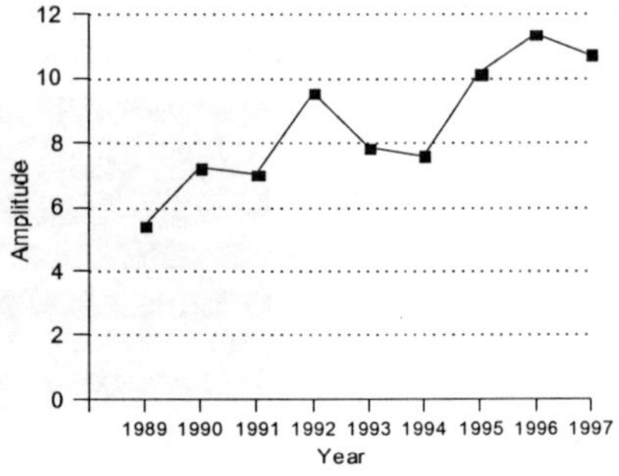

Figure 7. Increase in first-term amplitude values during a nine-year period.

4 Discussion and conclusions

Different land cover types possess distinctive seasonal greenness patterns that are represented as differential additive, amplitude, and phase values in the terms of the Fourier analysis. Fourier analysis of time-series data provides an unbiased, replicable method for assessing seasonal and interannual variability in sensitive landscapes, and offers considerable promise as a tool for monitoring directional landscape change. Interannual change in land cover condition is manifested as changes in these phase and amplitude values.

Changes in seasonal ***amplitude*** (phase unchanged) may indicate changes in land use/cover type (e.g., changes in a crop type or changes in natural vegetation), or in vegetation condition from drought, flooding, or overgrazing. Changes in ***phase*** (amplitude unchanged) may indicate changes in time of maximum greenness, which in turn may occur as a result of changes in the onset of greenness, changes in planting time, or changes in the time of harvest. Changes in ***both*** amplitude and phase over time would be indicative of major and significant changes in land surface condition. Such radical change could result from postfire regeneration, changes in land management (crop rotation, enrollment in conservation programs), loss of vegetation by natural or anthropogenic disturbance, or changes in regional climate itself driving changes in the vegetation. For example, land management practices such as a two-year crop-fallow rotation pattern should be manifested as high loadings on a bi-annual amplitude term.

Fourier analysis can be extremely useful for highlighting and isolating modes of variability in time series NDVI image data. A strong annual cycle clearly exists in the seasonal change in vegetation greenness as quantified by the NDVI, but of perhaps more interest are cycles that may exist both within and across the annual cycle. For example, the lag between vegetation and air temperature or between vegetation and precipitation can be particularly useful in determining the vegetation response to climate forcing. Fourier analysis allows us to quantify the amplitude and phase of seasonal and intra-seasonal cycles, and provides this information in a form that allows for easily inter-comparison between vegetation and climate. In an alternative scenario, we may wish to suppress the first-term (seasonal) component of vegetation greenness that represents phenology in order to highlight intraseasonal effects on vegetation state (e.g., precipitation events). Conversely, we might wish to suppress high-frequency information (from cloud cover, processing noise, or intraseasonal effects) in order to produce a more clean seasonal curve for land cover mapping purposes.

5 Acknowledgments

This research was conducted at the Kansas Applied Remote Sensing (KARS) Program (Edward A. Martinko, Director). The research described in this paper was funded in part by the National Institute for Global Environmental Change (NIGEC),

SouthCentral Regional Center at Tulane University (David Sailor, Director), through the U.S. Department of Energy (Cooperative Agreement No. DE-FC03-90ER61010). Any opinions, findings, and conclusions or recommendations expressed in this publication are those of the authors and do not necessarily reflect the views of the DOE.

References

1. S.E. Egbert, K.P. Price, M.D. Nellis, and R. Lee, Developing a land cover modeling protocol for the High Plains using multiseasonal Thematic Mapper imagery, *Proc. 1995 ASPRS/ACSM Ann. Mtng*, Charlotte, N.C. pp. 836-845,1995.
2. T.R. Loveland, J.W. Merchant, J. Brown, D. Ohlen, B.C. Reed, P. Olson, and J. Hutchinson, Seasonal land-cover regions of the United States, *Ann. Assoc. Amer. Geog.*, vol. 85, no. 2, pp. 339-355, 1995
3. J.R. Eastman and M. Fulk, Long sequence time series evaluation using standardized principal components, *Photogrammetric Eng. and Remote Sens*, vol. 59, pp.1307-1312, 1993.
4. J. Briggs and M.D. Nellis, Seasonal variation of heterogeneity in the tallgrass prairie: a quantitative measure using remote sensing, *Photogrammetric Eng. and Remote Sens.*, vol. 57 pp.407-411, 1999.
5. B.C. Reed, J. Brown, D. VanderZee, T.R. Loveland, J.W. Merchant, and D. Ohlen, Measuring phenological variability from satellite imagery, *Jour. Veg. Sci.,* vol. 5, pp. 703-714, 1994.
6. S.A. Samson, Two indices to characterize temporal patterns in the spectral response of vegetation, *Photogrammetric Eng. and Remote Sens.* vol. 59 no. 4, pp. 511-517, 1993.
7. D. Lloyd, A phenological classification of terrestrial vegetation cover using shortwave vegetation index imagery, *Int. Jour. Remote Sens.* vol. 11, no. 12, pp. 2269-2279,1990.
8. R.A. Schowengerdt, *Remote Sensing: Models and Methods for Image Processing*. San Diego: Academic, 1997.
9. J. Jensen, *Introductory Digital Image Processing*, 2nd Edition, New Jersey:Prentice-Hall, 1996.
10. S. Azzali and M. Menenti, Mapping vegetation-soil-climate complexes in southern Africa using temporal Fourier analysis using NOAA-AVHRR-NDVI data, *Int. Jour. Rem. Sens.*, vol. 21, no. 5, pp. 973-996, 2000.
11. J.C. Eidenshink, The 1990 conterminous U.S. AVHRR data set, *Photogrammetric Eng. and Remote Sens.*, vol. 58, no. 6, pp. 809-813, 1992.
12. J.C. Davis, *Statistics and Data Analysis in Geology*, 2nd Ed. John Wiley and Sons, 1986.

MULTI-TEMPORAL ANALYSIS USING NOAA-PATHFINDER AVHRR LAND IMAGERY FOR THE STUDY OF THE LAND COVER DYNAMIC IN THE MEDITERRANEAN BASIN

J. A. SOBRINO, J. CUENCA, J. EL KHARRAZ, M. GÓMEZ, J. C. JIMENEZ, N. RAISSOUNI, and G. SORIA

Global Change Unit, Dpt. Thermodynamics, University of Valencia, C/ Dr. Moliner 50, 46100 Burjassot, Spain.

E-mail:.sobrino@uv.es

A multi-temporal analysis using the Pathfinder AVHRR Land (PAL) data has been carried out to map and monitor changes in the biophysical characteristics of land cover. The PAL data covers a period from July 1981 throughout the present. The Mediterranean basin was selected as the area of study due to its high environmental diversity. The study has been carried out by combining both the information in the visible/near-infrared bands in terms of Normalized Difference Vegetation Index (NDVI) and in the thermal-infrared bands in terms of Land Surface Temperature (LST). First, we have adapted theoretical algorithms to obtain LST and emissivity (ε) from PAL data. Then, we have analyzed the space-temporal dynamics of these parameters. Finally, a method for land cover dynamic monitoring is presented. The method named *Vector of Land Cover Dynamic* (VLCD), analyses the evolution of LST and NDVI for the months of April and July. This permits characterizing the temporal trajectories, the separation and the distribution of the different zones according to the evolution of the vegetative systems.

1 Introduction

They have been numerous investigations in the field of cartography and analysis on land cover changes considering only the temporal analysis of Normalized Difference Vegetation Index (NDVI) [1][2]. However, few papers introduced the Land Surface Temperature (LST) parameter obtained from satellite data to study the land cover changes [3][4]. One of the motives is the difficulty of obtaining this parameter. Nevertheless, at present it is possible to determine LST within a precision of 1.3 K. Land cover classification and change detection using temporal series of LST and NDVI requires a good knowledge of the relationships between these biophysical parameters in different ecological conditions. In this paper we present a method named *Vector of Land Cover Dynamic* (VLCD) developed for the multi-temporal monitoring in the whole of the Mediterranean Basin

2 Algorithms

LST retrieval from satellite data is mainly influenced by the atmosphere and surface emissivity. If both effects are not corrected for the error in LST may be very large (up to 12 K). In order to obtain accurate estimations of LST, we have applied the following split-window algorithm [4] with an error of estimation of 1.3 K

$$LST = T_4 + 1.4\,(T_4 - T_5) + 0.32\,(T_4 - T_5)^2 + 0.83 + (57 - 5W)\,(1 - \varepsilon) - (161 - 30W)\,\Delta\varepsilon$$

T_4 and T_5 are the brightness temperatures measured in AVHRR Channels 4 and 5 respectively, ε is the average effective emissivity in both channels, $\Delta\varepsilon$ is the spectral variation of emissivity and W is the total amount of atmospheric water vapor (g cm^{-2}) obtained applying the Split-Window Covariance-Variance Ratio (SWCVR) method [5]; $W = 0.26 - 14.253\,(\cos\theta \ln R_{54}) - 11.649\,(\cos\theta \ln R_{54})^2$, where θ is the satellite observation angle and R_{54} is the ratio of the spatial covariance and the variance of image brightness temperatures in the split-window Channels 4 and 5. The estimated error is 0.5 gm^{-2}. To obtain ε we have applied the NDVI Threshold Method ($NDVI^{THM}$). This method integrates a wide spectral data set of bare soil reflectivity measurements in the band of 0.4-14 μm and uses different approaches in function of NDVI. Now, we proceed to explain the methodology in function of each NDVI value interval:

i) $0.2 \leq NDVI \leq 0.5$

It is the usual case that the surface at pixel scale in remote sensing is heterogeneous and rough. The emissivity for that surface can be obtained using a theoretical approach, which models a given surface by considering it as being made up of a mixture of bare soil and vegetation [6]

$$\varepsilon_i = \varepsilon_{vi} P_v + \varepsilon_{si} (1 - P_v) + C_i \quad (i = 4 \text{ or } 5)$$

where ε_{si} and ε_{vi} are, respectively, the emissivities of bare soil and vegetation for AVHRR Channel i (i= 4 or 5), P_v is the proportion of vegetation, and C_i is a term that depends on the surface characteristics and which takes into account the internal reflections (cavity effects), (i. e., $C_i \approx 0$ for homogeneous and flat surfaces).

In order to estimate the C_i parameter, different types of vegetated structures: legume, vineyard, orange grove and forest have been considered. From the geometrical model given in Sobrino et al., [6], C_i values for these typical vegetated structures are obtained and are approximated by $C_i = a_i + b_i P_v$. For AVHRR Channels 4 and 5, the values obtained for C_i are; $C_4 = 0.014 - 0.010\,P_v$ and $C_5 = 0.018 - 0.014\,P_v$. The values of ε_{v4}, ε_{s4}, ε_{v5} and ε_{s5} corresponding to Channels 4 and 5 are obtained from the emissivity data of different samples, provided by Dr. J. Salisbury (personal communication). Thus, the mean effective emissivity of two channels (ε) and the channel emissivity difference ($\Delta\varepsilon$) are respectively; $\varepsilon = 0.971 + 0.018\,P_v$ and $\Delta\varepsilon = 0.006\,(1 - P_v)$, where P_v can be derived according to Carlson and Ripley [7] from NDVI image using the following formulation

$$P_v = \frac{(NDVI - NDVI_{min})^2}{(NDVI_{max} - NDVI_{min})^2} = \frac{(NDVI - 0.2)^2}{0.09}$$

where $NDVI_{min}$=0.2 and $NDVI_{max}$=0.5 correspond to NDVI values for bare soil and full vegetation, respectively.

ii) NDVI < 0.2

The pixel is considered as soil with sparse vegetation and even bare soil (P_v=0). Using the spectral data of bare soils provided by Salisbury, with ε greater than 0.94, two linear relationships have been obtained by correlating the mean emissivity and the channel emissivity difference to the AVHRR Channel 1 reflectivity (ρ_1), $\varepsilon = 0.980 - 0.042\rho_1$ and $\Delta\varepsilon = -0.003 - 0.029\rho_1$ with an error of 0.004 for both equations.

iii) NDVI > 0.5

In the last case, the pixels are considered as fully vegetated and $\varepsilon_{v4} = \varepsilon_{v5} = 0.985$ are considered. Thus $\varepsilon = \varepsilon_4 = \varepsilon_5 = 0.985 + C_i$, with $C_i = 0.004$.

3 Multi-temporal analysis

3.1 Study area and satellite data

The study area is the *Mediterranean Basin*, located between 23.75° N and 46.25° N latitude and between 36.25° E and 17.5° W longitude. This particular geographical position gives the area great bioclimatic diversity. The climate is variable according to region and marked by strong annual and inter-annual irregularity. The area is also subjected to a significant problem with regard to desertification. The Pathfinder AVHRR Land (PAL) has been used to monitor this area. NOAA and NASA are the sponsors of the PAL project. Approximately 20 years of global AVHRR data are now available. The AVHRR data used are the Global Area Coverage GAC [8] with a resolution at nadir view of 4 Km. We have used images from July 1981 through August 1994. The data are accessible via FTP from the Goddard Distributed Active Archive Center

3.2 *Zones of interest*

In order to avoid monitoring land cover dynamics per pixel, which is time consuming and difficult to interpret, we have analyzed the evolution of varied landscapes of the study area. The different zones of interest were defined using as mapping criterion the image of the Total Average of NDVI (TANDVI) for the whole period (see Fig. 1). Following this procedure we defined a total of six zones (Table 1). The TANDVI interval in each zone is 0.1, in order to yield zones that match the different climatic zones in the study area.

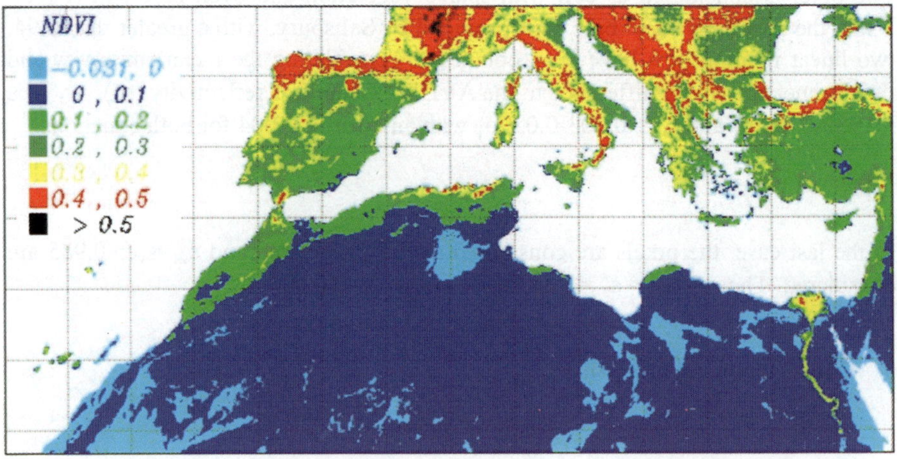

Figure 1. Image of the Total Average of NDVI (TANDVI).

Table 1. Zones of the study area and their main characteristics.

TANDVI	Characteristics	Name
< 0	Deserts. Limited opportunities for the development of the human activities. i.e., Sahara desert, Hejaz.	zone 1
0 -0.1	Zones of desert influence in Africa, desert zones in Spain and center Turkey. i.e., Pre-Sahara, Almería, Syria deserts.	zone 2
0.1 -0.2	Dry lands. Low precipitation. High temperatures. Low level of water retention. Sensitive to the effects of the climatic change and the loss of the biodiversity. i.e., Plains of the Iberian Peninsula and Turkey.	zone 3
0.2 -0.3	Low precipitation. Able to sustain vegetation. Sensitive to the effects of the climatic change and the loss of the biodiversity. i.e., Northern Africa coastal zones.	zone 4
0.3 -0.4		zone 5

3.3 Monthly evolution

Fig. 2a shows the results of the monthly evolution in terms of the NDVI in each of the defined zones. The arid zones (1 and 2) show small variations; with a maximum of 0.05 during spring and a minimum of –0.03 and 0 respectively during summer. Zones 3 and 4 reach the minimum in November (0.1 for zone 3 and 0.15 for zone 4), zones 5 and 6 presents their minimum in December (0.2). For these last zones, the months of June and July show a maximum NDVI (0.4-0.6) during summer. Fig. 2b is similar to Fig. 2a but for LST evolution. The maximum (> 320 K) is produced during the summer and the minimum (<290 K) in winter. Thermal amplitude in the study area, between zone 1 and zone 6, is of about 23 K during July and 14 K during January. To distinguish between zones is better to work with summer data (July). Finally, Fig. 2c is similar to the previous ones but for ε evolution. Thus, zone 6 reaches a maximum of 0.988 during summer and a minimum of 0.976 during winter. Zone 5 shows a similar behavior to the previous one with a maximum of 0.982 during summer and a minimum of 0.976 during winter. Zone 4 presents a maximum of 0.976 during spring and a minimum of 0.974 around the end of summer. Zone 3 presents an opposite behavior that zone 4, a maximum of 0.976 is produced during winter and a minimum of 0.973 is produced during summer. The arid zones (1 and 2) show a similar evolution to the previous zone, but with low values; a maximum of 0.973 in December and a minimum of 0.968 in July.

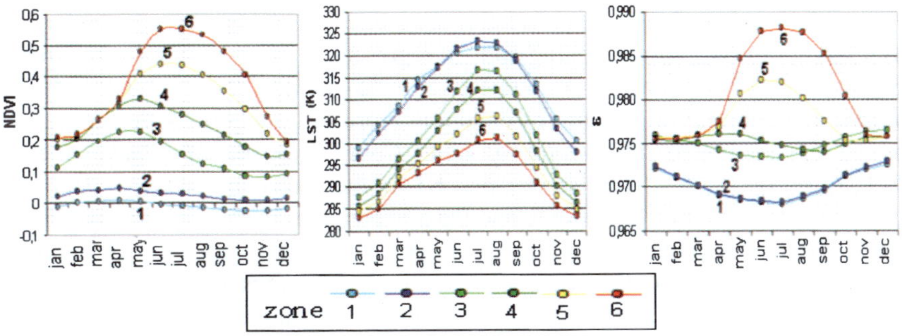

Figure 2. Plot of the monthly evolution of (a) NDVI, (b) LST and (c) ε in each zone.

3.4 Annual evolution

Fig. 3a shows the results of the evolution in terms of NDVI for the whole period. For 1982 and 1983 all the zones show the minimum values, a decrease is observed during the year 1988 for zone 6 and during 1992 for zones 3 and 4. Arid zones (1 and 2) show a constant behavior during the whole period. Fig. 3b is similar to Fig. 3a but for LST evolution. It is seen that the evolution hierarchy is conserved throughout the years. The arid zones (1 and 2) show high values of LST of about

312 K. The thermal amplitude in the Mediterranean basin between arid zones and those with abundant vegetation is of about 17 K. In Fig. 3b, the difference between zones 3, 4, 5, 6 is between 2 to 3 K, while this difference is greater between zones 1 and 2 and zone 3 (8 K). Finally, Fig. 3c is similar to the previous ones but for ε evolution. Here, zones 3 and 4 are the most stable during the whole period, the rest show some peaks, but in general the stable behavior of the zones is conserved throughout the period of study.

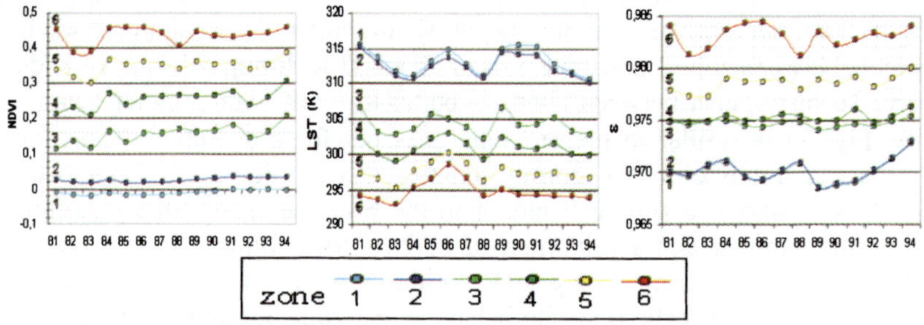

Figure 3. Plot of the annual evolution of (a) NDVI, (b) LST and (c) ε in each zone.

4 VLCD Method

Even though in the previous paragraphs is represented the evolution in the time of LST and NDVI separately. Our objective is to analyze the evolution of both parameters conjointly to the aim of obtaining more information. With this end, we defined the method of the *Vector of Land Cover Dynamic Monitoring* method (VLCD) in Fig. 4. This figure shows the LST on the y axis and NDVI(x100) on the x axis. The definition of the module and the arctangent of the VLCD are given in the figure. Both parameters introduce thermal amplitude and NDVI amplitude corresponding to the months of April and July. Fig. 5 represents for each zone and for the whole period (13 years), the module and the arctangent of the corresponding VLCD. The zones are well separated and distributed in three regions: ***vegetation*** (zones 5 and 6, arctangent<70º), ***semi-arid*** (zones 3 and 4, arctangent>70º and module>12.5), and finally ***arid*** (zones 1 and 2, arctangent>70º and module>12.5).

The VLCD method has the advantage of characterizing the temporal trajectories and the separation of the different zones according to the evolution of the vegetative systems. Fig. 5 permits defining four evolution trajectories: AB, showing the evolution of zones 1 and 2 in the arid region; BC and CD, the evolution of zones 3 and 4 in the semi-arid region; and finally DE, showing the evolution of zones 5 and 6 in the vegetation region. An evolution throughout AB, CD and DE with a convergence toward the point B, D and E respectively, would mean a positive evolution of the zone. The case defined by the path BC, would correspond to an evolution from arid to semi-arid zone. Otherwise, due to different causes i.e.; drought, desertification, forest fires, climatic changes, etc. the evolution would correspond logically to a negative evolution of the zones.

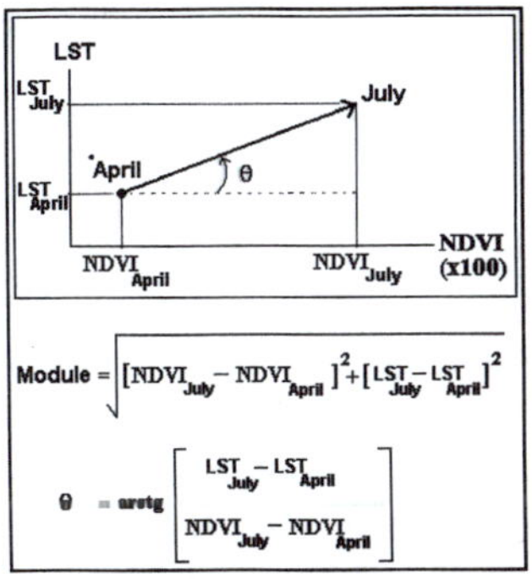

Figure 4. Definition of the VLCD.

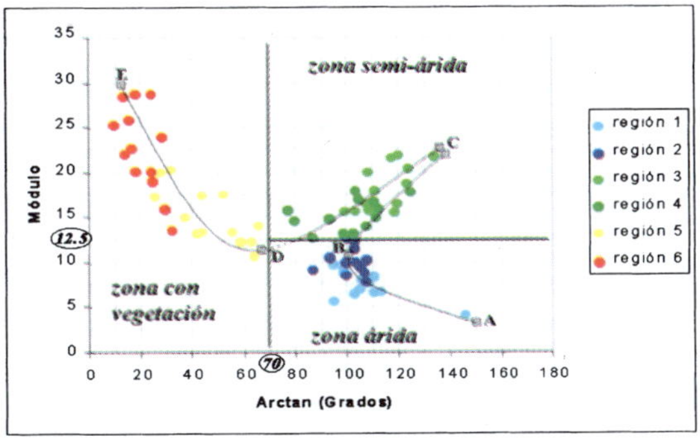

Figure 5. Representation of the VLCD for the study area.

5 Conclusions and perspectives

A preliminary study on land cover dynamic monitoring in the Mediterranean basin has been carried out. A method for the multi-temporal analysis called *Vector of Land Cover Dynamic* (VLCD) was presented. This method analyses the evolution of the slope and the module of the line which extremities are defined by LST and NDVI in April and July. The method permits characterizing the temporal trajectories and the evolution of the vegetative systems. Finally, among the future perspectives, we are interested; to include more years of PAL data to this study, to provide a classification map of the Mediterranean basin using the VLCD method and to apply the method to other future sensors data such as AATSR, MODIS, etc.

6 Acknowledgements

This research was supported in part by the European Union (WATERMED, Project No. ICA3-ct-1999-00015). The data used by the authors in this study include data produced through funding from the Earth Observing System Pathfinder Program of NASA's Mission to Planet Earth in cooperation with National Oceanic and Atmospheric Administration. The data were provided by the Earth Observing

System Data and Information System (EOSDIS), Distributed Active Archive Center at Goddard Space Flight Center which archives, manages, and distributes this data set.

References

1. C. J. Tucker, J. R. Townshend, T. E. Goff "African land-cover classification using satellite data", 1985, *Science*, 227, 369-375.
2. T. R. Loveland, J. W. Merchant, D. O. Ohlen, J. Brown "Development of a land-cover database for the conterminous, *U. S.*", 1991, *Photo Eng Remote Sens*, 57, 1453-1463.
3. E. F. Lambin, D. Ehrlich, "The surface temperature-vegetation index space for land cover and land-cover change analysis", 1996, *IJRS*, 17, 463-487.
4. J. A. Sobrino, N. Raissouni "Toward remote sensing methods for land cover dynamic monitoring. Application to Morocco", 2000, *IJRS, vol. 21, 2, 353-366.*
5. J. A. Sobrino, N. Raissouni, J. Simarro, F. Nerry, F. Petitcolin "Atmospheric Water vapor content over land surfaces derived from The AVHRR Data. Application to the Iberian Peninsula", 1999, *IEEE, Trans Geo Remote Sens*, 37, 1425-1434.
6. J. A. Sobrino, V. Caselles, F. Becker "Significance of remotely sensed thermal infrared measurements obtained over a citrus orchard", 1999, *ISPRS Photogramm. Remote Sens.*, 44, 343-354.ç
7. T. N. Carlson, D. A. Ripley "On the relation between NDVI, fractional vegetation cover, and leaf area index", 1997, *Remote Sens. Environ.* 62:241-252
8. J. R. Townshend "Global data sets for land applications from the Advanced Very High Resolution Radiometer", 1994, *IJRS*, 15, 3319-3332.

TRAJECTORIES OF LAND-COVER-CHANGE IN INTENSIVE AGRICULTURAL REGIONS : A SHORT-TIME PREDICTIVE APPROACH WITH THE DEMPSTER-SHAFER THEORY

L. HUBERT-MOY, M. BILLOT AND S. CORGNE

COSTEL LETG UMR 6554, Université Rennes 2, 6 avenue Gaston Berger ,35043 Rennes France

E-mail: laurence.hubert@uhb.fr

Y. QUETE

Géosciences - Université de Rennes 1, Bâtiment n° 15, Campus de Beaulieu, 35042 Rennes France

E-mail: Yves.Quete@univ-rennes1.fr

The Dempster-Shafer theory of evidence is often used to fusion data from different sources. This study intends to point out the interest of this theory to predict land cover at a local scale and for a short-time scale. To illustrate an application which aims to define the probability to recover corn in fields, one example of the two study areas where this method has been applied is exposed. Land cover changes and their related driven factors are determined on a small watershed located in an intensive agricultural region by experts' knowledge and remote sensing data. The first outcomes show that it is a tool well appropriate to definite uncertainty which, associated to the results, makes the predictions more realistic.

1 Introduction

Intensive land use influences downstream water quality, especially through chemical application and effluent spreading which involve a non-point source pollution. Water quality degradation is so high in some regions that delivering drinkable water has become a problem in areas where intensive agricultural practices are conducted. In this context, definition of land use and land cover trajectories (with a minimal duration of 4-5 years) and short-time predictive land allocation maps are needed both to modify land cover and agricultural practices in vulnerable areas towards pollutant fluxes and to assess actions of environmental policies.

Change detection in intensive watersheds is not an easy task, as spatio-temporal land use and land cover variations can be very high and are multiple driving factors dependant. In some cases uncertainty level can be significant, as far as meteorological constraints or economical and/or political events like subsidy allocation for certain crops are concerned.

Therefore, uncertainty in data has to be considered and quantified in modelling land use and land cover changes. Model predictions cannot be more accurate than data which produce them. Moreover, our frequently imperfect understanding of process functions makes the choice of variables incorporated in the land use-land cover modelling system a difficult task.

Unlike methods generally used, the Dempster-Shafer theory, also named the theory of evidence, introduces uncertainty in modelling, allows the expression of ignorance in the body of knowledge, and states that belief for a hypothesis is not necessarily the complement of its negation. This theory is usually applied to fusion data from different sources including remote sensing images with the main objective to improve classification in remote sensing in integrating multiple data sources (Lee and al, 1987; Le Hegarat-Mascle *et al*, 2000). The Dempster-Shafer theory is also frequently used to highlight land use and land cover changes from satellite images or aerial photographs and other data (Leduc *et al*, 2000; Leduc *et al*, 2001).

The objective of this study is to evaluate its interest in producing land use-land cover prediction in relation both with water quality and farmers' knowledge and motivation with the theory of evidence.

2 Study site and data

Two case studies on the modelling of land cover with a short-time predictive approach have been carried out in Britanny, on two watersheds where exceeding european standard nitrogene rates are registered for more than fifteen years.

One is a set of 14 small catchment areas called "Drains de Rennes 1" (25 km^2) and located in the northern-eastern part of the region.. These catchments, which are drained to supply drinkable water to the urban area of Rennes (300,000 habitants), constitute an important historical patrimony, since they have been collected water for the end of the 19th century. The watersheds form an homogeneous breeding area where an intensive dairy production combined with wet and warm automns produces significant amonts of nitrogene before winter infiltration of water. In this area, nitrogene inputs are still too high, especially on bare soils in winter, following and preceding corn implementation.

By contrast, the second case study is a larger watershed , the Yar, (65 km^2) located along the northern coast, where euthophisation occurs in the shape of bloom algae in the spring. This phenomenon, very injurious for tourism and more and more costly to discharge from coasts, is largely due to exceeding fertilization. Five different farming production types have been defined on this watershed.

Our experience in modelling integrated environmental system at the small catchment areas scale suggests that data requirements are highly system dependant. The hypothesis on land cover prediction have been set on each studied

area in relation with the specific local preoccupations : corn presence in crop rotations for one, state of bare soils in winter for the other (additional hypothesis could be defined in the future). The key variables of the two studied systems and their interactions, the hierarchisation of these variables and the assignment of basic probabilities on the selected indicators have been determined by experts in agricultural issues.
The series of pre-processed images used are different in the two case studies : Infrared aerial photographs at high resolution for one (one mission in summer and one mission in winter per year during 4 years since 1996), 10 satellites images SPOT and IRS-LISS III (2 per year during 5 years since 1996) for the other.
In both cases modelling land use and land cover changes is carried out in a G.I.S. through (**Figure 1**):

1. Change detection from image processing like vector change analysis or classifications before definition of land use and cover change trajectories.
2. Change prediction from fusion of land cover trajectories and selected indicators of driving factors of change with the rule of Dempster-Shafer.

This paper focuses mainly on the first study area to clarify the subject, and only results achieved on this site are exposed; Result assessment on the watershed of the Yar is in progress.

3 The Dempster-Shafer theory

The theory of evidence proposed by Dempster has been developed by Shafer in 1976 and basic concepts of this theory have often been exposed (i.e. in Richards, 1993 and in Srinivasan *et al,* 1990). Detailed applications of the Dempster-Shafer theory can be found in Eastman (1997) and in Mertikas and Zervakis (2001).
The Dempster-Shafer theory is a probabilistic approach, which starts by assuming a domain of reference called frame of discernment Ω, composed with a set of mutually exclusive hypotheses. The frame of discernment corresponds to a hierarchical structure and includes all possible combinations between the sets of hypotheses. For example for the three Hypotheses {A, B, C}, the possible combinations are [A], [B], [C], [A, B], [A, C], [B, C], [A, B, C] (**Figure 2**). The non-singleton hypotheses represent uncertainty. The commitment between these hypotheses is expressed through different functions :

- First, mass functions are defined, such each subset of Ω is associated with a basic probability assignment also called mass function. For example, *m* (A) represents the strenght of some evidence which supports the hypothesis A. One mass between 0 and 1 is assigned for each combination of the set of hypotheses. The sum of all basic probability assignments or mass always equals to 1, and *m* for the empty set is 0. Thus :

$$\sum_{A \subseteq \Omega} m(A) = 1$$

$$m_\Omega(\emptyset) = 0$$

- Then Dempster evidence functions, belief and plausibility, are derived from mass functions. For example, Belief function *Bel* (A) summarizes all the reasons to support A.The belief function corresponds to the sum of masses of belief of the subsets which involve A. *Bel* of a subset A of Ω is defined by :

$$Bel(A) = \sum_{E \in \Omega \text{ such that } E \subseteq A} m(E)$$

Where E represents all the subsets of A; If E = 1, then E = A.

Plausible belief *Pl* (A) expresses how much we should believe in A if all currently unknown facts were to support A. Plausibility represents the degree to which an hypothesis cannot be disbelieved. Thus :

$$Pl(A) = \sum_{E \in \Omega \text{ such that } E \cap A \neq \emptyset} m(E)$$

M (A), *Bel*(A) and *Pl*(A) have one value in the interval [0,1]. The true belief in A lies somewhere in the interval [*Bel*(A)-*Pl*(A)] which represents uncertainty (**Figure 3**). An area of high interest can be defined in the lower part of this interval insofar as it corresponds to the values for which the degree of uncertainty in establishing the presence of the hypothesis A is the lowest.

When mass functions are defined, they are combined according to the Dempster rule, also called Dempster's orthogonal sum, to aggregate probability statements from different sources of evidence, as follows :

$$Bel(A) = Bel_1(A) \oplus Bel_2(A)$$

In other words, if we are given two basic probability assignments *m*1 and *m*2 from two evidence sources to support the hypothesis A, they combine into a third basic probability assignment *m*1+*m*2 defined as :

$$m1{+}m2\,(A) = \frac{\sum m1\,(X) * m2\,(Y) \text{ for all X,Y with intersection A}}{\sum m1\,(X) * m2\,(Y) \text{ for all X,Y with empty intersection}}$$

if $\sum m1\,(X) * m2\,(Y) = 0$ for $X \cap Y = \emptyset$, then the equation becomes :

$$m1{+}m2\,(A) = \sum m1\,(X) * m2\,(Y) \text{ for } X \cap Y \neq \emptyset$$

All basic probability assignments are aggregated for all subsets of Ω, before final belief, plausibility and belief interval can be calculated for each hypothesis. Thus, uncertainty is measured at different levels of the full hierarchy of hypotheses.
When comparing with traditional probability theory, the main attribute of this theory is the ability to assign non-additive belief to subsets of Ω and therefore to express the degree of ignorance in a decision rule in handling uncertainty that involves ignorance (Srinivasan and Richards, 1990). In bayesian probability theory, only singleton subsets of Ω are considered and as they supposed to be exhaustive, they are affected one single value without taking in account uncertainty. With the theory of evidence, as a result of being able to assign belief and uncertainty values to set it is possible to suspend judgment.
When considering application of this theory, some critical points can be raised : Firstly, the choice of pieces of evidence and their relationship with the hypotheses lead to several questions (i.e. Are they direct or indirect pieces of evidence? Which hypothesis (es) do they support? How do they support one or several hypotheses?). In fact, these choices appear to be highly experts' knowledge dependant and therefore they should always be validated, even though it is not always easy to validate decisions derived from subject judgment or empirical data. Secondly, conflicts between sources of evidence are bad-managed when data are combined by normalization. This problem is all the more important as the number of data to fusion is high. Thirdly, the choice of criterion decision for a given hypothesis is subjective too, and should be clearly justified, that is not evident in any case.

4 Change prediction based on the Dempster Shafer evidence theory

The main research question which is addressed in this study is : In which locations will land cover changes likely to occur ? Answering to this question requires firstly the designation of the biophysical and anthropic data which contribute the most to the explanation of land cover changes and secondly the determination of the land cover change trajectories.
On the site of the Drains de Rennes 1 the driving factors of changes for corn allocation in the fields have been designated by experts from the Chambre d'Agriculture d'Ille et Vilaine. Considering the independance of sources, the

selected factors are past observed crop rotations, field size, distance from farm buidings, and protection perimeters set up by local authority, as it is the city of Rennes. Past observed crop rotations have been determined by crossing successive classifications of aerial photographs with the classical maximum likelihood method which gives a very good precision with such documents at this observation scale (Hubert-Moy *et al*, 2001).

Building of land use and land cover trajectories depends on the homogenity of the studied system : as one trajectory is sufficient to describe uniform systems (Drains de Rennes 1), several trajectories are needed for each farming production system in more diverse areas. For the first, trajectory definition strategy is only based on the field unit, whereas farm units are also considered for the second. Therefore, only one frame discernment is needed for one site, whereas several are required for the second, accordingly to farmers' behaviour with regard to their production system.

The frame of discernment for the site Drains de Rennes 1 comprises the following three hypotheses : [Fields with corn], [Fields without corn], [Fields with corn, Fields without corn].

Each of the two first hypotheses are supported by two sources of evidence, while uncertainty is affected to the third (**Figure 4**). Sources of evidence have been validated with past observed data on corn location.

Mass images are produced from sources of evidence through fuzzy functions. The shape of membership function is chosen according to the evidence image distribution and expert's knowledge (i.e. the control points of the fuzzy function related to the distance from farm buildings are respectively 0; 1,000m; 3,000m; and 5,000m as the probability of finding fields with corn is low under a distance inferior to 1,000 m, increases between 1,000 m and 3,000m, and finally decreases over 5,000 m). Ignorance is represented in one of the four sources of evidence : The distance from farm building is unknown for some fields, because of a recent land management and of several farm building locations out of the study area.

Probability images have been aggregated with the Dempster Shafer rule to produce belief, plausibility and interval belief images.

Maps related to the hypothesis [Fields with corn] show that strong belief (more than 0.7) for this hypothesis concerns a rather limited number of fields (**Figure 5a**), whereas a rather low plausibility (less than 0.4) is associated with numerous fields (**Figure 5b**). This means that uncertainty related with corn presence is quite important on this site. The interval belief image allows to locate the fields where uncertainty is rather low. The fields of concerned are almost located outside of the protection perimeters, which is quite encouraging for water quality.

These maps have been validated in crossing them with the classification of aerial photographs which have been registered last winter (**Figure 6**). It constitutes a first assessment as fields with no or few vegetation cover at the end of december will be for the most part of them covered with corn, thus some of them will correspond to young meadows or to fields with wheat which have been sawn later in the end of winter. Therefore this first validation has to be completed with data registered during the next summer. Regardless to this limit, the results show a quite good correlation between expected fields with corn with observed bare soils. Almost all fields which did not have any or few vegetation cover on january 2001 have been depicted with the model (Site A, Figure 6) : 87% of predicted corn fields on the belief map for classes over 0..5 correspond to bare fields, though some fields where there will be corn are not identified as bare fields on the winter classification, as farmers are encouraged to cover fields with vegetation between two corn successions; 75% of the fields classified as bare fields in winter correspond to predicted corn fields with a plausibility less than 0.4. The relative under-estimation of bare fields by the model can be easily explained by climatic conditions : precipitations were exceptionnaly high last winter, and soils were too saturated to sow wheat in the fields. Therefore in some places fields with corn will be this year more numerous as expected (Site B in Figure 6), as fields with wheat will be far less represented as usual. This explains that some fields which have been expected with vegetation covering last winter (wheat and meadows) will be sawn with corn in reality.

Nevertherless, these results point out that the progressive reduction of corn spatial extent from the protected perimeters (which have been set at the scale of the collecting water area) near the river up to the watershed limits which should have gone on next year will probably not occurr due to climatic conditions. The same phenomenon is visible on the other study site : Land use and land cover trajectories have been defined for the five last years, and projection of observed tendancies has been produced too for one trajectory related to the bare soils in winter on this watershed. The preliminary results point out that the significative reduction of bare soils between two annual crops which has been observed from 1996 should continue, but not with the same intensity due to exeeding precipitations. This shows that prediction has to be updated as new parameters like climatic conditions or economical events modify agricultural pratices.

5 Conclusions

Based on the obtained results, we illustrate the interest to use the Dempster-Shafer's theory of evidence to produce short-time predictions in land cover.

Uncertainty associated to the results makes the predictions more realistic and political decisions based upon them will be better able to account for the possible risk of failure.Though, uncertainty is reduced as new data are integrated in the

system, predictions becoming stronger and more explicit when long series of data are processed.
We also demonstrate that this model will operate as well as on homogeneous areas which can directly be analysed at the watershed scale or on areas where different types of farming productions and a large diversity of biophysical factors are encountered. Thus it can be considered as an auxiliary and assessment tool for the environmental action programmes which aim to reconquer water quality.

Figure 1. Structure of land use and land cover modelling

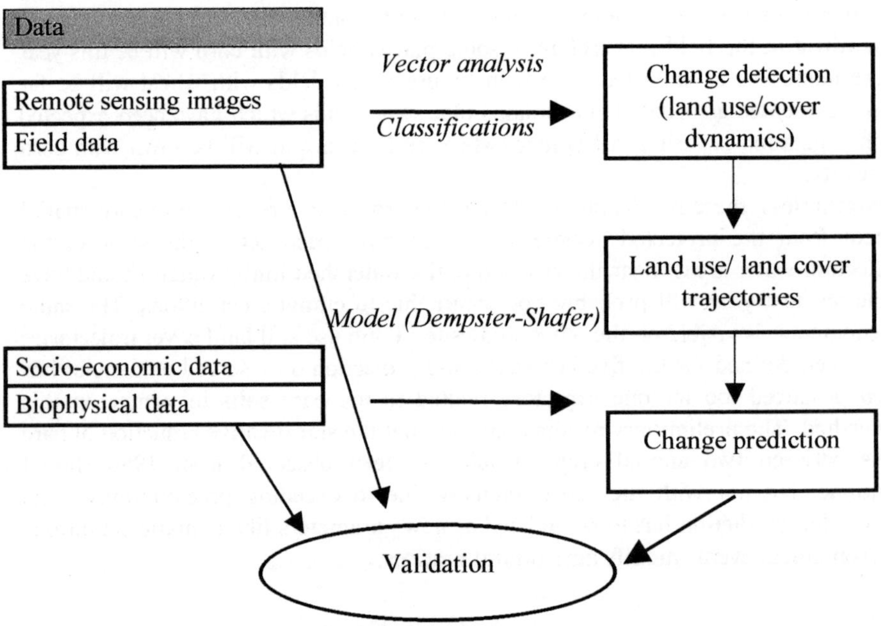

Figure 2. The hypothesis frame of discernment

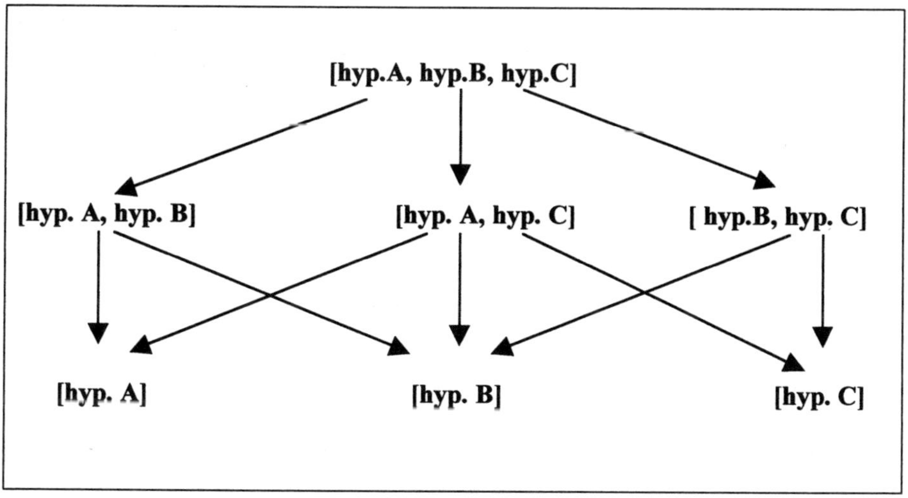

Figure 3. Measures of Belief, Plausibility and Uncertainty : Representation for an hypothesis A in the interval unit

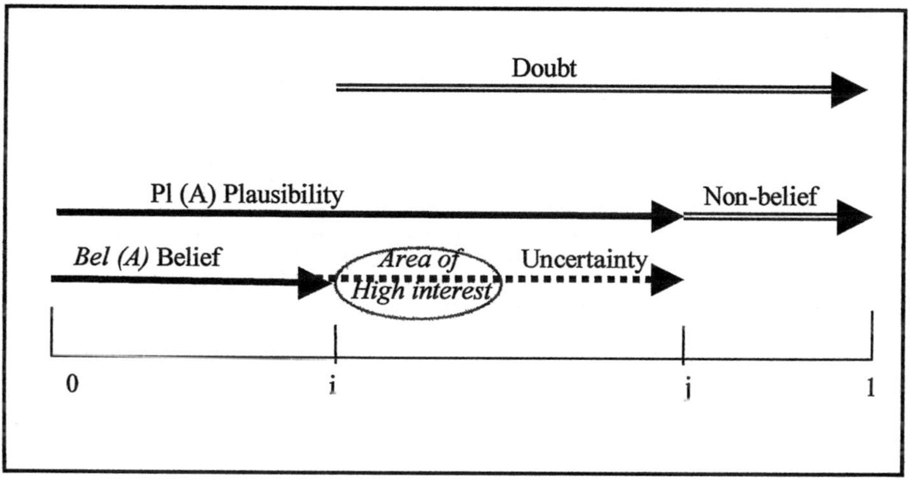

Figure 4. Data fusion with the rule of Dempster-Shafer (Example on the site "Drains de Rennes1)

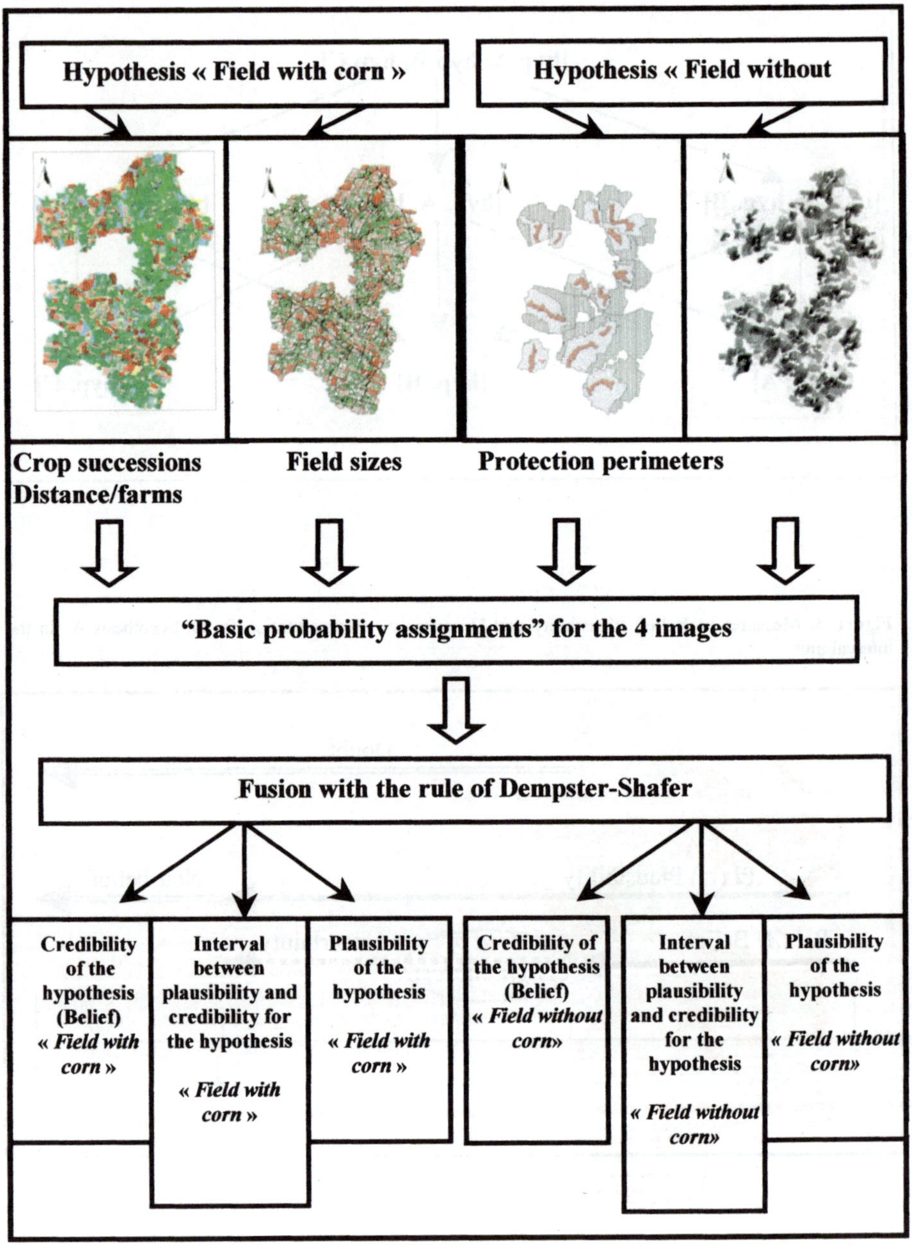

Figure 5 . Classification of Belief image (a) and Plausibility image (b) for the hypothesis "Corn"

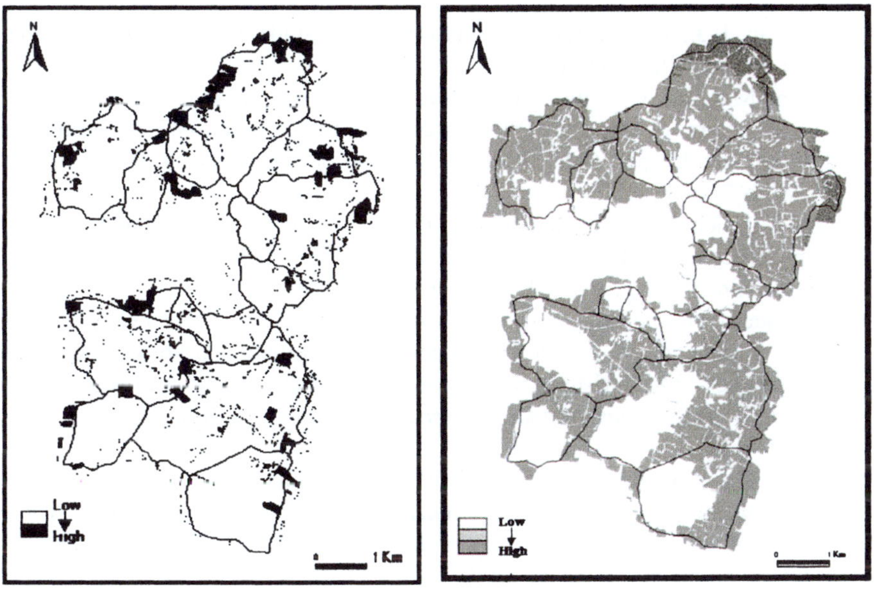

Figure 6. Classification of vegetation cover in winter 2000/01

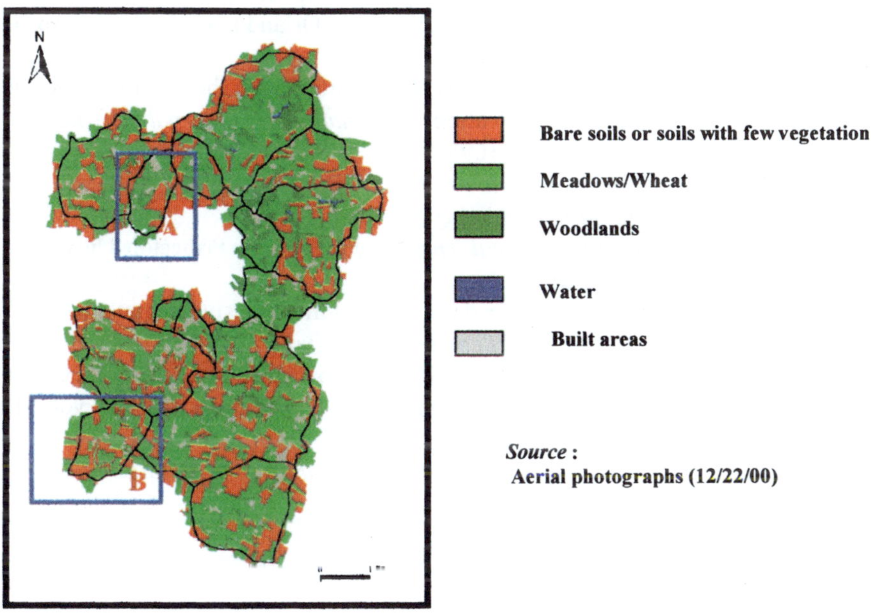

References

1. Eastman, R.J., 1997, Idrisi for Windows User's guide, version 2.0, Worcester, Idrisi Project, Clark University.
2. Hubert-Moy, L., Cotonnec, A., Le Du, L., Chardin, A., and Perez, P., 2001, A comparison of classification procedures of remotely sensed data applied on different landscape units, *Remote Sensing of Environment,* ed. Elsevier, vol.75, n°2, p.174-187.
3. Lee, T., Richards, J.A., and Swain, P.H., 1987, Probabilistic and Evidential Approaches for Multisource Data Analysis, *IEEE Transactions on Geoscience and Remote Sensing*, Vol. GE-25, N°3, pp. 283-293.
4. Le Hégarat-Mascle, S., Quesney, A., Vidal-Madjar, D., Taconey O., Normand M. and Loumagne C., 2000, Land cover discrimination from multitemporal ERS images and multispectral Landsat images: a study case in an agricultural area in France, *International Journal Of Remote Sensing*, vol. 21, N°3, pp.435 -456.
5. Leduc, F., Solaiman, B., and Cavayas, F., 2000, A fuzzy and evidential approach for temporal changes interpretation in a forestry context using cartographic and satellite data", *Géomatique 2000*, Palais des Congrès, Montréal-Canada, 8-10 mars 2000.
6. Leduc, F., Solaiman, B., and Cavayas, F., 2001, Combination of fuzzy sets and Dempster-Shafer theories in forest map updating using multispectral data, *Sensor Fusion*, SPIE, Vol. 4385, pp.323-334, Orlando, USA, 18-20 April, 2001.
7. Mertikas, P., and Zervakis M.E., 2001, Exemplifying the theory of evidence in remote sensing image classification, *International Journal Of Remote Sensing*, vol. 22, N°6, pp.1081 –1095.
8. Richards, J.A., 1993, *Remote Sensing Digital Image Analysis: an introduction*, 2nd ed., Springer-Verlag, Berlin.
9. Shafer, G.A., 1976, *Mathematical Theory of Evidence*, Princeton University Press, Princeton, New Jersey.
10. Singh, A., 1989, Digital change detection techniques using remotely sensed data", *International Journal Of Remote Sensing*, vol. 10, N°6, pp.989-1003.
11. Srinivasan, A., and Richards J.A., 1990, Knowledge-based techniques for multi-source classification, *International Journal Of Remote Sensing*, vol. 11, N°3, pp.505-525.
12. Zadeh, L.A., 1965, Fuzzy sets, *Information and control*, vol. 8, pp.338-353.

Classification and land use evolution in the South Milan Agricultural Park

STEFANO BOCCHI, DARIO BELLINGERI, ANDREA GALLI

Department of Crop Science - State University of Milan ,via Celoria, 2 20133 Milan Italy

E-mail: stefano.bocchi@unimi.it - dariobel@hotmail.com – Galli@popcsi.unian.it

The results of a study on a Metropolitan Park land use evolution by using multitemporal remote sensing techniques are presented. A set of 5 Landsat images was used to classify the crops with a combination of NDVI and Principal Components method. A simple Area Stability Indicator was calculated by difference between the results obtained with 1991 and 1999 classifications, for many municipalities located in the study area. A significant increase of the more profitable crops (i.e.corn) was observed with a corresponding decrease of winter crops and grassland, indicating a trend of intensification. The methodology appears easily reproducible and feasible for the Institution that will need to update the database for a efficient landscape monitoring

1 Introduction

South Milan Agricultural Park (SMAP) is a protected, green area belting the City of Milan with a total of about 46,000 ha grouping in 61 municipalities where agriculture is the main land use. The Park aims at planning, managing and monitoring both natural and artificial resources and giving the people living inside and outside the City a better environment [1]. With the project SITPAS (a Park Geographic Information System) the Park is collecting and arranging in a Data-Base many information related to the agricultural activities.

A major requirement of any agricultural information system for crop identification, area and production is the regular input of updated land cover information. Such information can be collected directly through interviews, ground surveys and local cartography, but it is better derived from satellite imagery which provides additional update information to the system at intro-annual, annual or multi-year intervals. The use of a such powerful technology is much more indicated where the land use is rapidly changing like in the case of SMAP, where the conflicts among sectors - agriculture development, urbanization, industrial plants, services - are increasing [2,3].

The aim of this study was to analyze the land use changes from 1991 up today by comparison of different classes of land cover for different homogeneous areas, by using LANDSAT images and NDVI trends. The ISA (Area Stability Indicator) was also calculated for different representative municipalities.

2 Materials and methods

2.1 *Study area*

The South Milan Agricultural Park (SMAP) has been existing since 1990 and is, in Italy, the first great park on a metropolitan scale. The SMAP forms half a ring south Milan, linking Ticino Park, located West, and Adda Park to the East. The objectives are to safeguard agricultural activities, cultivation and forests, to protect natural locations, to recover or preserve landscape, environment and countryside, and to inform and guide farmers and citizens in respecting environmental resources [1].

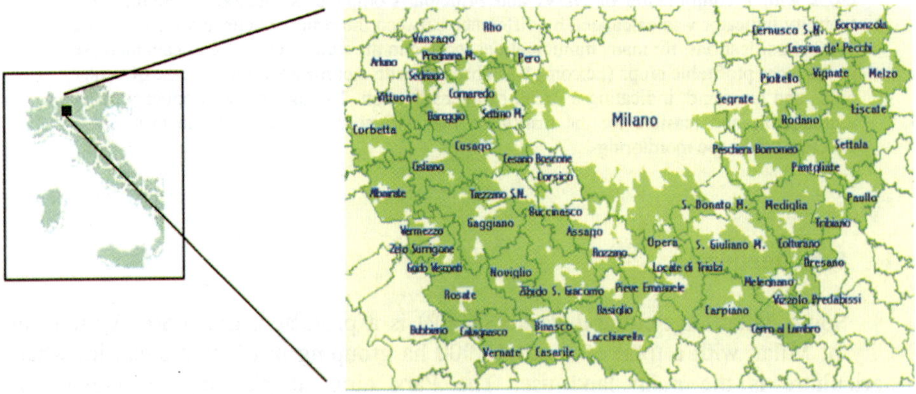

Figure 1. Study area South Milan Agricultural Park

2.2 *Data set*

Database is composed by two passages of Landsat-5 TM in 1991 (18 June and 5 August), just after the institution of the Park, and three passages in 1999, namely 23 May and 26 July (Landsat-5 TM), and 6 October (Landsat-7 ETM+), for describe the actual situation. The choice of the dates, moreover, was based upon the knowledge of the phenology of the main crop grown in the Park [2,3].

2.3 *Methods*

Geometric registration was first performed to make data comparable and overlayable; sixteen Ground Control Points were selected both on topographic map and on satellite images. After the second degree polynomial warping and the nearest neighbor resampling, geometric accuracy obtained gave root mean square errors less than 0.5 pixel.

The methodology for the classification and for the evalutation of land use change is represented in fig 2.

Figure. 2. Theoretical scheme of the classification and land use change evaluation process.

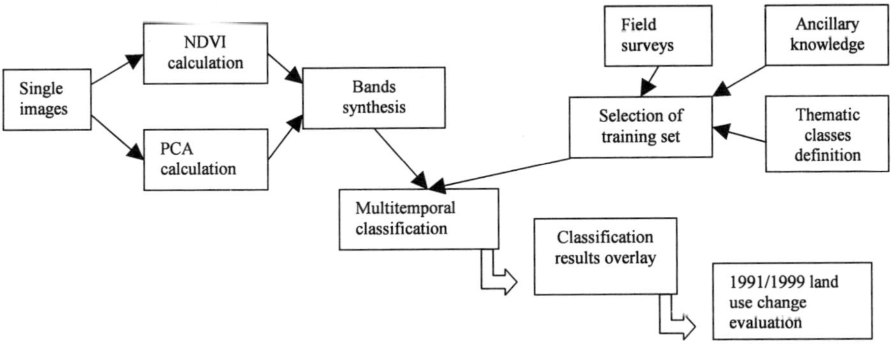

3 Results

3.1 Classifications

The algorithm of classification chosen was the Maximum Likelihood, with threshold at 90%; the land use classification was performed separately on the two data set of 1991 and 1999. In this paper the focus is on the situation of 1999, when the ground cover types are the following:

Table 1. Information classes, number of pixels and description, relatively to 1999 case.

CLASSES	N° of points of TRAINING SET	DESCRIPTION
Water	87	Artificial lakes, rivers
Residential urban	80	-
Industrial urban	90	Factories, generally high reflectance surfaces
Woodland	106	Residual woods, poplar-woods
Grassland	144	Meadows, water meadows
Flooded rice	132	-
Autumn/winter crops	102	Barley, wheat
Spring/summer crops	85	Corn (silage)
Spring/summer crops 2	143	Corn (grain), soybean
Second harvest	90	Barley-corn; raygrass-corn

The major information provided by the three images was synthesize using the Normalized Difference Vegetation Index (NDVI) and the first two Principal Components of every image [4,5]; in fact, some information class shows a lack of separability using only the NDVI (see fig. 3 and table 2).

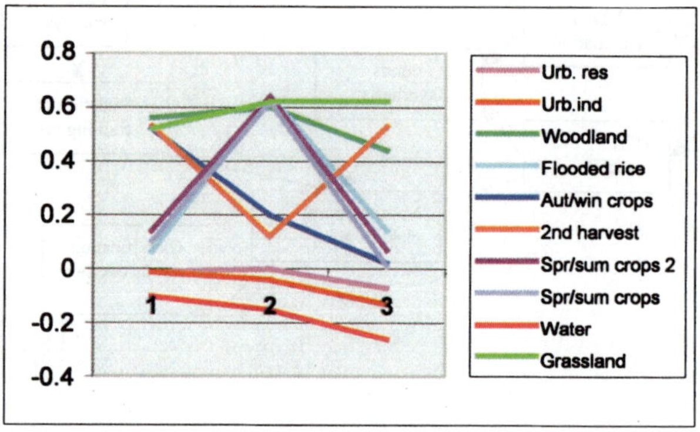

Figure 3. Mean values of NDVI of the training set (1:May; 2:July; 3: October)

Table 2. Principal components values

	MAY	MAY	JULY	JULY	OCTOBER	OCTOBER
	1st comp.	2nd comp	1st comp	2nd comp	1st comp	2nd comp
Water	-89	-38	-15	98	-87	-37
Grassland	-5	39	-16	-31	-1	66
Urb.res.	-1	-28	40	23	-68	-30
Urb. ind.	70	-26	112	-45	143	-31
Woodland	-32	29	-23	13	-36	13
Flooded rice	-66	-22	-30	4	17	2
Aut/win crops	-20	40	62	-36	24	-17
2nd harvest	-19	41	72	-35	-6	50
Spr/sum crops 2	42	-7	-29	-10	-19	-16
Spr/sum crops	47	-11	-28	-2	58	-12

Table 3. Scheme of the spectral behavior in the three dates (green arrow: high vegetative response; red arrow: absence of vegetation; yellow arrow: intermediate behavior)

CLASSES	23-may	26-jul	06-oct
woodland	↑	↑	↑
flooded rice	↔	↑	↔
grassland	↑	↑	↑
autumn/winter crops	↑	↓	↓
spring/summer crops	↓	↑	↓
spring/summer crops 2	↓	↑	↑
2nd harvest	↑	↓	↑

Focusing the attention on the vegetated classes, table 3 shows the important contribution of multi-temporal images in the distinction between the different types of cultivation, characterized by different phenological cycle [3].

The results of the 1999 classification is shown in figure 4; classification accuracy was evaluated with a contingency table (table 4), and it resulted greater than 90% for every information class, except for 2nd harvest (with high variability in spectral response).

Table 4. 1999 land use classification contingency table (pixel)

Classes	Water	Grass.	Wood.	Flooded rice	Aut/ win	2nd harvest	Spr/ sum	Spr/ sum 2	Urb. Ind.	Urb. Res.	Total	Acc. (%)
Unclassified	0	1	1	1	1	4	2	4	3	0	17	-
Water	82	0	0	0	0	0	0	0	0	0	82	98.8
Grassland	0	140	0	0	0	16	0	1	0	0	157	93.9
Woodland	0	0	113	0	0	0	0	0	0	0	113	91.9
Flooded rice	0	0	0	138	0	0	0	0	0	0	138	96.7
Aut/win crops	0	0	0	0	98	2	0	1	0	3	104	95.2
2nd harvest	0	8	5	0	0	98	0	0	0	0	111	81.7
Spr/sum crops	0	0	4	1	0	0	139	5	0	0	149	98.5
Spr/sum crops 2	0	0	0	3	3	0	0	102	0	0	108	90.3
Urb. Ind.	1	0	0	0	1	0	0	0	99	4	105	93.4
Urb. Res.	0	0	0	0	0	0	0	0	4	128	132	94.8
Total	83	149	123	143	103	120	141	113	106	135	1216	93

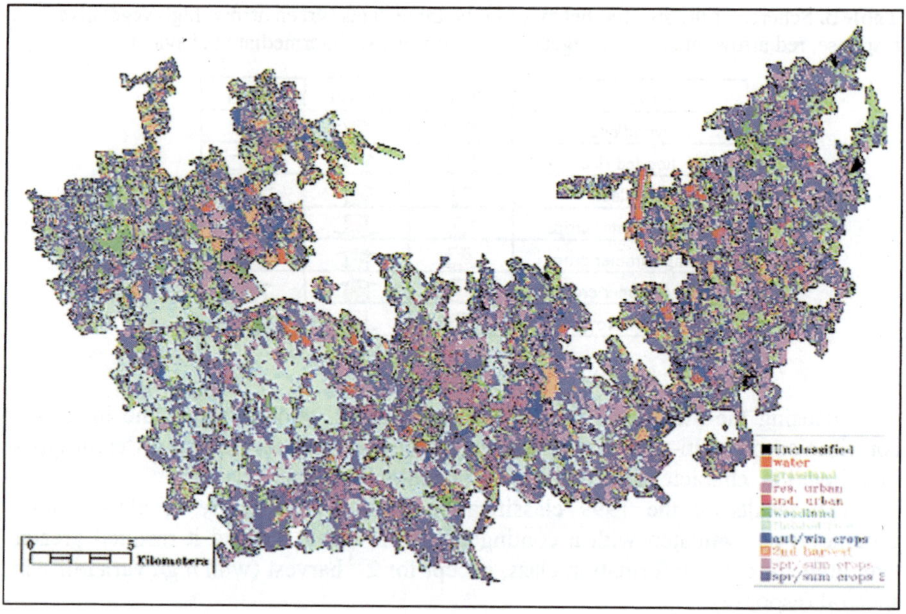

Figure 4. Multitemporal land use classification (1999)

With the same methodologies for the 1999 case, the classification of land use relative to 1991 was obtained; in this case, having only two images, it was not possible to discriminate the areas interested by second harvesting, and to distinguish between the two types of spring/summer crops.

3.2 Land-use change

The evaluation of the land use changes between 1991 and 1999 was performed overlaying the two thematic maps, exploiting the fact that every pixel represents the same object in the different dates.

Figure 5 synthesizes these changes, in terms of percentage of surface; it shows the stability of water, the little increase of both residential and industrial urban, while the increase of woodland is mainly due to poplar cultivation; the greatest percentage of change is clearly in the agricultural areas.

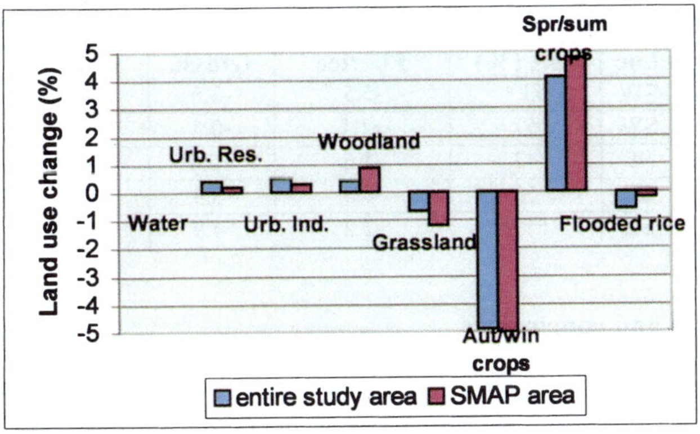

Figure 5. Land use change (calculated as "1999 less 1991"), for the entire image (40x30 Km) and for the area of the Park

The evaluation of land use change, moreover, was performed on some sub-areas, representatives of the different agricultural and cropping systems adopted in the Park.. Here the evolution is shown, only for the four classes of agricultural use, of the Zibido San Giacomo area, where rice is the main, predominant crop. Land use change is appreciable from the surface percentages (figure 6), and from the crossing of the two classifications, performed for every pixel and shown in figure 7: histograms in light blue, green, dark blue and purple represent unchanged points, and yellow histograms are all the possible dynamics. The sum of all the pixel belonging to the colored diagonal, in this case the 53.4 %, can be identified as an index of the stability of the area (I.S.A.). For other sub areas, see table 5.

Figures 6/7. Surface % at Zibido and representation of Indicator of Stability (I.S.A.)

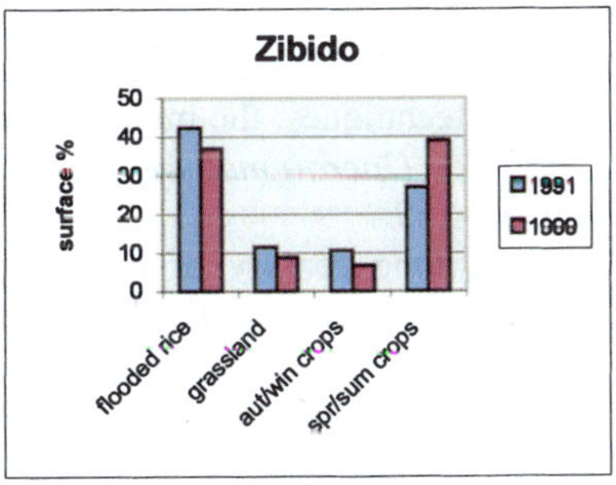

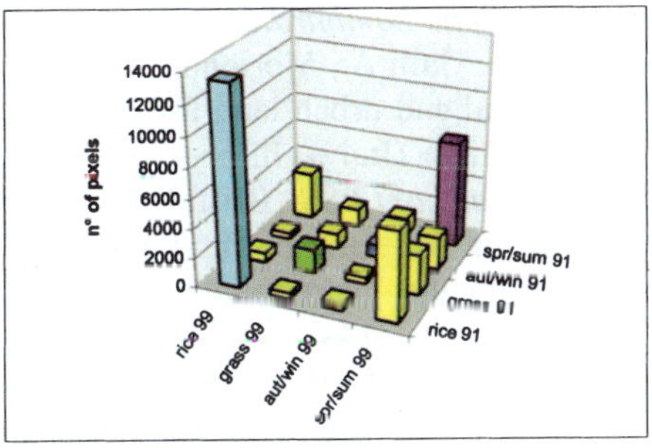

Table 5. Summary of I.S.A. and % of land use change in agricultural classes for 5 sub-areas

Area	Loc	ISA (%)	Fl. rice	Grassl.	Aut/win	Spr/sum
Zibido	SW	53,4	-5.5	-2.6	-3.8	+12.1
Gudo	SW	57	+1.1	-0.8	-1	+1.5
Melzo	NE	47	-1.6	+0.8	-9.8	+11.4
Dresano	SE	56	-0.5	-5.4	-8.2	+16.4
Melegnano	SE	51,7	-2.2	-4.9	-2.8	+12.3

4 Discussion and conclusions

The methodology allowed to obtain the land use classification of the area of SMAP relatively to 1991 and 1999. The two thematic maps have a partial different quality, in terms of information classes and overall accuracy: this is due to the different quantity of satellite imagery, and to the different quality of ancillary data.

Multi-temporal features have been used not only to obtain the two classifications separately, after a suitable synthesizing of the information, but also to evaluate the evolution between the situation of land use in the Park in 1991 and 1999: the little increase in classes such as residential and industrial urban denotes that a request to edification still exists. Moreover, the strong increase of more profitable and productive cultivation like corn, with the relative decrease of autumn/winter crops, means that the agriculture in this area is even more intensive, to the detriment of traditional cultures like, for example, irrigated meadows and *marcite*. Rice, on the contrary, appears more stable, being interested only by a little decrease. Finally, the methodology used in this works is reproducible, and offers the possibility of constant updating of the thematic maps.

5 References

1. AA.VV., Istituzione del Parco Agricolo Sud Milano, in *Bollettino Ufficiale della Regione Lombardia* n° 17, 27/4/1990
2. Bocchi S., Galli A., Moriondo A., Rossi B., Tomasoni R. Changes detection in agricultural and urban land use by remote sensing techniques: the metropolitan area of Milan (Italy). In *proc. of Remote Sens. For Operational Application, 15th Annual Conf.RSS,* Bristol sept., (1989) pp. 63 – 69.
3. Gomarasca M., Brivio M., Pagnoni F., Galli A, One century of lande use changes in the metropolitan area of Milan (Italy). *International Journal of Remote Sensing,* **14**-2 (1993) pp. 211-233.
4. Michener D. and Houhoulis F., Detection of vegetation changes in a forest ecosystem. *Photogrammetric Eng. And Remote Sensing,* **63**-12 (1997), pp. 1363-1374.
5. Sunar F, An analysis in changes in a multi date data set: a case study in the Ikitelly area, Istambul, Turkey. *Int J. of R. S.*, **19**-2, (1998), pp. 225-235.

LAND USE CHANGES IN SOUTHERN LAZIO, ITALY, BY SATELLITE DATA MULTITEMPORAL ANALYSIS

R. CASACCHIA (*), ,R. SALVATORI (*), A. GRIGNETTI (*), S. MANDRONE (*)
(*) CNR-IIA Viale Marx, 15 – 00137 Roma Tel: 0686090522 Fax 0686090360
rcasak@gioconda.polar.rm.cnr.it

A. CAPRARO (**)
(**) Geologist– Aquino (FR)
antogeol@libero.it

M. P. BOGLIOLO (***)
(***) ISPESL-DIPIA Via di Fontana Candida 1-00040 Monteporzio Catone (RM)

In this paper images from different satellite sensors were analysed to detect the major land use changes in two test areas of southern Lazio, Italy. Land modifications were analysed by means of three different indices (NDVI, PVI, SAVI). The goal of this paper is to test, in areas were these investigations were rarely performed, different methodologies to monitor urban growth, impact of new infrastructure, and the relationship between soil and vegetation. Multitemporal analysis was carried out on Landsat 5 TM, Landsat 7 ETM and IRS-LISS III multispectral images. Good results are performed by means of quantitative multitemporal comparisons between NDVI, PVI and SAVI indices.

1 Introduction

In the last decades several regions of the Italian territory underwent profound and rapid changes due to the development of the local economy. After the expansion of the large urban and industrial settlement occurred from the fifties to the seventies, a second phase was undertaken characterized by a intense growth of small residential and industrial areas external to the city limits. This process mainly concerned small and medium towns producing a dramatic environmental impact whose effects have caused several difficulties in the management of these areas. For this reason local administrations begin to be aware of the necessity of better understanding the pressure that local economies produce on the territory, using instruments that allow synoptic and multitemporal analyses. In this context remote sensing plays an important role due to the experience of the scientific community in the application of multispectral image processing techniques in studying land use change. Also the increased spatial resolution of the new satellite sensors, along with the historical data series, makes possible the monitoring of different environments and the analysis of features of specific interest. However it has to be taken into account that the expertise of local administration personnel is not always adequate to fully manage the entire process from visual inspection of raw data to the production of the final thematic information. The role of scientific institutions is therefore to provide the necessary tool to understand the meaning of satellite-derived thematic maps.

The goal of this study is to apply different processing techniques to multispectral satellite images in order to determine the most suitable algorithm to detect land cover changes in selected areas of southern Lazio.

2 Test areas and satellite data processing

The investigated areas were selected for co-existence of urban, agricultural and protected areas (highly fragmented territory); availability of multitemporal images acquired by different satellite sensors; availability of ground-truth information both referred to spectroradiometric data and to the author's direct knowledge.

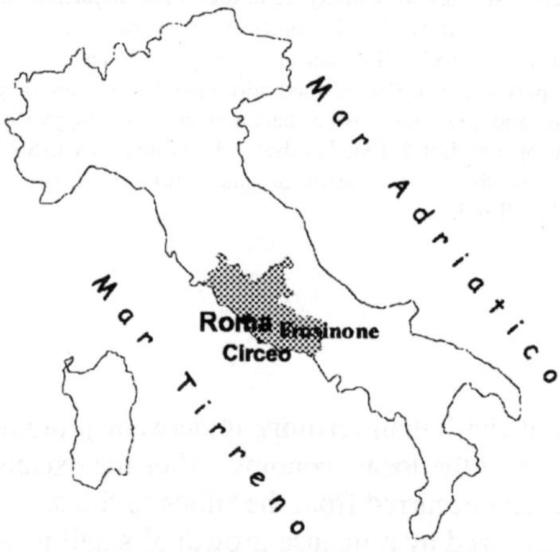

Figure 1 Location mao of the investigated sites

The chosen areas are the Circeo region and the Province of Frosinone, located respectively along the southern coast and the southern inland region of Lazio (Figure 1).

The Circeo region includes the eastern boundary on the Circeo National Park, the small town of San Felice and several villages spread out in the country side (surface area of about 25200 ha). In the last two decades an intense urban growth took place in the eastern part of this region, due to trade and resort activities. The western part, where the National Park is located, was quite well preserved. Data processing of this region concerned a Landsat 5 Thematic Mapper scene (path/row 190/31), acquired on August 11, 1986 and a Landsat 7 ETM scene acquired on September 24, 1999.

The province of Frosinone includes the city and the surrounding country side, the Lepini range in the southern part and the Ernici range in the northern part; it covers an area of about 57000 ha. The territory is extremely fragmented and characterised by urban and industrial settlements, agricultural and rural areas and natural forest vegetation.

In the last 15 years the entire province recorded a significative urban and industrial increase, that resulted in a remarkable change in land use. One of the major events that determined such environmental impact was the construction of the new Roma-Napoli railway (TAV project), that began in early nineties. The satellite images used for this area were Landsat 5 Thematic Mapper (path/row 190/31) scene acquired on August 11, 1986 and a IRSS LISS-III (path/row 31/40) scene acquired on August 4, 1998.

To the purpose of multitemporal analysis, environmental indices were computed from images calibrated into reflectance. Reflectance data were obtained by the empirical line method [1], using field spectroradiometric data acquired on invariant targets (soil and asphalt) by a portable instrument operating in the wavelength range 350-2500 nm. Landsat 7 ETM and IRSS LISS-III data were then normalised with respect to TM. Based on reflectance images the Normalised Difference Vegetation Index (NDVI), the Perpendicular Vegetation Index (PVI) and the Soil Adjusted Vegetation Index (SAVI) were computed. Vegetation indices allow to quantitatively and qualitatively estimate the distribution and the amount of the vegetation cover [2] and can be used for discriminating among vegetated areas, bare soils, anthropic and urban areas.

NDVI can be easily computed from multispectral satellite images as follows:

$$NDVI = NIR - R / NIR + R$$

where R and NIR are respectively the spectral channels in the red and in the near-infrared (i.e. bands 3 and 4 of Landsat and IRS/LISS sensors) [3]. High values of NDVI correspond to high biomass density.

PVI (Perpendicular Vegetation Index) can be extracted as follows:

$$PVI = a\rho_{IR} - b\rho_R$$

where a and b are coefficients of the line of soil, while ρ_{IR} and ρ_R are respectively reflectances in bands 4 and 3PVI<0 is characteristic of water surface, low PVI values (near to zero) correspond to soil and urban areas, while higher PVI values refer to vegetation [4]. This index allow to estimate the effects of soil on the spectral response of vegetation and can be used to monitor vegetation growth. The computation of this index requires that soil surfaces are recognised in the satellite data.

SAVI (Soil Adjusted Vegetation Index) is generally used to reduce soils brightness effects, assuming a linear relationship between soil reflectance in red and near-infrared bands. SAVI is computed by:

$$SAVI = \frac{\rho_{IR} - \rho_R}{\rho_{IR} + \rho_R + L} \cdot (1 + L)$$

where L is a correction factor depending on vegetation density usually assumed to be = 0.5 while ρ_{IR} and ρ_R are respectively reflectances in the 4 and 3 bands [4].

3 Results

In Figures 2 and 3 a multitemporal comparison of results obtained by NDVI, PVI and SAVI are shown, respectively for Circeo and Frosinone test areas. Histograms show, as percentage over the total surface area, the coverage for main land cover units.

A general trend can be recognised for all the indices: in the last decade land use changes are quite evident, with a transformation of natural vegetation areas into urban (buildings), bare soils (that represents degraded areas) and cultivated areas. In fact the percentage of vegetated areas decreased while percentage of bare soils and anthropic areas increased.

In particular Circeo NDVI graph (figure 2.a) shows the effectiveness of NDVI index in quantitative evaluation and discrimination of surfaces coverage. By multitemporal comparison is quite easy to follow the growth of urban areas (NDVI values from 0.13 to 0.25), agricultural soils (values from 0.25 to 0.38) and bare soils (range 0.38-0.51) versus a corresponding decrease of vegetated areas (NDVI >0.51). The computed increase of anthropic areas is over 12 percent. NDVI processed images are very detailed, enabling an accurate discrimination (at Landsat resolution) of urban areas and buildings. This comes out very useful for environmental applications, in case of local studies.

In the Frosinone area (figure 3.a) it is possible to notice that after the completion of TAV project, an intense land use change occurred, resulting in a sharp decrease of the vegetated and cultivated surfaces respect to a strong increase of urban and bare soil surfaces.

Circeo PVI graph (figure 2.b) also shows the ability of this index to describe the general trend of target area, but it is less accurate and representative than NDVI. Intervals 0.09-0.18 and 0.18-0.27 represent urban areas, bare soils or agricultural land, while PVI > 0.27 should represent vegetated areas. In PVI processed image part of urban areas is included in open wood class (range 0.36-0.45): this explains the high percentage of vegetation classes. Moreover, using this index is quite difficult to discriminate between urban areas, bare soils and agricultural land.

In the Frosinone area PVI provides a high detail in discriminating vegetation units; for this reason data interpretation may reveal some difficulties. From figure 3.b it is quite evident the expansion of the urban features (range 0.12-0.32) and of the surrounding degraded surfaces (range 0.32-0.38) that determined a substantial decrease of the agricultural and naturally vegetated areas (PVI > 0.38). However in some cases PVI values of urban areas and outcropping limestone (Mt. Lepini and Mt. Ernici) are in the same range.

SAVI processed images allow a better and more precise identification of land cover units than PVI images, because the soil contribution is considered in computing this index. In Circeo area appreciable results (figure 2.c) are obtained in discriminating between urban areas and buildings (range 0.08-0.21),

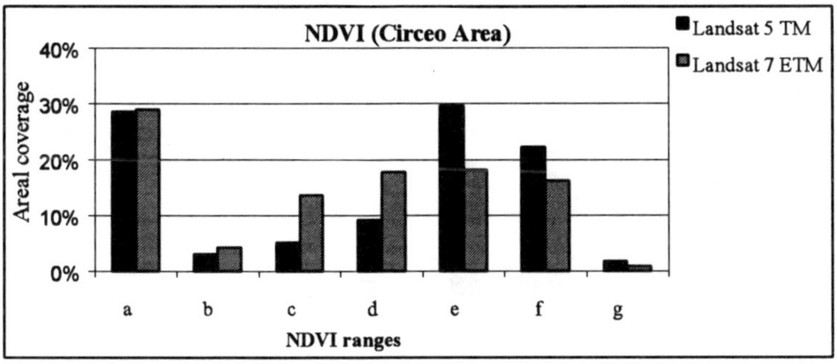

Figura 2a **a**=Sea water(<0.13), **b**=Urban (0.13-0.25), **c**=Agricultural soils (0.25-0.38), **d**=Bare soils (0.38-0.51), **e**=Herbaceous formation (0.51-0.64), **f**=Open wood (0.64-0.79), **g**=Forests (>0.79)

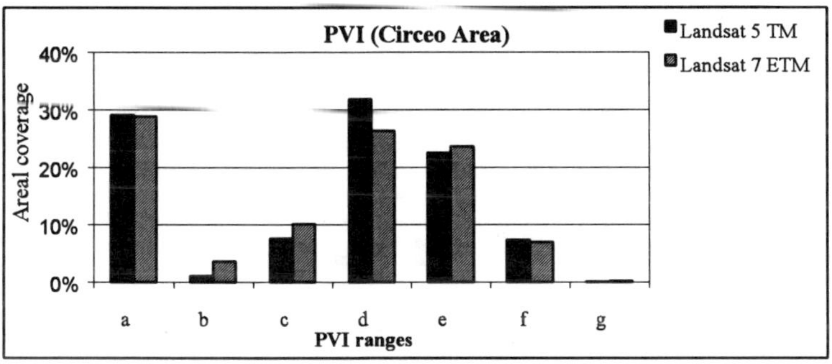

Figura 2b **a**=Sea water(<0.09), **b**=Urban (0.09-0.18), **c**=Bare soils (0.18-0.27), **d**=Herbaceous formation (0.27-0.36), **e**=Open wood (0.36-0.45), **f**=Forests (0.45-0.63), **g**=Forests (0.63-0.72)

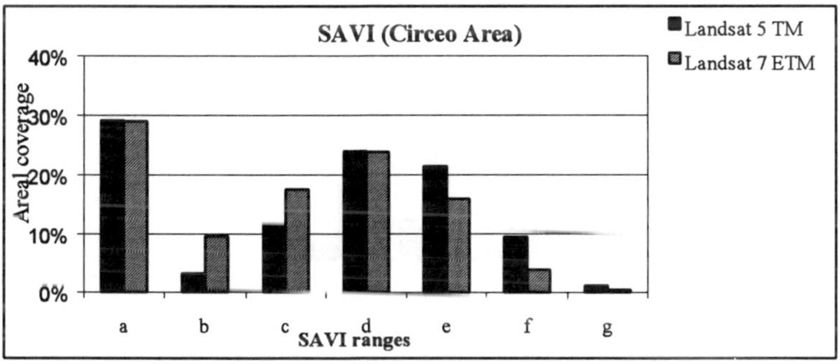

Figura 2c **a**=Sea water(<0.08), **b**=Urban (0.08-0.21), **c**=Anthropic areas (0.21-0.32), **d**=Bare soils (0.32-0.45), **e**=Herbaceous formation (0.45-0.57), **f**=Open wood (0.57-0.69), **g**=Forests (0.69-0.81)

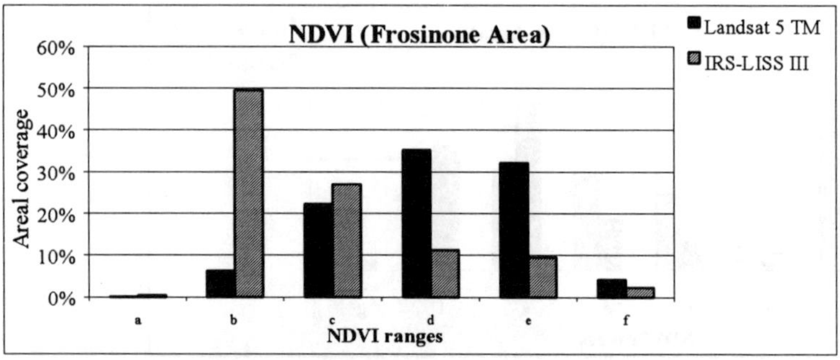

Figura 3a **a**=Other (< 0.11), **b**=Urban (0.11-0.20), **c**=Bare soils (0.20-0.33), **d**=Agricultural soils (0.33-0.45), **e**=Mixed vegetation, **f**=Forest (0.60)

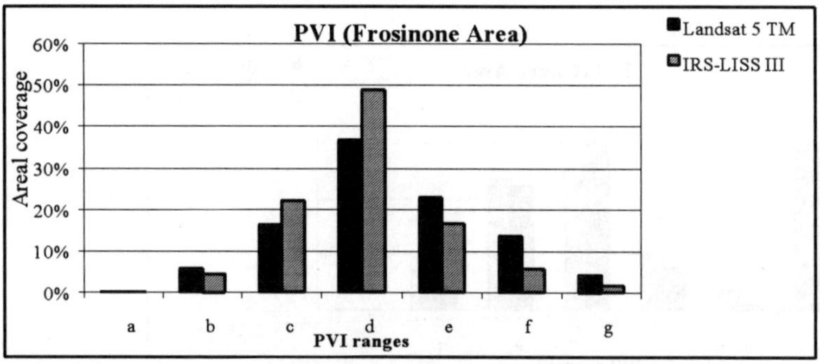

Figura 3b **a**=Other (< 0.12), **b**=Urban (0.12-0.27), **c**= Anthropic areas (0.27-0.32), **d**=Bare soils (0.32-0.38), **e**=Herbaceous formation (0.38-0.42), **f**=Forest (0.42-0.48), **g**=Forest (>0.48)

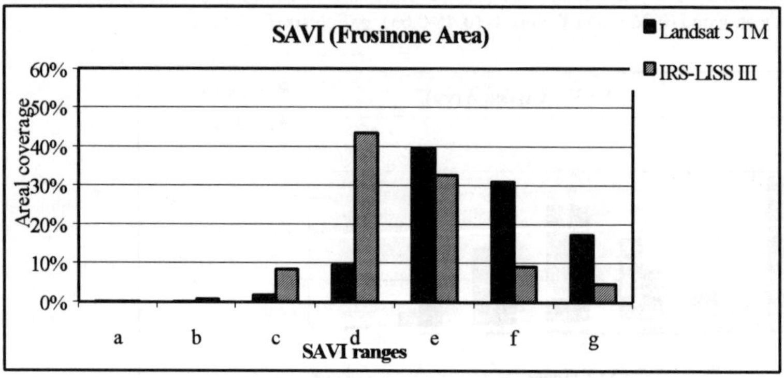

Figura 3c **a**=Other (< 0.005), **b**=Urban (0.005-0.12), **c**=Bare soils (0.12-0.24), **d**=Mixed vegetation (0.24-0.36), **e**=Mixed vegetation (0.36-0.49), **f**=Mixed vegetation (0.49-0.59), **g**=Forest (>0.59)

anthropic and agricultural lands (range 0.21-0.32), bare soil areas (range 0.32-0.45) and vegetation coverage units (values from 0.45 to 0.81). However in SAVI image a overestimation is registered in bare soil areas extension: surface area of about 24% is registered, more large than the real extension (ranging about from 10% to 18% respect the whole target area).

In Frosinone area SAVI (figure 3.c) provides good results in discriminating the vegetated surfaces with respect to any other surface material. However, SAVI index values shows a class referred to forested area and three classes all referred to mixed vegetation in which can be recognised the contribution of both bare and agricultural, soil. For this reason image maps produced by SAVI are less fragmented than those obtained by PVI and can be easily used to estimate the extension of the forested areas.

4 Conclusions

In this paper results of multitemporal studies using satellite images processed by environmental indices are presented. NDVI, PVI and SAVI could be an interesting alternatives to classification procedures in land cover changes identification. Using these indices a quantitative evaluation of land cover units extension and their changes in time can be readily carried out with appreciable results.

Table 1. Main characteristics of used image processing technique and their applications

	FEATURES	PROCESSING CHARACTERISTICS	APPLICATIONS
PVI	Highly specific for vegetation studies.	Large computing-time; operator decision process is required for soil line extraction	Preliminary land use analysis (discrimination vegetated/non vegetated areas)
SAVI	Specific for vegetation studies. It is not well-suitable for refined environmental analysis	Easy processing; quite reduced computing-time	Discrimination soil/vegetation
NDVI	Quantitative index, well suitable for environmental analysis. Very accurate discrimination of different surface covers.	Easy and rapid processing	Environmental analysis; quantitative multitemporal comparisons; land cover change detection; environmental evolution trend identification.

In particular NDVI allowed a good discrimination of the different surface covers, with results very similar to ground truth surveys and can be consider a reliable environmental index, useful to evaluate environmental impact due to land use changes. In Table 1 the main characteristic of used techniques are summarised.

All the image processing techniques discussed in this work can be used to produce land use maps rapidly and with relatively low cost investment. This may represent an advantage for local administrations interested in multitemporal monitoring of environmental changes for land management and decision making.

5 Acknowledgement

This work was performed under the research contract CNR/ISPESL Ric.N.92/97.

References

1. Smith G.S. and Milton E.J., The use of the empirical line method to calibrate remotely sensed data to reflectance, *Int. Jour. Remote Sensing* **20** (1999) pp. 2653-2662.
2. Mas J.F., Monitoring land-cover changes: a comparison of change detection technique, *Int. Jour. Remote Sensing* **20** (1999) pp.139-152.
3. Lawrence R.L. and Ripple W.J., Calculating change curves for multitemporal satellite imagery: Mount St. Helen, *Remote Sens. Environ.* **67** (1999) pp. 309-319.
4. Bannari A., Morin D. and Bonn F., A Review of Vegetation Indeces, *Remote Sens. Reviews* **13** (1995) pp.95-120.
5. Chavez P.S. and Kwarteng, A. Y., Extracting Spectral Contrast in Landsat Thematic Mapper Image Data Using Selective Principal Component Analysis, *Photogrammetric Engineering and Remote Sensing* **55** (1989) pp. 339-349.
6. Fung T. and LeDrew E., Application of Principal Components Analysis to Change Detection, *Photogrammetric Engineering and Remote Sensing* **53** (1987) pp. 1649-1658.

USING REMOTE SENSING TO UPDATE GEOGRAPHIC INFORMATION SYSTEM

P. SBARDELLA
Datasiel S.p.A., Via Merano 22, Genova, ITALY
E-mail: p.sbardella@datasiel.net

R. BARICHELLO
C.A.A.R. (Regione Liguria), Loc. Pallodola, Sarzana ITALY
E-mail: agmcaar@regione.liguria.it

Updating Environmental Geographic Information Systems is a complex problem, due to the need to balance best frequency and detail of process repetition A general suggested approach is the data and model integration, using both satellite image and field survey as input data in the Regione Liguria G.I.S. environment. The suggested method provides an annual multi-temporal analysis of satellite imagery effective for land-cover change detection, to obtain an updated and validated "Forest Land Vegetation Map" for the whole Ligurian Territory (over 500,000 ha). This approach is based on subsequent approximations to define the best ratio between the effectiveness and the simplicity of the operational methodology.

1 Introduction

The update of the Regione Liguria Geographic Information System using remote sensing applications is a very important goal, in which many operative tasks of the Regional Government converge to.

Many cartographic layers in the Regione Liguria G.I.S. often need updates, with variable frequency of repetition, according to the themes and the applications (information needs, scale, budget).

The goal of this project is to take existing geographic information system, field survey and satellite image processing technologies and develop problem–oriented methods and services for assessing and monitoring land–cover change (especially in forestry and environmental fields).

2 Background

This project, born as a pilot project, is part of the S.P.I.R.L. Program (Servizio Previsione Incendi boschivi Regione Liguria) (see reference [3]) and its main goal is to define an operational methodology, both effective and simple, to support the dynamic implementation process of the "Carta dei combustibili" (a prevision map for fire primer in wood–land).

These prevision maps need meteorological and vegetation high quality data.

Detailed field survey will assure the initial update of the Regional Vegetation Map. This project would assure a support method for the next annual updates.

3 Methodology

The suggested method provides an annual multi–temporal analysis of satellite imagery, by specific vegetation indexes (like NDVI or similar) effective for land–cover change detection (see reference [2]).

Landsat 7 TM data offer a good choice in term of confidence, geometric and radiometric quality and price, but the frequently cloud–cover on the Ligurian region in the topic seasonal period (May, June) cannot exclude future experimentation using different data (ERS or ENVISAT).

Before the multi–temporal analysis it is necessary to carry out the atmospheric corrections and the phyto–climatic calibrations of the satellite data. To obtain a good result in the image–to-image comparisons (index-to-index comparisons), it is required to correct the aberrant factors (like the atmospheric influence) and the divergence factors (like the differences of the vegetation "mean" phenologic development) present at any acquisition date.

3.1 Atmospheric Corrections and Phyto-Climatic Calibrations

Experimental uses of agro-meteo-climatic data (mainly humidity and temperature data), obtained by the Regione Liguria monitoring network, are carried out both in the specific methodology for the "Carta dei Combustibili" definition and in this project.

Wide availability of data and the complexity of the problem suggested an approach founded on subsequent approximations.

The first approximation level consists in a simple atmospheric correction technique named "haze removal" (see reference [6]). The large approximation of this level is in the direct extrapolation of the correction values (known only for same points in the Ligurian Sea) for each pixel of the image.

A second approximation level provides the extrapolation of the previous correction values using the spatial distribution of humidity data obtained by the regional Agro-Meteorological Service.

The next approximation level aim is carrying out a complex atmospheric correction, using supplemental meteo-climatic data and correlation analysis, both local (in the regional monitoring network nodes) and spatial (through now-casting and forecasting maps produced by the regional Agro-Meteorological Service every day).

The Phyto-Climatic Calibrations use the specific field survey data (other than the previous ones) and the same above-mentioned correlation analysis.

Except for the first approximation level, for all the others the image data registration is required. The applied resample algorithm is the Nearest Neighbor (see reference [4]), the reference band is the band 8 (panchromatic), the initial Ground Control Point data set is standardized.

3.2 Synthetic Indexes production and their multi-temporal comparisons

Synthetic index extraction from Landsat TM imagery aims to standardize a high correlation level between spectral variation in the imagery and vegetation change.

At present the choice of the NDVI (Normalized Difference Vegetation Index) (see reference [5]) is suggested because of a large references availability, but we cannot exclude future experimentation using different indexes relating to the expected results.

However, index-to-index standard comparison also requires that variations (caused by factors like, e.g., seasonal influence) should be pinpointed and, if possible, excluded.

The quality of the above-mentioned corrections and calibrations can be evaluated empirically in test-areas, where the vegetation structure, at the two corresponding satellite acquisition dates, is well known.

In this project phase the final product is a "Index Values Difference Map" derived with a simple difference operation between index values of the second year and that of the previous one.

3.3 Detection and measurement of change in forest land vegetation

The "Index Values Difference Map" become the scheduled support tool for assessing and monitoring land-cover change only if it is studied within the Regione Liguria G.I.S. (see reference [7]).

The comparison between the land-cover layers (up-to-date at the first satellite acquisition date approximately) and this new Map is carried out using spatial filtering and topological analysis (see reference [1]).

The comparison yields geo-statistics and spatial distribution of data in terms of potential relationship of cause and effect in land-cover change detection.

Workflow goes on with the Index Difference Values classification to achieve "change intensity" classes, according to the vegetational structure. The classification is carried out using a simple "density slicing" algorithm. The choice of the class limits is founded on the previous statistical results.

The goal of this classification is to improve the support quality for detecting land-cover change.

For any vegetation change detection it is analyzed the possible cause-effect relationship with fires, landslides and further instances, both recent and past, stored in the Regione Liguria G.I.S.

Both ancillary data (aerial or high-resolution satellite imagery) and specific field survey can resolve an interpretative doubt, to reach the G.I.S. update (for land-cover or different thematic layers) and the improvement of the Methodology.

4 Project state-of-the-art

The initial step for the development of this project is a comparative evaluation of data sources and methods to be performed on any types of vegetation change in a test-area in East Liguria, for which there is a high-quality data availability, up-to-date at 1999 (see table 1 and Figure 1).

Table 1. Data sets available for the test-area.

Theme	Acq. Date	Scale/Res.	Data Format
Land-Cover (rev.)	1999	1:25,000	vector
Forest Map (rev.)	1999	1:25,000	vector
Landsat TM7 covers	1999, 2000	15 – 30 m	raster (MS+PAN)
Landsat TM5 covers	1992	30 m	raster (MS)
Fires in wood-land	1999, 2000	1:10,000	vector
Landslide	1999, 2000	1:10,000	vector
Aerial Ortho-photo	1999	1 m	raster (color)
Digital Terrain Model	1999	1:10,000	grid, vector
Ikonos (data fusion)	2000	1 m	raster (MS+PAN)
Topographic Maps	variable	1:5,000	vector, raster
Topographic Maps	variable	1:10,000	vector, raster
Topographic Maps	variable	1:25,000	raster
Further aerial photo	variable	variable	variable
Further Env. Layers	variable	variable	vector

Test-area (about 50,000 ha, covered by the 193/029 Landsat frame) contains also five stations of the Regione Liguria current agro-meteo-climatic monitoring network.

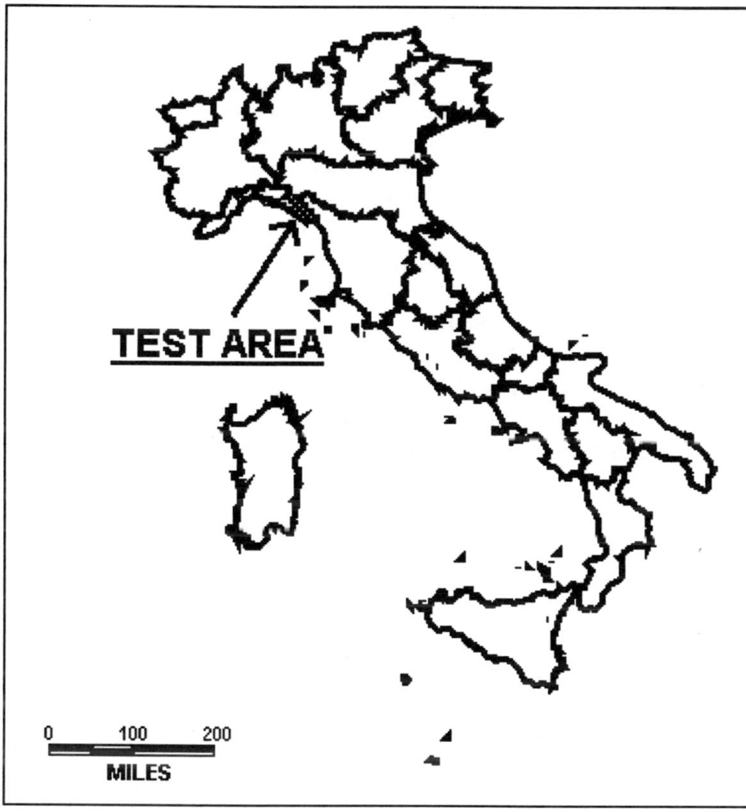

Figure 1. Test-area in East Liguria (Italy).

Thus the period for the land-cover change detection is between 1999 and 2000.

With the first approximation level of the atmospheric correction and without phyto-climatic calibration, it was yielded an "Index Values Difference Map" that was geo-coded with the Nearest Neighbor resample algorithm.

The imagery registration was performed using both Aerial Ortho-photos and Topographic Maps (in Gauss-Boaga coordinate system) as Control Image and Band 8 (Landsat7 Panchromatic Band, with 15 meters pixel-resolution) as Input Image. The resultant RMS (with "Affine" transformation, a linear model that corrects rotation, translation, skew, and scale in X and Y) is approximately 6.5 meters for both the 1999 and the 2000 satellite images. The registration of the others bands (30 meters pixel-resolution multi-spectral bands 1-7,9), for any acquisition date, was performed using the same Panchromatic Band GCP, dividing input image coordinates by 2 and applying a 15 meters Nearest Neighbor resample algorithm.

Despite the extreme simplifications due to the first approximation level, a correlation appears between the Map values and the regional G.I.S. thematic data.

Analyzing relationship between vegetation change detection quality and recent fires stored in the Regione Liguria G.I.S, both fire properties (typology, spatial dimension, date) and vegetation properties (structure, stratification, phenologic phase) seem to play a primary role in the response variability. Thus is necessary to improve the quality of data stored in the regional G.I.S.: for instance, the fire typology it is not available yet. Besides, the problematic availability of cloud-free Landsat images in the topic seasonal period (June, for our application type) can increase the multi-temporal analysis complexity (hill-shade and season-climatic conditions divergence between the first satellite acquisition and the second one).

Table 2 shows principal fire case-studies occured between the first and the second Landsat acquisition (13/09/1999 and 29/07/2000 respectively). The mentioned Regione Liguria Land Cover Classes are: AM = "Angiosperma Mesofila" wood, AC = "Angiosperma Termofila" wood, A = "Arbusteto l.s.".

Table 2. Shematic comparisons between fire case-studies.

Fire Date	Land Cover	Area (ha)	Detection Quality
19/01/2000	AM	8	Low
19/01/2000	AM	3	Low
04/05/2000	AM	8	Medium-low
08/06/2000	AC	10	High
03/07/2000	A	9	Medium-high

An experimental method was carried out with different Vegetation Indexes, like IR index (Normalized Difference between Band 4 and Band 5), Simple Vegetation Index (ratio between Band 3 and Band 4) or Normalized Band 7 (see reference [2]), obtaining a general results convergence.

In any case, Index Difference Values classifications (carried out using simple "density slicing" algorithms to improve the support quality for detecting land-cover change) achieve "change intensity" maps with an high complexity interpretative level.

Both these results and the experimental method (with an higher complexity level, included Phyto-Climatic Calibration) are currently under evaluation.

However, the main goal of this project is to define an operational methodology in which effectiveness and simplicity are in balance. For this reason the integration of this methodology with mono-temporal satellite analysis and photo-interpretation is scheduled.

5 Conclusion

Multi-temporal analysis using Landsat7 data, will play an important role in monitoring land-cover changes within the regional G.I.S., but requires an improved input data-set (availability of "good" satellite acquisition dates and exhaustive Geographic Knowledge Base more wide then the actually used ones).

The Phyto-Climatic Calibration of satellite imagery, its suitability limits and its integration within a multi-temporal analysis are open problems. A wide availability of agro-meteo-climatic data (Humidity, Temperature and many other data) is offered by the Regione Liguria monitoring network (over 50 monitoring points).

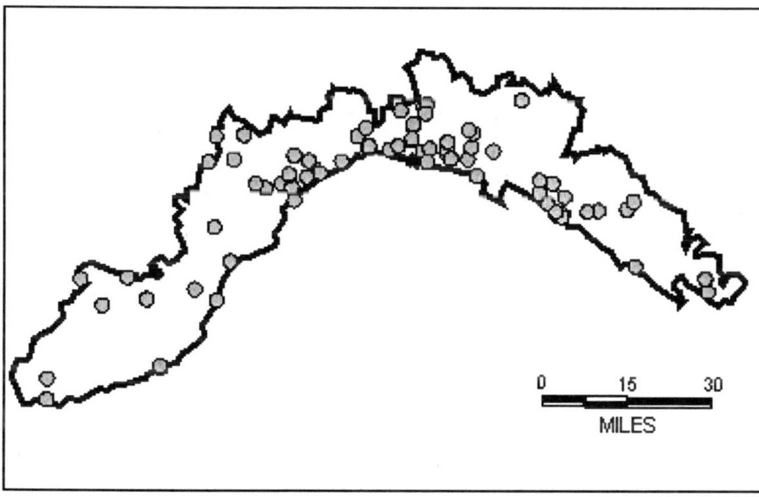

Figure 2. Spatial distribution of the Regione Liguria current monitoring network.

Mono-temporal satellite analysis and photo-interpretation can be a fairly good integration of the proposed methodology.

This project is based on the idea that the best approach to update any Geographic Knowledge Base is a well-balanced integration among image processing technologies, geographic information systems and field survey.

Moreover the final scopes and the end-user operative needs must be considered as main goals. This is the guide thread of the Project.

The quantitative results and the new results will be exposed in a next paper.

References

1. Fung T. and LeDrew E., The determination of optimal threshold levels for change detection using various accuracy indices, *Photogrammetric Engineering & Remote Sensing*, Vol. **54**, No. 10 (Oct. 1988), pp. 1449-1454.
2. Green K., Kempka D. and Lackey L., Using Remote Sensing to Detect and Monitor Land-Cover and Land-Use Change, *Photogrammetric Engineering & Remote Sensing*, Vol. **60**, No. 3 (March 1994), pp. 331-337.
3. Regione Liguria – Centro di Agrometeorologia Applicata Regionale, *Programma S.P.I.R.L. (Servizio Previsione Incendi Boschivi Regione Liguria)*, Regione Liguria (Sept. 2000).
4. Richards J.A., *Remote Sensing Digital Image Analysis*, 2nd edition. (Berlin: Springer-Verlag, 1993).
5. Rouse J.W., Haas R.H., Shell J.A., Deering D.W. and Harlan J.C., Monitoring the vernal advancement of retrogradation of natural vegetation, *Final Report*, Type III, NASA/GSFC, Greenbelt, MD (1974), pp. 371.
6. Sbardella P., *Carta dell'Uso e della Copertura del Suolo della Regione Liguria – Relazione tecnica finale*, Regione Liguria (1994).
7. Sbardella P., *Osservazioni e limiti della cartografia di Land Cover CORINE della Regione Liguria*, Regione Liguria (1995).

Author index